T0174762

A Guide to Outcome Modeling
In Radiotherapy and Oncology
Listening to the Data

Series in Medical Physics and Biomedical Engineering

Series Editors: John G. Webster, E. Russell Ritenour, Slavik Tabakov,
and Kwan-Hoong Ng

Series in Medical Physics and Biomedical Engineering

A Guide to Outcome Modeling In Radiotherapy and Oncology
Listening to the Data

Issam El Naqa
Associate Professor, University of Michigan
Ann Arbor, Michigan

CRC Press
Taylor & Francis Group
Boca Raton London New York

CRC Press is an imprint of the
Taylor & Francis Group, an **informa** business

CRC Press
Taylor & Francis Group
6000 Broken Sound Parkway NW, Suite 300
Boca Raton, FL 33487-2742

First issued in paperback 2020

© 2018 by Taylor & Francis Group, LLC
CRC Press is an imprint of Taylor & Francis Group, an Informa business

No claim to original U.S. Government works

Version Date: 20171211

ISBN-13: 978-0-367-57208-2 (pbk)
ISBN-13: 978-1-4987-6805-4 (hbk)

This book contains information obtained from authentic and highly regarded sources. Reasonable efforts have been made to publish reliable data and information, but the author and publisher cannot assume responsibility for the validity of all materials or the consequences of their use. The authors and publishers have attempted to trace the copyright holders of all material reproduced in this publication and apologize to copyright holders if permission to publish in this form has not been obtained. If any copyright material has not been acknowledged please write and let us know so we may rectify in any future reprint.

Except as permitted under U.S. Copyright Law, no part of this book may be reprinted, reproduced, transmitted, or utilized in any form by any electronic, mechanical, or other means, now known or hereafter invented, including photocopying, microfilming, and recording, or in any information storage or retrieval system, without written permission from the publishers.

For permission to photocopy or use material electronically from this work, please access www.copyright.com (http://www.copyright.com/) or contact the Copyright Clearance Center, Inc. (CCC), 222 Rosewood Drive, Danvers, MA 01923, 978-750-8400. CCC is a not-for-profit organization that provides license and registration for a variety of users. For organizations that have been granted a photocopy license by the CCC, a separate system of payment has been arranged.

Trademark Notice: Product or corporate names may be trademarks or registered trademarks, and are used only for identification and explanation without intent to infringe.

Library of Congress Cataloging-in-Publication Data

Names: El Naqa, Issam, editor.
Title: A guide to outcome modeling in radiotherapy and oncology : listening to the data / edited by Issam El Naqa.
Other titles: Series in medical physics and biomedical engineering.
Description: Boca Raton, FL : CRC Press, Taylor & Francis Group, [2018] | Series: Series in medical physics and biomedical engineering | Includes bibliographical references and index.
Identifiers: LCCN 2017036390 | ISBN 9781498768054 (hardback ; alk. paper) | ISBN 1498768059 (hardback ; alk. paper) | ISBN 9781498768061 (e-book) | ISBN 1498768067 (e-book)
Subjects: LCSH: Cancer--Radiotherapy--Data processing. | Oncology--Data processing.
Classification: LCC RC271.R3 G85 2018 | DDC 616.99/40642--dc23
LC record available at https://lccn.loc.gov/2017036390

Visit the Taylor & Francis Web site at
http://www.taylorandfrancis.com

and the CRC Press Web site at
http://www.crcpress.com

To my parents (Rola & Mustafa), brothers (Rami & Rabie), my wife Rana and daughters (Layla & Lamees).

Contents

SECTION II **Top-down Modeling Approaches**

CHAPTER 7 ▪ Analytical and mechanistic modeling   65

VITALI MOISEENKO, PhD, JIMM GRIMM, PhD, JAMES D. MURPHY, PhD, DAVID J. CARLSON, PhD, and ISSAM EL NAQA, PhD

CHAPTER 8 ▪ Data driven approaches I: conventional statistical inference
             methods, including linear and logistic regression   85

TIZIANA RANCATI and CLAUDIO FIORINO

CHAPTER 9 ▪ Data driven approaches II: Machine Learning 129

SARAH GULLIFORD, PHD

SECTION III Bottom-up Modeling Approaches

CHAPTER 10 ▪ Stochastic multi-scale modeling of biological effects induced by ionizing radiation   147

WERNER FRIEDLAND and PAVEL KUNDRÁT

CHAPTER 11 ▪ Multi-scale modeling approaches: application in chemo– and immuno–therapies   181

ISSAM EL NAQA

Section IV **Example Applications in Oncology**

Chapter 12 ▪ Outcome modeling in treatment planning 197

X. Sharon Qi, Mariana Guerrero, and X. Allen Li

MATTHEW J SCHIPPER, PHD

J. SCHUEMANN, A.L. MCNAMARA, and C. GRASSBERGER

Chapter 15 ▪ Modeling response to oncological surgery   259

J. Jesús Naveja, Leonardo Zapata-Fonseca, and Flavio F. Contreras-Torres

Chapter 16 ▪ Tools for the precision medicine era: developing highly adaptive and personalized treatment recommendations using SMARTs   283

Elizabeth F. Krakow and Erica E. M. Moodie

Foreword First edition

The goal of medicine is to improve the lives of our patients as much as possible while not causing harm. In my over 30 years as an attending radiation oncologist, I have tried to do this through reading, attending lectures, and thinking deeply about my successes and failures among the thousands of patients I have treated during my career. But how to organize all of this information in a way that is most beneficial to our patients? There is well-worn joke in medicine about the "physician" (pick your favorite specialty to tease) who has seen one patient, two patients, or three patients with a particular problem, and who spouts, with great authority, "In my experience", "In my series", or "In case after case after case". Likewise, the standard saying in medicine, "Three anecdotes? That's data!" Surely, in this era of "Big Data" we can do better for our patients than thinking of the last time we saw a patient who was similar to the one in front of us, and hoping to treat this new patient the same way, if our prior treatment was successful, or differently, if treatment was unsuccessful.

The answer is yes! This informative multi-author book edited by Dr. Issam El Naqa gives us insights into how we can organize our data by using models that can predict, and, thereby improve the outcome of treatment. Chapters are devoted to the input, such as incorporating clinical, imaging, dosimetric data, as well as modern radiation biology, and how to use those data to build models that will improve the physician's ability to balance efficacy and toxicity far more accurately than relying on pure clinical judgement. This book is coming at a particularly good time. The tsunami of data to which we now have access overwhelms the physician's ability for incorporation into a treatment plan, but is too valuable to ignore. At the University of Michigan, we are using these models to help us make treatment decisions. Even when we do not yet feel that the data are strong enough to build a robust model for predicting outcome, these models help to organize our approach and aim us toward the data we need to gather to improve our clinical decision making.

An unusual feature of this book is brilliance of the editor. Multi-author books are often uneven, with gaps and repetitions resulting from the lack of unifying force. Dr. El Naqa's remarkable breadth of knowledge gives this book a unity rarely found in such texts and greatly increases its value.

Big Data are here to stay... and will only get Bigger. This book can help us turn data into knowledge, and, perhaps, wisdom.

Theodore S. Lawrence, MD, PhD, FASTRO, FASCO
Isadore Lampe Professor and Chair
Department of Radiation Oncology
University of Michigan

Foreword First edition

Over the past 50 years, several books have been written on the subject of outcome modeling in radiation oncology. They illustrate a continuous progress in our understanding of effects of radiation treatment on cancer and normal tissues, and progress in methodology of analyzing ever increasing pool of clinical and experimental data. This book, edited by Dr. El Naqa, clearly indicates that the field of outcome modeling is changing rapidly and in a qualitative way. That is, several book chapters cover subject matters that did not even exist not long ago. For example, book chapters on machine learning, radiomics, or radiogenomics, present new and exciting areas of active research that may lead to improved outcomes of cancer treatment. Chapter 1 is an excellent introduction to the subject of outcome modeling, and the last Chapter 16 discusses the challenges and opportunities in the era of personalized medicine. The remaining 14 chapters are divided into four logical sections on sources of data, top-down modeling approaches, bottom-up modeling approaches, and example applications in oncology. All chapters are well-written and present in-depth discussion of state-of-the-art in the respective areas. The book is very well edited and should be on a must-read list for anyone who is interested in outcome modeling research. For someone who has been doing outcome modeling for over 30 years, the book also well illustrates one of my favourite quotes attributed to Albert Einstein "The more I learn, the more I realize how much I don't know." It rings very true for me personally, and this book will be a valuable addition to my library. However, it also applies to the entire field of outcome modeling. Finding or developing the "final" and "true" model of relevant outcomes in oncology is still as challenging, and as exciting, as ever before. All 300 hundred or so pages of this book well illustrate both, the challenges and the excitement of new approaches and possible solutions that may improve the outcomes of treatment for patients with cancer.

Andrzej Niemierko, Ph.D.
Associate Professor of Radiation Oncology
Institution Massachusetts General Hospital
Radiation Oncology Department
Harvard University, Massachusetts General Hospital 10
Emerson #122, Boston MA 02114
email: aniemierko@partners.org Phone 617/724-9527

Preface

OUTCOME modeling in radiation oncology and the broader oncology field plays an important part in understanding response, designing clinical trials, and personalizing patients' treatment. It attempts to provide an approximate representation of a very complex cancer environment within an individual patient or population of patients. This representation would allow for predicting response to known therapeutic interventions and further extrapolate this treatment effect beyond current standards. Modeling of outcomes has accompanied cancer treatments since its early days. Initially, this was based on cognitive understanding of the disease, accumulated experiences and simplistic understanding of observed effects clinically or experimentally. This was based primarily on using *in vitro* assays, which constituted the basis for developing *analytical* or sometimes refereed to as *mechanistic* models of response, which have been applied widely for predicting response and designing early clinical trials over the past century. However, due to the inherent complexity and heterogeneity of biological processes, these traditional modeling methods have fallen short of providing sufficient predictive power when applied prospectively to personalize treatment regimens. Therefore, more modern approaches are investigating more data-driven, advanced informatics, and systems engineering techniques that are able to integrate physical and biological information and to adapt intra-treatment changes and optimize post-treatment outcomes using *top-down* approaches based on big data complex system analyses or *bottom-up* techniques based on first principles and multi-scale approaches.

The original idea of this book was conceived by the editor's own (re)search for a comprehensive resource on outcome modeling and was encouraged by discussions with colleagues and trainees whom where looking into materials that could help them get their initial footing into this mystifying field of developing outcomes models; as part of their practice to understand already existing ones or to learn the field to become a new contributor with new fresh and unexplored ideas. The world of outcome modeling is too vast to be covered in single volume and every one working in the field of oncology whether directly or indirectly is a likely contributor. Who is in the field of oncology is not curious or is not tempted to predict what will happen with a treated patient? However, in this book we confine ourselves to the computational aspects that would use existing information (knowledge and/or data) as a resource and feed this resource into a suitable mathematical framework that could relate the stream of current information to a futuristic clinical endpoint such as local control, toxicity, or survival. A book of this nature can not be written by a single individual with specialized expertise, but it draws on the knowledge and wisdom of many experts from many domains including but not limited to clinical practitioners, biologists, physicists, computer scientists, statisticians, bioinformaticians, all are more than happy to call themselves *M*odelers who continue to learn from each other. Personally, I've been fortunate to gain my first steps into outcome modeling from Dr. Joseph O. Deasy and continue to explore its deep horizons with Dr. Randall T. Ten Haken, to whom I'm very grateful for their mentorship, guidance, and friendship. I would like also to acknowledge our clinical colleagues, particularly Dr. Jeffrey D. Bradley and the team at Washington University in St. Louis including Dr. Andrew Hope and Dr. Patricia Lindsay; and the team at McGill University led by Dr. Jan

Seuntjens and Dr. Carolyn Freeman; and the many colleagues and collaborators that I had the privilege to work with on many of the studies cited in this book. I would like also to acknowledge my trainees' invaluable contributions from them I learned possibly as much as I taught them. Ultimately, the goal of such a textbook is to provide a didactic path to outcome modeling in oncology as a promising tool as well as fulfilling a duty to serve our cancer patients and improve their care and quality of life through advanced education and research.

The book is an open invitation to the world of outcome modeling in oncology and its esteemed modelers, where basic science meets data science, art is embedded in Greek letters and Arabic numerals, skill is highlighted in computer coding, intuition is expressed in mathematical formulas, and prediction aims to give cancer patients and their caregivers new light and genuine hope in their treatment choices.

In a data-centric world, the first section of the book aims to provide the reader with overview information about the different and heterogeneous data sources and resources that could be all or partly used in model design and building. These data could be clinical or preclinical, and could span basic patient and/or treatment characteristics, to medical imaging and their features (radiomics) into more molecular biological markers (genomics, proteomics, metabolomics, etc.) constituting the so called pan-omics of oncology, generated over a treatment period that can span a few days to several weeks or months.

The second and the third sections of the book, aim to introduce the two main categories of outcome modeling according to their design methodology, namely; top-down and bottom approaches. The *top-down* approaches are presented in section II, starting with analytical (mechanistic models) through conventional statistical approaches into advanced machine learning techniques, which are currently the cornerstone of the Big Data era. Section III presents *bottom-up* approaches using first principles and multi-scale techniques with numerical simulations based on Monte Carlo with application in radiotherapy and non-Monte Carlo approaches with application in chemotherapy and immunotherapy. Description of softwares and tools for development and evaluation was intentionally embedded into most of these chapters to make it more accessible to learn hands-on by example and also experiment with many of the featured modeling approaches.

The fourth and last section of the books presents a diverse selection of common applications of outcome modeling from a wide variety of areas: treatment planning in radiotherapy, utility-based and biomarker applications, particle therapy modeling, oncological surgery, and design of adaptive and SMART clinical trials. These applications aim to serve as demonstrations of the potential power of outcome modeling in personalizing medicine in oncology, as well as samples to highlight technical issues and challenges that may arise during the processes of model building and evaluation when applied to real-life clinical problems.

Contributors

David J. Carlson, PhD
Department of Radiation Oncology,
Yale University, New Haven, CT, USA

James Coates, PhD
Department of Oncology,
University of Oxford, Oxford, UK

Flavio F. Contreras-Torres, PhD
Centro del Agua para America Latina y el
  Caribe.
Tecnologico de Monterrey, Monterrey,
  Mexico

Robert P. Coppes, PhD
Departments of Radiation Oncology and
  Cell Biology,
University Medical Center Groningen,
  Groningen, the Netherlands

Nicholas J. DeNunzio, MD, PhD
Department of Radiation Oncology,
University of Rochester, 601 Elmwood Ave.
  Box 647, Rochester, NY, USA

Issam El Naqa, PhD
Department of Radiation Oncology,
University of Michigan, Ann Arbor, MI,
  USA

Claudio Fiorino, PhD
Medical Physics Department,
San Raffaele Scientific Institute, Milan,
  Italy

Werner Friedland, PhD
Helmholtz Zentrum Munchen - German
  Research Center for Environmental
  Health (GmbH),
Institute of Radiation Protection,
  Neuherberg, Germany

C. Grassberger, PhD
Departement of Radiation Oncology,
Massachusetts General Hospital & Harvard
  Medical School, Boston, MA, USA

Jimm Grimm, PhD
Department of Radiation Oncology,
Johns Hopkins University, Baltimore, MD,
  USA

Mariana Guerrero, PhD
Department of Radiation Oncology,
University of Maryland School of Medicine,
  Baltimore, MD 21201, USA

Sarah Gulliford, PhD
The Institute of Cancer Research,
Sutton, Surrey SM2 5NG, UK

Sarah L. Kerns, PhD
Department of Radiation Oncology,
University of Rochester, Rochester, NY,
  USA

Elizabeth F. Krakow, PhD
Division of Clinical Research, Fred
  Hutchinson Cancer Research Center and
  Department of Medical Oncology,
University of Washington, Washington,
  USA

Pavel Kundrat, PhD
Helmholtz Zentrum Munchen - German
  Research Center for Environmental
  Health (GmbH),
Institute of Radiation Protection,
  Neuherberg, Germany

X. Allen Li, PhD
Department of Radiation Oncology,
Medical College of Wisconsin, Milwaukee,
  WI 53226, USA

Yi Luo, PhD
Department of Radiation Oncology,
University of Michigan, Ann Arbor, MI,
USA

A.L. McNamara, PhD
Department of Radiation Oncology,
Massachusetts General Hospital & Harvard
Medical School, Boston, MA, USA

Michael T Milano, MD, PhD
Department of Radiation Oncology,
University of Rochester, 601 Elmwood Ave.
Box 647, Rochester, NY, USA

Vitali Moiseenko, PhD
Department of Radiation Medicine and
Applied Sciences,
University of California San Diego, La
Jolla, CA, USA

Erica E. M. Moodie, PhD
Department of Epidemiology, Biostatistics
& Occupational Health,
McGill University, Quebec, Canada

James D. Murphy, PhD
Department of Radiation Medicine and
Applied Sciences,
University of California San Diego, La
Jolla, CA, USA

J. Jesus Naveja, PhD
PECEM. Facultad de Medicina,
Universidad Nacional Autonoma de Mexico,
Mexico City, Mexico

X. Sharon Qi, PhD
Department of Radiation Oncology,
UCLA school of Medicine, Los Angeles, CA
90095, USA

Tiziana Rancati, PhD
Prostate Cancer Program,
Fondazione IRCCS Istituto Nazionale dei
Tumori, Milan, Italy

Barry S. Rosenstein, PhD
Departments of Radiation Oncology and
Genetics and Genomic Sciences,
Icahn School of Medicine at Mount Sinai,
New York, NY, USA

Matthew J Schipper, PhD
Departments of Radiation Oncology and
Biostatistics,
University of Michigan, MI, USA

J. Schuemann, PhD
Departement of Radiation Oncology,
Massachusetts General Hospital & Harvard
Medical School, Boston, MA, USA

Corey Speers, MD PhD
Department of Radiation Oncology,
University of Michigan, Ann Arbor, MI,
USA

Randall K. Ten Haken, PhD
Department of Radiation Oncology,
University of Michigan, Ann Arbor, MI,
USA

Catharine M.L. West, PhD
Division of Cancer Sciences, Manchester
Academic Health Science Centre,
Christie Hospital NHS Trust,
University of Manchester, Manchester, UK

Peter van Luijk, PhD
Department of Radiation Oncology,
University Medical Center Groningen,
Groningen, the Netherlands

Leonardo Zapata-Fonseca, PhD
PECEM. Facultad de Medicina,
Universidad Nacional Autonoma de Mexico,
04600, Mexico City, Mexico

I

Multiple Sources of Data

CHAPTER 1

Introduction to data sources and outcome models

Issam El Naqa

Randall K. Ten Haken

CONTENTS

ABSTRACT

Outcome modeling plays a pivotal role in oncology. This includes understanding response to different therapeutic cancer agents, personalization of treatment and designing of future clinical trials. Although application of outcome models has accompanied oncology practices since its inception, it also evolved from simple hand calculations of doses based on experiences and simplified understanding of cancer behavior into more advanced computer simulation models driven by tremendous growth in patient-specific data and an acute desire to have more accurate predictions of response. This chapter reviews basic definitions used in computational modeling and main data resources, and provides an overview of the increasing role of outcome models in the field of oncology.

Keywords: outcome models, oncology, data, applications.

1.1 INTRODUCTION TO OUTCOME MODELING

M̲ODELING is a process of generating an abstract representation of a subject. In the context of outcomes, this process entails identifying the underlying relationship between observed clinical outcomes and the different types of data that generated them. Models in their inherent nature are approximations of relationships between the observations and the data, otherwise, they would be considered as laws of nature such as Newton's laws of motion, for instance. This notion was perhaps best articulated by George Box, a renowned statistician, who famously stated: "All models are wrong but some are useful" [1]. This statement should be kept as a reminder that models are often idealized simplifications of reality and may not be the "truth" sought after [2]. However, their usefulness should be judged depending on their ability to solve the problem at hand. In case of outcome models, this would be translated into the ability of such models to *sufficiently* predict a patient's response to a certain treatment regimen given known *relevant* characteristics of the patient, the disease, and the treatment itself. The question of *sufficiency* is tied to the desired predictive power of the model and its ability to differentiate between high and low risk patient outcomes. On the other hand, the *relevancy* of the model characteristics is related to the data types employed and the minimum number of variables needed in a model to make it predictive. This is succinctly summarized in another famous quote from Albert Einstein: "Everything should be made as simple as possible, but no simpler." This is an important principle in modeling known as *parsimony* or the *Occam's razor*, which advocates that among all the models representing available observations, the simplest one is the most plausible one [3]. However, this should be cautiously applied and is actually contradicted in the case of mechanistic or systems biology approaches, where it is implied that by incorporating more assumptions into a model (not necessarily for the purpose of improving the predictive power) will allow the learning and extracting of more information from the data at the possible expense of risking more uncertainty [4].

The process of outcome modeling, following early attempts to systematically apply therapeutic agents at the end of the 19th century, has continuously evolved from trial and error approaches with back of the envelope hand calculations into the application of advanced statistical and mathematical techniques. Modeling has long been recognized as a useful tool to synthesize existing knowledge, test new treatment scenarios that may not have been observed in clinical trials, and compare different treatment regimens. This role has been further invigorated by advances in patient-specific information, biotechnology, imaging, and computational capabilities. In the era of big data and personalized or precision medicine, it is expected that outcome modeling will be at the center of these initiatives paving the way to better understanding of cancer biology and its response to treatment. It is currently spearheading the process of development of appropriate clinical decision support systems for cancer management across the spectrum of available treatment modalities to achieve an improved quality of care for cancer patients [5].

1.2 MODEL DEFINITION

The Merriam-Webster Dictionary defines a model "as a system of postulates, data, and inferences presented as a mathematical description of an entity or state of affairs" while the Oxford Dictionary defines it as "a simplified description, especially a mathematical one, of a system or process, to assist calculations and predictions." In the context of computational modeling, the National Institute of Biomedical Imaging and Bioengineering (NIBIB) defines it as the use of mathematics, physics and computer science to study the behavior of complex systems by computer simulation to make predictions about what will happen in the real system that is being studied in response to changing conditions. For the purpose of outcome modeling in oncology, we modify this definition as follows:

Definition 1.1 *Outcome modeling is the general process of using mathematics, statistics, physics, biology, and computer science to characterize the behavior of tissue response to a treatment (radiation, drug, surgery, etc.) such as tumor eradication (control) or side effects (toxicities) via approximate computer simulations in order to make predictions about what will happen in the patient under changing therapy conditions.*

Outcome models can help to understand the underlying biology of cancer response to treatment. They can approximate the treatment environment and be used to develop decision support tools for oncologists that can provide guidance for treatment or design of future clinical trials as depicted in Figure 1.1.

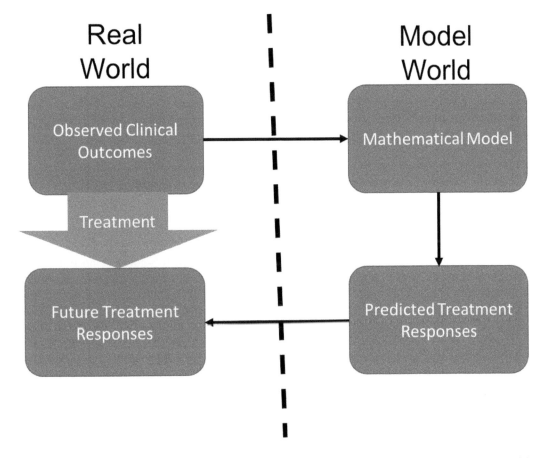

Figure 1.1 The process of outcome modeling as an approximation of the real world by mathematical constructs that would allow future prediction of response and comparison with future treatment responses.

1.3 TYPES OF OUTCOME MODELS

Outcome models can be generally categorized: (1) based on their objective into predictive or prognostic; (2) based on their methodology into top-down or bottom-up approaches; and (3) based on their assumptions into analytical and data-driven models.

1.3.1 Prognostic versus predictive models

A model could be categorized into being a predictive or prognostic model depending on whether the treatment parameters are considered as part of the model construction or not, respectively. A *prognostic model* can inform about a patient's likely cancer outcome (*e.g.*, disease recurrence, disease progression, death) independent of the treatment received. For instance, an image-based nomogram using Cox regression was developed for locally advanced cervical cancer from tumor metabolic activities, tumor volume, and highest level of lymph node (LN) involvement [6]. The model outperformed classical FIGO staging in terms of predicting different clinical endpoints (recurrence-free survival (RFS), disease-specific survival (DSS), and overall survival (OS)). A *predictive model* informs about the likely benefit from a therapeutic intervention. For instance, the Quantitative Analyses of Normal Tissue Effects in the Clinic (QUANTEC) effort presented several predictive models of radiation-induced normal tissue toxicities with the purpose of limiting the amount of harmful radiation exposure received by uninvolved tissue under different treatment scenarios according to these model predictions [7]. Moreover, the website: PredictCancer.org lists several models for predicting response to chemoradiation in lung, rectum, head and neck cancers [8].

Outcome modeling tends to lend itself primarily to the predictive modeling category, where the main focus is to select or modify treatment regimens according to outcome model predictions.

1.3.2 Top-down versus bottom-up models

Outcome modeling schemes in oncology could be divided into top-down or bottom-up approaches as depicted in Figure 1.2 [9]. *Top-down* approaches start from the observed clinical outcome and attempt to identify the relevant variables that could explain the phenomena. They follow from generalizing the notion of integrative statistical modeling where treatment outcomes can be optimized through using complex data analyses (e.g., machine learning methods) [10]. For instance, different machine learning techniques were contrasted for predicting tumor response post-radiotherapy in lung cancer [11]. *Bottom-up* approaches start from first principles of physics, chemistry, and biology to model cellular damage temporally and spatially up to the observed clinical phenomena (*e.g.*, multi-scale modeling) [12]. Typically, bottom-up approaches would apply advanced numerical methods such as Monte-Carlo (MC) to estimate the molecular spectrum of radiation or drug damage in clustered and non-clustered DNA lesions ($Gbp^{-1}Gy^{-1}$). For instance, the temporal and spatial evolution of the effects from ionizing radiation can be divided into three phases: physical, chemical, and biological. Different available MC codes (see Chapter 10) aim to emulate these phases along the molecular and cellular scales to varying extents [13].

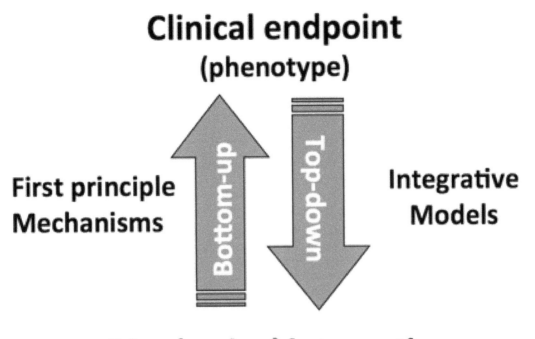

Figure 1.2 Outcomes modeling approaches in oncology could be divided into: top-down (starting from the observed clinical outcome and attempting to identify the relevant variables that could explain the phenomena) or bottom-up (starting from basic principles to the observed clinical outcome in a multi-scale fashion [9].

1.3.3 Analytical versus data-driven models

Outcome models could be divided based on utilizing existing knowledge into analytical and data-driven models, respectively. *Analytical* models, also known as mechanistic or phenomenological modeling approaches, attempt to predict outcomes or treatment effects based on reductionist simplifications or biological effects. They include basic mechanisms of action that agree with experimental results and are thus thought to be partially theory-based. For instance, in radiotherapy, the linear-quadratic (LQ) model is a prime example of this category that has witnessed significant success in radiation oncology and has been frequently used in designing clinical trials for several decades [14]. The LQ model can be interpreted in terms of DNA track structures, which is represented by two exponential components: a linear component resulting from single-track events and a quadratic component arising from two-track events as discussed in Chapter 10.

Data-driven modeling, also known as statistical modeling techniques, does not rely only on mechanistic principles related to disease manifestation but rather on empirical combinations of relevant variables. In this context, the observed treatment outcome is considered as the result of functional mapping of several input variables through a statistical learning process that aims to estimate dependencies from data [15]. Among the most commonly used mathematical methods for outcome modeling are functions with sigmoidal shapes (S-curved) such as logistic regression, which also have nice numerical stability properties. The results of this type of approach are expressed in the model parameters, which are chosen in a stepwise fashion to define the abscissa of the regression model. However, it is the user's responsibility to determine whether interaction terms or higher order variables should be added. Penalty techniques based on ridge (L2-norm) or Lasso (L1-norm) methods could aid in the process by eliminating least relevant variables and imposing sparsity conditions as discussed in Chapter 8. An alternative solution to ameliorate this problem is offered by applying machine learning methods.

Machine learning techniques are a class of artificial intelligence (e.g., neural networks, support vector machines, deep neural networks, decision trees, random forests, etc.), which are able to emulate human intelligence by learning the surrounding environment from the given input data and can detect nonlinear complex patterns in such data. Based on the human-machine interaction, there are two common types of learning: supervised and unsupervised. Supervised learning is used when the endpoints of the treatments such as tumor control or toxicity grades are known; these endpoints are provided by experienced oncologists following institutional or National Cancer Institute (NCI) criteria, for instance. Supervised learning is the most commonly used learning method in outcomes modeling as further discussed in Chapter 9 [16].

It is worth mentioning that methods based on Bayesian statistics can act as a bridge between pure analytical and data-driven approaches by incorporating prior knowledge into the modeling process.

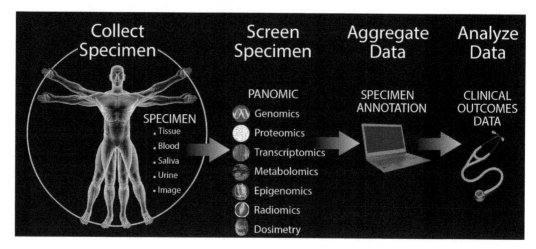

Figure 1.3 The human body is a valuable resource for varying solid and fluid types of specimens, which can yield different omic (genomics, proteomics, transcriptomics, metabolomics, radiomics) predictive biomarkers, in addition to dosimetric and clinical factors used in radiotherapy that would undergo major processes of annotation, curation, and preparation before being applied into outcome modeling of treatment outcomes (*e.g.*, tumor response, toxicity) [17].

1.4   TYPES OF DATA USED IN OUTCOME MODELS

There are numerous data types that may impact response to therapy starting from traditional clinical data (*e.g.*, patient demographics, cancer stage, tumor volume, histology, co-morbidities, weight loss, adjuvant therapies, etc.), and therapy related data (*e.g.*, total dose, dose amount, scheduling, etc). Data types include biological markers (genetic variations, gene and protein expressions, etc.) and imaging-based metrics (intensity, volume, heterogeneity metrics). In recent years there have been exponential growth in patient-specific information generated from biotechnology high-throughput data (genomics, proteomics, transcriptomics, metabolomics) and quantitative imaging (radiomics) data together forming this *Big data* or *panOmics* [18] lake of useful knowledge for outcome modeling as discussed in subsequent chapters and depicted in Figure 1.3 [17].

1.5   THE FIVE STEPS TOWARDS BUILDING AN OUTCOME MODEL

The process of successful model would require multiple steps. This could be summarized in five main steps of Exploration, Reconstruction, Explanation, Validation, and Utilization summarized in the acronym EREVU. The process is partly science with well established theoretically proven procedures and partly art relying on proper intuition and common sense heuristics as will be become clearer throughout the book. This pyramid of success for outcome modeling process is depicted in Figure 1.4 and explained briefly below.

1. **Exploration.** The tremendous growth of patient specific information and plateaued cancer incidences, constitutes a challenge known as the *Pan- versus p-OMICs dilemma*, where the relationship between a large number of variables needs to be explained from a limited patient sample sizes [9]. The literature is rich with different statistical and information theoretic techniques for exploratory data analysis as originally presented by Tukey [19] and later expanded by Burnham and Anderson [2]. These methods include statistical resampling such as cross-validation and bootstrapping [20], model selection techniques based on information criteria (Akaiki information criteria (AIC), Bayesian information criteria (BIC), etc)[2], and dimensionality reduction and variable selection techniques (principal component analysis (PCA), linear discriminant analysis (LDA), etc) [21]. Many of these methods will be described in subsequent chapters on outcome model development.

2. **Reconstruction.** Model building depends on the nature of the selected model type (*cf.* section 1.3). For instance, in the case of data-driven approaches using statistical learning methods, the model reconstruction process involves selecting the most relevant variables and the appropriate model parameters for building the optimal parsimonious model that could perform well on out-of-sample data [22]. Two primary challenges exist when it comes to constructing robust models and attempting to apply them to a wider population: *under-fitting*, whereby a model is not robust enough to apply to a wider population, and *over-fitting*, whereby the true signal is mistaken for noise and is being fit instead. In addition to traditional multiple comparison corrections such as Bonferroni adjustments or permutation testing [23], more advanced statistical techniques, can be also used to determine if the model over-fit or under-fit a given signal; as discussed, methods to accomplish such tasks may include statistical resampling (cross-validation or bootstrapping) or information theory approaches [24].

3. **Explanation.** The ability to interpret a model in an intuitive manner is an essential

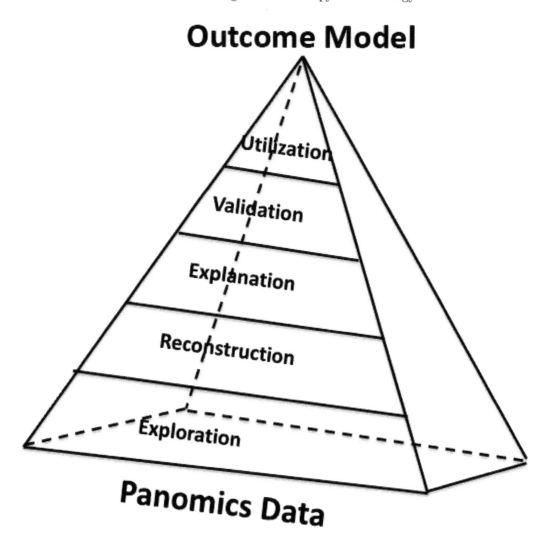

Figure 1.4 The outcome modeling pyramid (EREVU).

component for understanding its underlying assumptions and for its acceptance by the clinical community. This is particularly true in the case of data-driven models. For instance, correlative models based on simple linear or logistic regression are more widely accepted and clinically used compared to machine learning methods despite the proven superiority of the latter due to the black box stigma. However, this trend is changing given the interest in precision medicine and the rise of big data analytics in recent years [25]. Methods based on generative approaches such as Bayesian networks, which use graph-based techniques, can act as an intermediary between the simplicity of linear/logistic regression methods and the complexity of more advanced machine learning based on deep learning, for instance.

4. **Validation.** The gold standard for validation remains independent evaluation of model performance on an unseen cohort. However, independent evaluation is not always practical due to the lack of availability of such data until close to clinical

implementation and is thus of limited use in the early stages of model building. The Transparent Reporting of a multivariable prediction model for Individual Prognosis Or Diagnosis (TRIPOD) recommendation developed a set of 22 items checklist for the reporting of studies developing, validating, or updating a prediction model [26]. The TRIPOD) recommendations emphasize the importance of internal validation as a necessary part of model development (TRIPOD level 2 validation). For model evaluation, it suggests reporting performance measures (with confidence intervals) of the proposed prediction model. Independent evaluation of the model prior- (TRIPOD level 3) or post-publication (TRIPOD level 4) is highly encouraged and are considered the highest desired validation levels.

5. **Utilization.** Once an outcome model has been developed and validated, it can become an important component of a more informed clinical decision support system or could be employed as a new tool for developing and designing more useful clinical trials. However, any implementation or use of an outcome model needs to be consistent with its original assumptions and the results should be interpreted in terms of its reported uncertainty to achieve its goals and avoid ill-informed predictions.

1.6 CONCLUSIONS

Outcome modeling will play a prominent role in oncology in the era of precision medicine and big data analytics. Its role in supporting clinical decisions and designing clinical trials will continue to thrive and invoke recent progress in computational data analytics. Better understanding of the outcomes model development process from early data gathering to prediction applications is important for successful implementation and for achieving its promised reward in improving patients' care and reducing possible waste in health care costs. Moreover, the development of successful outcome models would also require the collaboration among all stake holders: modelers, physicians, statisticians, bioinformaticians, physicists, biologists, and computer scientists. These concepts are represented throughout the rest of this book with the ultimate goal of reducing ineffective trial and error decision making while simultaneously achieving better outcomes for all cancer patients.

CHAPTER 2

Clinical data in outcome models

Nicholas J. DeNunzio, MD, PhD

Sarah L. Kerns, PhD

Michael T Milano, MD, PhD

CONTENTS

ABSTRACT

Clinical data has historically represented the initial resource for outcome modeling in oncology. In this chapter, we review the impact of this resource on modeling outcomes and we specifically draw on different examples from radiotherapy modeling of radiation-induced toxicities in cancer patients. We present several representative examples starting by the seminal work of Emami and Lyman and how essential guidelines for radiotherapy treatment evolved from round table opinion-based models into becoming more data-centric recommendations in the modern radiotherapy era.

Keywords: clinical data, outcome modeling, radiotherapy, QUANTEC.

2.1 INTRODUCTION

For several decades, there has been interest in better predicting tumor control probability (TCP) and normal tissue complication probability (NTCP) risks after therapeutic radiation, with significant effort focused on predictive dosimetric parameters for the latter. Several landmark efforts have worked towards this end. The classic Emami and Lyman paper [27] was published in 1991, prior to the adoption of 3D conformal radiation; from decades-worth of published data and clinical expertise of colleagues, the authors published a comprehensive table with information on whole and partial organ tolerance doses for a wide variety of organs. During the 1990s and 2000s, numerous studies analyzed associations between dose/volume parameters and normal tissue complication risks. Much of this was reviewed by Milano et al in 2007 and 2008 [28, 29]. In 2010, the QUANTEC (quantitative analysis of normal tissue effects in the clinic) review articles, representing a task force jointly supported by the American Society of Radiation Oncology (ASTRO) and the American Association of Physicists in Medicine (AAPM), summarized the available data on dose-volume metrics predictive of NTCP risks, in an effort to update and refine the estimates provided by Emami [30, 31]. A similar joint effort, hypofractionated radiotherapy tissue effects in the clinic (HYTEC), has focused on dosimetric measures impacting NTCP as well as TCP after hypofractionated stereotactic/image-guided radiotherapy. Another ongoing endeavor, pediatric normal tissue effects in the clinic PENTEC, will focus on NTCP risks in the pediatric population.

These aforementioned cooperative efforts, and all of the published data that were incorporated into the published reviews and pooled analyses, elucidate the importance of dose-volume metrics in predicting clinical outcomes, whether NTCP or TCP. Clinicians have become reliant on published dosimetric guidelines to help inform their treatment planning decisions. In fact, variation in normal tissue radiosensitivity has been attributed to a combination of deterministic and stochastic factors. These include treatment type (surgery, chemotherapy, etc.), radiotherapy delivery parameters, and non-genetic patient factors such as age, smoking status, and diabetes [32]. Nevertheless, our understanding of NTCP and TCP remains superficial, even with robust and reproducible models. For example with NTCP predictions, there are several limitations; including: (1) the recommended dosimetric constraints generally do not account for regional variation in organs (*i.e.*, susceptibility of different parts of an organ); (2) differences in how institutions define targets and normal tissues, as well as dosimetric planning algorithms, can impact the calculated dose distribution and therefore the calculated probability risks; (3) reliance on single dosimetric points, generally from a 2-dimensional dose volume histogram, may cloud better understanding of the true 3-dimensional dose distribution [33]. Despite these limitations, commonly used published evidence-based dose-volume metrics represent practical constraints that are useful to clinicians in day-to-day practice. The tables from Emami, QUANTEC, cooperative groups studies, and other efforts (such as AAPM TG 101) [34] can often be found hanging on the walls in offices and dosimetry/planning rooms, providing clinicians with concrete guidelines to help inform them of anticipated outcomes after radiotherapy. However, missing from these tables are clinical and biological factors potentially affecting NTCP and TCP modeling, factors that feasibly weigh into treatment planning decisions. This largely reflects a paucity of data, data inconsistencies, and the complexities associated with incorporating multifaceted patient- and tumor-related factors into dosimetric guidelines.

Radiotherapy interacts, on the cellular level, with DNA and cell membranes, eliciting a cascade of biological reactions. These include the classically termed *acute effects* occurring in the midst of or shortly after radiation, likely resulting from cellular depopulation and inflammation, and *late effects* occurring months after radiotherapy, likely resulting from fibrosis and microvascular devascularization. Particularly severe acute reactions may ultimately result in "consequential late effects" perhaps due to stem cell depletion. The extent of injury in normal tissue is likely dependent on host (patient) factors, the underlying milieu of the irradiated tissues, and environmental factors, as well as interactions between these factors. Similarly, the extent of response to target/tumor cells (*i.e.*, TCP) is likely dependent on host and tumor factors. However, the extent that any individual biologic factor contributes to radiation response and any potential interplay with other variables is not fully understood. These interactions are summarized in Figure 2.1.

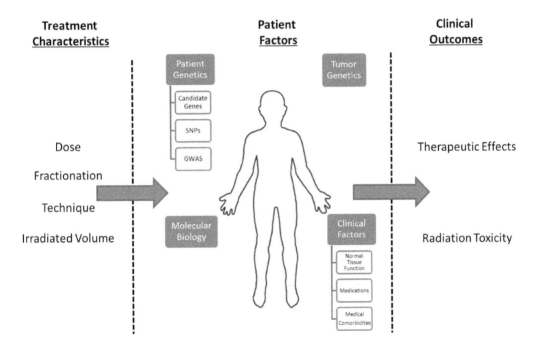

Figure 2.1 The radiotherapy interaction chain from treatment to outcomes.

Simple factors (perhaps with complex etiologies) potentially affecting outcomes include gender, race and age. Pre-existing medical conditions may also influence normal tissue radiotherapy response (*e.g.*, different patients with different levels of tissue/organ reserve are likely to respond differently to radiation). These underlying medical conditions are not necessarily binary states (*i.e.*, with or without disease state), but rather most likely a continuum in the severity and nature of the underlying comorbidities. Biomarkers that are quantitatively or qualitatively correlated with NTCP and TCP may be considered a surrogate for an underlying biologic state potentially affecting outcomes, although the underlying biologic mechanism of action may or may not be clear.

There are also relatively rare known genetic factors (*e.g.*, ataxia telangiectasia mutation) that predispose patients to greater risks of radiation toxicity, and relatively poorly characterized genetic susceptibility to normal tissue response (again, likely representing a spectrum). Perhaps there are other unknown or poorly characterized host/genetic factors affecting NTCP and TCP. This chapter will focus on some of the common clinical factors that may affect NTCP risks in patients undergoing stereotactic body radiation therapy (SBRT) in concert with dosimetric parameters and an ever evolving understanding of the molecular underpinnings of tumor and normal tissue responses. Before we specifically address biologic factors impacting toxicity after SBRT, we will summarize (1) published data on the effect of collagen vascular disease (CVD) on toxicity risks, which represents a classic model of host biology affecting NTCP; and (2) the use of genetic markers to predict toxicity risks, which represents a burgeoning field with great promise. While specific data on CVD or genetic markers predicting NTCP after SBRT are lacking, we present the data here as concepts that would potentially be relevant to SBRT patients.

2.2   COLLAGEN VASCULAR DISEASE

CVD is a general term that was initially used to refer to a variety of pathologic entities mainly rooted in the connective and vascular tissues. However, we now have more specific designations for these disease processes, such as rheumatoid arthritis (RA), systemic lupus erythematosus (SLE), scleroderma, dermatomyositis, ankylosing spondylitis, polyarteritis nodosa, psoriatic arthritis, and undifferentiated systemic rheumatic disease. Traditionally, it has been thought that the presence of CVD may predispose a patient to increased toxicity after receiving radiation therapy. This hypothesis was largely driven by early low-level evidence in the form of case reports documenting severe instances of toxicity [35], rather than reflecting the more general spectrum of CVD entities. In addition, radiation techniques have changed substantially over the decades during which many of these studies have taken place.

Most of what is published on CVD is from patients treated with conventional radiation. From a retrospective analysis from Massachusetts General Hospital, examining acute and late effects of radiotherapy in 209 patients with CVD (62% with RA), the authors concluded that RA was not associated with an increased risk of late toxicity but that non-RA CVD was [36].

The associations of scleroderma and active SLE with debilitating late fibrosis have been recognized for decades, warranting caution in patients with breast and head and neck cancers for whom the skin dose is relatively high. However, overly cautious withholding of radiation for patients with SLE is a concern [37]. Clearly, better understanding what drives individual responses in acute and chronic toxicity, and therefore being able to assess a patient's risk prior to treatment, is of interest and likely to be of value.

A small cadre of retrospective studies published in the late 2000s from the Mayo Clinic and the University of Michigan addressed these questions using data collected from the mid-1980s to the mid-2000s. These analyses more closely examined disease entities outside of RA, notably SLE [38] and scleroderma [39], or an aggregate overview of multiple CVD disease entities [40, 41]. Their results suggest that normal tissues other than skin and subcutaneous tissue do not appear to be as susceptible to severe late complications in patients with scleroderma or SLE. In addition, there appears to be some dose-volume threshold for skin and subcutaneous tissue below which late toxicity is unlikely to occur, somewhere between the relatively low exposure in those patients treated in the Mayo Clinic studies and high-dose exposure in patients with breast and head and neck malignancies. However, this is not well-characterized in published studies. Perhaps unsurprisingly, acute toxicities were associated more with curative-intent treatment in patients with SLE [38], likely reflective of higher prescribed doses with curative-intent treatment. These findings were generally not observed in the companion scleroderma study (except for increased acute toxicity associated with curative treatment), although higher dose cutoffs were used in their analyses. To better understand how CVD severity, as reflected by the number of organ systems involved, may impact risk of severe chronic toxicity from radiation therapy, the SLE (21 consecutive patients) and scleroderma (20 consecutive patients) data sets from the Mayo Clinic were analyzed in aggregate [40]. Although there was no statistically significant association between CVD severity and risk of chronic complications of radiotherapy when individually considering SLE or scleroderma, the aggregate analysis showed a significantly higher rate of toxicity in patients with high-severity CVD (upper 50% of organ systems involved) versus those with low-severity disease ($p = 0.006$).

The effect of systemic medication use for CVD on radiation-induced toxicities was examined as part of a 3:1 match-control comparison study by Lin *et al.* [41], of 73 patients with CVD (RA, SLE, scleroderma, dermatomyositis/polymyositis, ankylosing spondylitis, polymyalgia rheumatic, Wegener granulomatosis, and mixed connective tissue disorders/other were included). Medication classes studied were varied and included non-steroidal anti-inflammatory drugs, statins, calcium channel blockers, antimalarials, oral cytotoxics, and infusional chemotherapies. Of these, infusional chemotherapy accounted for significantly increased risks of severe acute and chronic toxicities; oral cytotoxic agents significantly increased the risk of acute toxicity.

Data on CVD's potential impact on toxicity risks after SBRT are scant. In a study of chest wall toxicity risks among 140 patients (with 146 lesions), connective tissue diseases were significantly (p=0.036 and 0.008 on univariate and multivariate analyses, respectively) associated with chest wall pain after SBRT, although only 3% of patients had connective tissue disease [42]. To our knowledge, CVD has not been well-addressed in other studies.

Despite the limitations of the published data on CVD's effect on radiation-related toxicity risk, including small study sizes, a wide variety of disease phenotypes, pathophysiologic mechanisms, and medications for these diseases, and despite more modern evidence showing tolerability of radiation in many CVD subgroups, CVD should be accounted for in modeling toxicity risks after SBRT, particularly with such high prescribed fractional doses. Ideally, this would be accompanied by approaches to better understand the biology behind normal tissue responses to radiation, and how CVD may affect that risk.

2.3   GENETIC STUDIES

In recent years, there has been a burgeoning effort to study and understand how genetic predispositions might predict radiation toxicity. As evidenced by one early study examining skin telangiectasia after radiotherapy, 81-90% of the patient-to-patient variation in severity of post-radiotherapy telangiectasias could be explained by patient-related factors [43]. Though genetic syndromes exist that are known to predispose patients to increased radiosensitivity (*e.g.*, ataxia telangiectasia, Nijmegen breakage syndrome, LIG4 syndrome [44], these are rare.

In an effort to better understand the genetic basis of radiation response, radiobiologic studies have been pursued on a variety of levels. These include investigations of: molecular signaling pathways thought to be responsible for affecting general mechanisms underlying acute and late sequela of radiotherapy; cellular radiosensitivity assays; and broader genome-based searches to identify candidate genes, groups of genes, genetic variants, and/or epigenetic modifications that may dictate how sensitive or resistant a given patient's tissues are to radiation-induced damage. Indeed, radiosensitivity is considered to be a phenotype that is regulated by many different genes, which yields significant complexity in what determines the response observed in any given person. Given its broader scope and the emphasis of this chapter, we will focus on genetic analyses, also called radiogenomics.

Recent studies have focused on correlates of single nucleotide polymorphisms (SNPs) to toxicity risks. SNPs are genome sequence variations that are present in a non-trivial portion (generally $\geq 1\%$) of the population. If residing in amino acid-coding or regulatory regions of proteins involved in the repair of tissue or DNA in response to radiation-induced damage, SNPs may (in part) account for the varied responses observed across individual patients. There may also be SNPs in genes that control normal tissue functions and determine the so-called functional reserve of tissues. While these would not necessarily be radiation-response genes, they may be important for susceptibility to radiation damage.

The initial approach to identifying SNPs relevant to radiobiologic processes involved candidate gene studies, which rely on previously established information about molecular pathways involved in the radiation response. Advantages to this approach include a focused scope of inquiry and relatively low cost. However, implicit in this strategy is the assumption that the components of the biological system under scrutiny are completely understood. Unfortunately, this does not seem to be the case in so far as normal tissue toxicity from exposure to radiotherapy is concerned. Early studies were also limited in the number of SNPs analyzed within a given gene, and in most studies, only one or a few of the hundreds to thousands of SNPs within a given gene were assayed. The limitations of early candidate gene studies are illustrated by the fact that findings from individual SNP studies could not be independently validated using a larger dataset of patients who underwent treatment for breast or prostate cancers, implying that any one SNP may not be that clinically meaningful and that a more comprehensive assessment of the genome may be required to enumerate multiple subtle contributions to normal tissue responses [45]. Despite these limitations of the candidate gene approach, there have been some notable successes, with SNPs in TNFα [46], HSBP1 [47, 48], TXNRD2 [49], XRCC1 [50], and ATM [51] showing replicated associations with radiation toxicity.

More recently, however, genome-wide association studies (GWAS) have been pursued due to the shortcomings noted above. These do not require advanced or presumed knowledge of the molecular players in a biological system and are able to examine a much larger pool of SNPs with varied allelic frequencies, penetrance, and population genetics dynamics. These characteristics increase sample size requirements, however, and make statistically significant associations difficult to establish. Consequently, the resources required to conduct these studies are more intensive, necessitating coordinated efforts in tissue banking, data generation (facilitated by relatively recent decreased costs in genome sequencing), and data sharing, such as that provided through the Radiogenomics Consortium, which is supported by the NCI/NIH (http://epi.grants.cancer.gov/radiogenomics). To our knowledge, all of the published data on SNPs, including GWAS, are from patients treated with conventional or hypofractionated radiotherapy protocols. We present the concept here in anticipation that these approaches will eventually be applied to SBRT.

A seminal GWAS examined the genetic basis of erectile dysfunction (ED) after radiotherapy in 79 patients with prostate cancer [52] in whom approximately 500,000 SNPs were analyzed. Unexpectedly, a SNP in the follicle stimulating hormone receptor gene was significantly associated with development of ED post-treatment. Four additional SNPs showed a trend towards association with ED. The risk allele for each of these top SNPs is common in populations of African ancestry but quite rare in those of European ancestry, underscoring the importance of clearly identifying the study population in GWAS. While this initial study was carried out in a single patient set of very small sample size and requires validation, several other radiogenomic GWAS have been published in larger patient populations following either a multi-cohort staged design or individual patient data meta-analysis approach where results in a single patient set are validated in additional patient sets. These studies have successfully identified risk SNPs in TANC1 [53], KDM3B, and DNAH5 [54]. Importantly, a radiogenomics GWAS of 3,588 patients with breast or prostate cancer provided evidence that many more common SNPs are associated with radiotherapy toxicity and remain to be discovered through additional, larger studies [55].

In addition to strictly genome-based analyses, a related but largely distinct line of inquiry (also coined *radiogenomics* or *radiomics*) correlates the results of imaging studies with individual clinical outcomes. Models have been developed to describe associations between imaging features and genetic aberrations in renal cell carcinoma, to identify the luminal B molecular subtype of breast cancer, and to evaluate genetic proxies portending survival in patients with non-small cell lung cancer (NSCLC) [56]. A study of 26 patients with NSCLC treated with resection (and therefore may not apply to unresectable patients treated with radiation) constructed a predictive model of survival first by correlating preoperative imaging and gene expression data and then, using a second publicly available gene data set with survival data, to see how well imaging findings predict survival. With 59-83% accuracy of genes predicted by imaging features, there is promise for stronger models to evaluate a patient's disease and potential to optimize therapeutic modalities.

Ultimately, it is the hope that better understanding the molecular mechanisms underlying the response to radiation-induced tissue damage and genetic variations that affect the observed phenotype in a given patient will allow for means to optimize treatment (*e.g.*, radiation dose and dose fractionation, radiation modality, use of radiation modifiers) to limit toxicity and maximize tumor control and patient survival [57].

2.4 BIOLOGICAL FACTORS IMPACTING TOXICITY AFTER SBRT

SBRT is defined by the American Society for Radiation Oncology (ASTRO) and American College of Radiology (ACR) as an external beam radiation therapy method that very precisely delivers a high dose of radiation to an extracranial target. SBRT is typically a complete course of therapy delivered in 1 to 5 sessions (fractions) (American College of Radiology 2014), and has been a widely adopted treatment modality by radiation oncologists with an estimated majority of physicians using SBRT when surveyed in 2010 and a majority of non-users planning to adopt this technology [58].

Among physicians using SBRT, the most common anatomic sites treated were the lung (89%), spine (68%), and liver (55%), with the lung experiencing the greatest adoption rate increase up through that time. Its popularity can be attributed to successful ablation of target lesions (often on the order of 80% or higher at one year regardless of anatomic site being studied), focal treatment with more limited toxicity to peripheral tissues receiving low dose radiation, short treatment durations, and applicability to patients who are non-operative candidates. However, treatment of lesions in the lung and liver come with a risk of toxicity to the chest wall along with the primary tissues themselves.

2.4.1 Chest wall toxicity after SBRT

Chest wall toxicity after SBRT can profoundly impact a patient's quality of life. Often, but not exclusively, defined as chest wall pain and/or rib fracture, chest wall toxicity has been difficult to assess in part because of the heterogeneity of signs or symptoms or anatomic region to which the term refers across available studies. There is a growing body of literature describing dosimetric predictors of chest wall toxicity [59, 60, 61, 62, 63]. There is some variability across studies, due in part to differences in definition of the chest wall (*e.g.*, ribs vs. 2 cm rind of tissue outside of the lung) as well as outcome (*e.g.*, pain, rib fracture).

Furthermore, treatment parameters do not provide for a comprehensive risk assessment model. Several authors have described biologic factors affecting chest wall toxicity risks. Examination of tumor locations and geometries and patient-related factors, many of which are particularly relevant to today's US population, have shown that BMI [64], tumor-chest wall distance [65, 66], patient age [67], female sex [67, 68], and osteoporosis in patients with lesions adjacent ($< 5mm$) to the chest wall [69] elevate risk of chest wall toxicity. Furthermore, diabetes mellitus has been found to contribute to risk of chest wall toxicity in patients with $BMI > 29$ [64]. Taken together, development of the ideal comprehensive risk assessment model for chest wall toxicity would integrate both aspects of treatment: treatment delivery and personal medical context (ie, medical comorbidities and/or genetic predispositions).

2.4.2 Radiation-induced lung toxicity (RILT) after SBRT

Radiation-induced lung toxicity (RILT), which includes radiation pneumonitis (RP) and pulmonary fibrosis, can range in severity from mild to life-threatening to lethal. RP can be either an acute or late sequela of radiotherapy, generally occurring months after radiation, but possibly as soon as one week after completing treatment. RP after SBRT seems to occur at a median of approximately 3-6 months after SBRT [70, 71, 72, 73]. Studies aimed at understanding risk of developing RP have focused to a large extent on dosimetric factors, particularly dose-volume parameters like V_x (where x is the dose received by some volume of lung tissue) and mean lung dose (MLD). However, much like the study of chest wall toxicity in response to radiation exposure, the heterogeneity of these study parameters (*e.g.*, $V_5, V_{10}, V_{15}, V_{20}, V_{30}, V_{65}$) and degree/grade of toxicity outcomes can be substantial, thereby complicating higher-level analysis of clinical risks.

Patient factors predisposing to development of RP have come under scrutiny as well, including (but not limited to) age, patient gender, smoking status (current vs former vs never), disease distribution within the lung, and the use and timing of chemotherapy in relation to receiving radiation. Pertaining to the relationship between SBRT and RILT, a pooled analysis of 88 studies was recently completed [74]. Thirteen of these studies analyzed smoking history, which did not appear to influence RILT. In the same analysis, gender (analyzed in 73 studies) and pathologic subtype (adenocarcinoma vs. others, analyzed in 53 studies) were not significant factors, while older age was significantly ($p = 0.045$) associated with a higher rate of grade 2+ RILT as were mean lung dose, V_{20}, and maximum tumor diameter.

In addition to the above patient factors, comorbid pulmonary disease, such as interstitial lung disease (ILD), may predispose to development of RP after SBRT. ILD broadly refers to inflammation in the lung parenchyma between alveoli. Left unchecked, it can lead to lung fibrosis and tissue destruction. As such, understanding how radiation may affect outcomes in patients with ILD, including whether this predisposes to RP development, has been of interest. In this context, a retrospective study examined patients with stage I lung cancer to determine factors that predict for development of severe RP [75]. Decreased FEV1, decreased FVC, increased V5, and increased MLD were all associated with grade 3+ RP on univariate analysis. FEV1 was associated with grade 3+ RP on multivariate analysis. However, others have shown that poor baseline pulmonary function may not increase the risk of RILT in the context of SBRT in patients with chronic obstructive pulmonary disease (COPD; [76]). When viewed in combination with COPD, significant smoking history, as reflected by a Brinkman Index (cigarettes/day x years) ≥ 400, has been determined to be a risk factor for prolonged minimal RP [77]. However, another study showed that patients with severe COPD do not appear to experience greater risks of lung toxicity after SBRT [78].

Importantly, and as a consequence of discovering risk factors for development of RP, predictive models have been developed based on expression biomarkers and pre-treatment imaging characteristics. For instance, pre-screening patients for risk of developing severe RP in response to SBRT has been pursued using pre-treatment CT scans to identify an interstitial pneumonitis shadow in conjunction with serum biomarkers Krebs von den Lungen-6 (KL-6) and surfactant protein-D (SP-D) [79]. The duration of this study included a period during which patients were treated with and without this screening procedure in place and showed a statistically significant reduction in the development of severe RP as a result of its implementation. Interestingly, and in contrast to the copious studies evaluating and establishing dose-volume parameters in assessing risk, the authors reported there was no such association in this study (though perhaps underpowered to detect such an association). This concept of dose-volume independence and toxicity has been postulated by others as well, to be at least responsible for some risk associated with radiation in the background of ILD [80]. The paradigm of using radiographic findings of ILD, irrespective of functional assessment (e.g., FEV1), to predict for outcome/risk in treating lung tumors with SBRT was also implemented by Bahig et al.[75] and has also been associated with increased volumes of irradiated lung (and, presumably, V_x parameters that are typically associated with increased risk of RP) [81].

In contrast to those with ILD, which seems to predispose a patient to developing RP, patients with emphysema appear to be at lower risk of developing RP based on a small recent retrospective study from Japan [82]. This work examined 40 patients treated for stage I non-small cell lung cancer using a dose of 75 Gy in 30 fractions (i.e., not SBRT), 17 of which were diagnosed with emphysema of varied severity while six had slight interstitial changes seen on CT imaging. The data showed a lower risk of RP in patients with severe emphysema (per modified Goddard's criteria) compared to moderate, mild, or no underlying disease. Further underscoring the variant radiobiology inherent in emphysematous lung tissue, albeit from a dosimetric perspective, is the study by Ochiai et al. [83]. Here, radiation plans for 72 patients undergoing SBRT for lung tumors were reviewed, 43 of which were noted to have emphysematous changes in the lung. This analysis showed that dosimetric parameters like heterogeneity and conformality indices can be influenced by the presence of underlying emphysema.

2.4.3 Radiation-induced liver damage (RILD) after SBRT

Radiation-induced liver damage (RILD) can result from the treatment of primary liver neo-plasms, liver metastases or targets within organs in close proximity to the liver (*e.g.*, right lung base, pancreas, and stomach). As there are no effective treatment options for RILD other than symptom control, great care must be taken to reduce the risk of RILD, making a comprehensive and predictive risk model particularly important.

Much work has been done to describe the dosimetric parameters predictive of RILD. Starting with the seminal Emami data [27], conventional fractionation to a mean liver dose of 30 Gy was thought to yield a 5% chance of RILD at five years out from treatment. A similar result was obtained by the University of Michigan about ten years later using the Lyman-Kutcher-Burman NTCP model [84].

However, given the liver's parallel architecture, the requirement for higher dose to achieve control of primary liver neoplasms, and the capability to escalate dose in a limited volume using SBRT, these metrics may not reflect dose-volume tolerance after treatment of a more limited volume of the liver. Several series have investigated development of RILD in rela-tion to SBRT and have implied not only that different metrics may be required to describe the toxicity complication risk but also that the biology of the target lesion(s) and patient medical comorbidities are likely to contribute to NTCP risks.

The potentially differing biology of intrahepatic tumors vs liver metastases has been underscored by several studies reviewed by Pan *et al.*in their QUANTEC paper on RILD [85]. Among several highlighted studies was a Dutch report following 25 patients undergo-ing SBRT for treatment of HCC (eight patients) or liver metastases (17 patients). Though two cases of RILD were observed in both cohorts, its severity was less in those treated for metastatic disease, in whom no grade 4-5 toxicity was observed [86], potentially indicating a difference in susceptibility between primary and metastatic liver disease. In addition, a pair of phase I studies was conducted at Princess Margaret Hospital, separately examining distinct patient cohorts. They showed a higher risk of toxicity in patients with primary intrahepatic lesions (HCC or intrahepatic cholangiocarcinoma) compared to those treated for liver metastases (from breast, colon, and other cancers) with similar median mean liver dose, dose-fractionation, and total dose received [87, 88].

Along with histologic considerations, additional patient risk factors have been consid-ered for RILD risks after three-dimensional conformal radiotherapy, including baseline liver function and background hepatic cirrhosis as reflected by the Child-Pugh classification [89]. In addition, associations of liver damage with hepatitis B viral carrier status or the re-gional distribution of chemotherapeutic agents have been postulated [90, 91]. After SBRT for small HCC, Child-Pugh classification B was a clinical factor associated with a signifi-cantly increased risk of developing at least grade 2 RILD in a multivariate analysis whereas associated parameters on univariate analysis also included normal liver volume and normal liver volumes receiving 15-60 Gy [92]. The positive association between RILD after SBRT and pre-treatment cirrhosis severity has also been described for patients with primary liver cancers [93].

2.5   BIG DATA

In an effort to mitigate many of the limitations of small retrospective studies, large-scale, or *big data*, efforts are being made to better utilize clinical, laboratory, and dosimetric data to develop (better) predictive models to make personalized radiotherapy a reality. Simply put, big data would better facilitate incorporation of biologic factors since there are likely many variables to study. Most available data (97%) have not been incorporated into assessment of prospective randomized trials conducted through cooperative groups and may be a rich source of information given the cost and limitations associated with these studies [94]. Studies utilizing big data may focus on clinical or basic scientific aspects of radiotherapy such as SBRT, cost effectiveness, or patient reported outcomes, though investigations are not limited to just these aspects of care. Furthermore, the best predictive models are likely to incorporate patient-specific data such as biomarkers or specimen characteristics. For genetic studies, consortia have been established domestically and internationally to curate data that may be used for large-scale studies. These include the US-based Gene-PARE and British RAPPER projects [95, 96, 97]. Importantly, in light of (genomic) data that may be accrued from different sources, guidelines have been published in an effort to improve the quality of reported data so that differences between sources/studies may be noted and accounted for [98]. A similar approach has been recommended for biomarker reporting in scientific journals using REMARK criteria [99, 100]. This approach will be crucial to apply to clinical data sets given the heterogeneity in radiation treatment plans and the manner in which outcomes are scored, as noted above.

2.6   CONCLUSIONS

Over the 25 years since the seminal Emami and Lyman paper was published, many strides have been made to better understand the factors that influence NTCP so that optimal treatment plans can be formulated with respect to clinical impact while also minimizing toxicities to the greatest extent possible. Further advances, however, will depend on a variety of factors and a multidisciplinary approach to achieve that obtains input from physicists, physicians, statisticians, and radiobiologists, among others. Certainly, large-scale retrospective and prospective trials that focus on integration of dosimetric data and patient factors, genetic and otherwise, will be valuable in strengthening existing clinical correlations, as well as bring new ones into view. Indeed, the increasing interest in and evidence of immunotherapy's utility in combating neoplastic disease may yield further complexity in determining NTCP for many tissues, particularly given the immune-mediated aspects of normal tissue damage and the enhancement of the immune response that underlies this approach to treatment. Additional advances in the development of various treatment modalities (e.g., proton therapy, intensity modulated proton therapy) in a field that has already made tremendous strides over the past decade, including the development of IMRT, will also contribute to the complexity in being able to determine the true tissue complication probability. This interdisciplinary approach will strengthen the foundation that QUANTEC has laid (and HYTEC and PENTEC are pursuing), providing a holistic understanding of how patient factors that lie outside the realm of quantitative assessment can impact objective measurements. Ideally, however, through radiomics, genomics and other molecular studies we will better understand the inner workings of normal tissue responses to radiation damage as well as the molecular effects of environmental exposures. This will thereby pave the way for thoughtful and deliberate manipulation of a very complex system to improve the therapeutic window of radiotherapy.

Imaging data: Radiomics

Issam El Naqa

CONTENTS

ABSTRACT

Imaging features could be exploited as a rich resource of patients-specific information for predicting outcomes and personalizing treatment as part of an emerging field called *radiomics*. In this chapter, we discuss the application of imaging-based approaches to predict outcomes from single and hybrid imaging modalities. We describe the different steps involved in radiomics analysis through examples.

Keywords: outcome models, imaging, radiomics.

3.1 INTRODUCTION

RADIOMICS is extraction of quantitative information from medical imaging modalities and relating this information to biological and clinical endpoints [101, 102, 103]. It is an essential part of synthesizing knowledge from imaging. Historically, images have been long recognized to be a rich resource of information; "a picture is worth a thousand words."

Radiomics could be thought of as consisting of two main steps: (1) extraction of relevant static and dynamic imaging features; and (2) incorporating these features into a mathematical model to predict outcomes as depicted in Figure 3.1. This chapter will focus on the process of feature extraction, while modeling approaches described in Part II of the book can be applied to predict outcomes once these images become mineable.

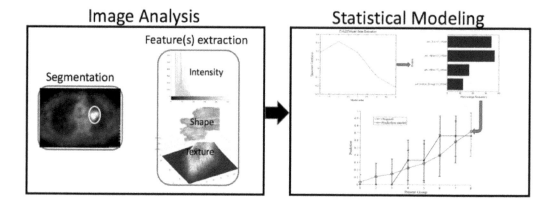

Figure 3.1 A general radiomics framework showing the two main steps of image analysis and modeling.

3.2 IMAGE FEATURES EXTRACTION

The features extracted from medical images could be divided into static (time invariant) and dynamic (time variant) features according to the nature of the acquisition protocol at the time of scanning, and into pre- or intra-treatment features according to the scanning time point [18].

3.2.1 Static image features

These image feature category is independent of time and could be divided into:

Standard uptake value (SUV) or Hounsfield Unit (HU) descriptors: SUV is a standard image quantification method particularly used in positron emission tomography(PET) analysis [104], likewise HU is used in computed tomography (CT). In this case, raw intensity values are converted into SUVs/HUs and statistical descriptors such as maximum, minimum, mean, standard deviation (SD), and coefficient of variation (CV) are extracted.

Total lesion glycolysis (TLG): This is also used in FDG-PET and is defined as the product of volume and mean SUV [105].

Intensity volume histogram (IVH): This is analogous to the dose volume histogram (DVH) widely used in radiotherapy treatment planning in reducing complicated 3D data into a single easier to interpret curve. Each point on the IVH defines the absolute or relative volume of the structure that exceeds a variable intensity threshold as a percentage of the maximum intensity [101]. This method would allow for extracting several metrics from images for outcome analysis such as I_x (minimum intensity to x% highest intensity volume), V_x (percentage volume having at least x% intensity value), and descriptive statistics (mean, minimum, maximum, standard deviation, etc.). We have reported the use of the IVH for predicting local control in lung cancer [106], where a combined metric from PET and CT image-based model provided a superior prediction power compared to commonly used dosimteric-based models of local treatment response. More details are provided in the example of Section 3.3.1.

Morphological features: These are generally geometrical shape attributes such as eccentricity (a measure of non-circularity), which is useful for describing tumor growth directionality; Euler number (the number of connected objects in a region minus the solidity (this is a measurement of convexity), which may be a characteristics of benign lesions [107, 108]. An interesting demonstration of this principle is that a shaped-based metric based on the deviation from an idealized ellipsoid structure (*i.e.*, eccentricity), was found to have strong association with survival in patients with sarcoma [109, 108].

Texture features: Texture in imaging refers to the relative distribution of intensity values within a given neighborhood. It integrates intensity with spatial information resulting in higher order histograms when compared to common first-order intensity histograms. It should be emphasized that texture metrics are independent of tumor position, orientation, size, and brightness, and take into account the local intensity-spatial distribution [110, 111]. This is a crucial advantage over direct (first-order) histogram metrics (*e.g.*, mean and standard deviation), which only measures intensity variability independent of the spatial distribution in the tumor microenvironment. Texture methods are broadly divided into three categories: statistical methods (*e.g.*, high-order statistics, co-occurrence matrices, moment invariants), model based methods (*e.g.*, Markov random fields, Gabor filter, wavelet transform) and structural methods (*e.g.*, topological descriptors, fractals) [112, 113]. Among these methods, statistical approaches based on the co-occurrence matrix and its variants such as the grey level co-occurrence matrix (GLCM), neighborhood gray tone difference matrix (NGTDM), run-length matrix (RLM), and grey level size-zone matrix (GLSZM) have been widely applied for characterizing tumor heterogeneity in images [114]. Four commonly used features from the GLCM include: energy, entropy, contrast, and homogeneity [110]. The NGTDM is thought to provide more human-like perception of texture such as: coarseness, contrast, busyness, and complexity. RLM and GLSZM emphasize regional effects. More details are provided in the example of Section 3.3.1.

3.2.2 Dynamic image features

The dynamic features are extracted from time-varying acquisitions such as dynamic PET or magnetic resonance imaging (MRI). These features are based on kinetic analysis using tissue compartment models and parameters related to transport and binding rates[115]. Recently, using kinetics approaches Thorwarth *et al.*published provocative data on the scatter of voxel-based measures of local perfusion and hypoxia in head and neck [116, 117]. Tumors showing wide spread in both characteristics showed less reoxygenation during RT and had worse local control. For instance, when a 3-compartment model is used, extracted parameters would include transfer constant (Ktrans), the extravascular-extracellular volume fraction (ve), and the blood volume (bv) [118]. A rather interesting approach to improve the robustness of such features is the use of advanced 4D iterative techniques [119]). Further improvement could be achieved by utilizing multi-resolution transformations (*e.g.*, wavelet transform) to stabilize kinetic parameter estimates spatially [120].

3.3   RADIOMICS EXAMPLES FROM DIFFERENT CANCER SITES

In the following, we will provide two representative cases of image-based outcome modeling and discuss the processes involved in such development. In one case, we will use separate extracted features from PET and CT for predicting tumor control in lung cancer. In the other case, fused extracted features from PET and MR are used to predict distant metastasis to the lung in soft-tissue sarcoma.

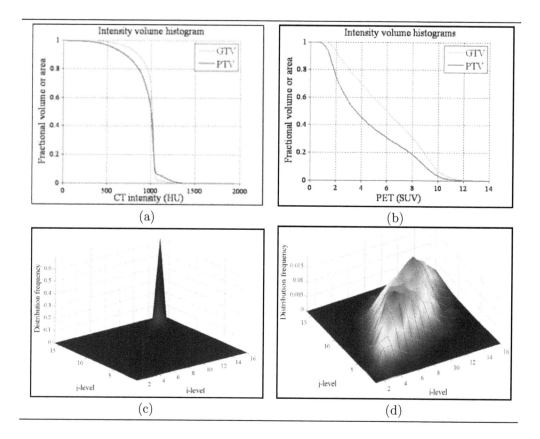

Figure 3.2   (a) Intensity Volume Histograms (IVH) of (b) CT and (b) PET, respectively. (c) and (d) are the texture maps of the corresponding region of interest for CT (intensity bins equal 100 HU) and PET (intensity bins equal 1 unit of SUV), respectively. Note the variability between CT and PET features: The PET-IVH and co-occurrence matrices show much greater heterogeneity for this patient. Importantly, patients vary widely in the amount of PET and CT gross disease image heterogeneity between patients.

3.3.1   Predicting local control in lung cancer using PET/CT

In a retrospective study of 30 non-small cell lung cancer (NSCLC) patients, thirty features were extracted from both PET and CT images with and without motion correction. The extracted features included tumor volume; SUV/HU measurements, such as mean, minimum, maximum, and the standard deviation; IVH metrics; and texture based features such as energy, contrast, local homogeneity, and entropy. The data corrected for motion artifacts based on a population-averaged probability spread function (PSF) using de-convolution methods derived from four 4D-CT data sets [121]. An example of such features in this case is shown in Figure 3.2.

Using modeling approaches based on statistical resampling and implemented in the DREES software [122], Figure 3.3 shows the results for predicting local failure, which consisted of a model of 2-parameters from features from both PET and CT based on intensity volume histograms provided the best.

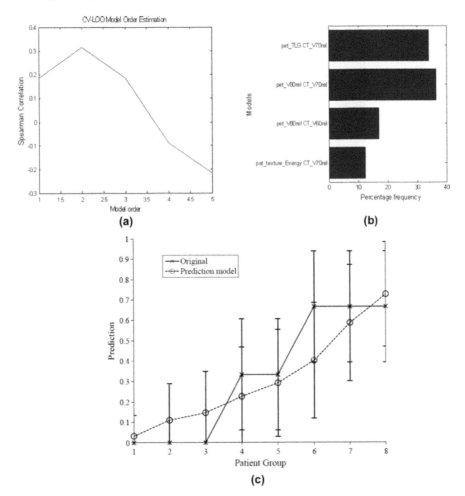

Figure 3.3 Image-based modeling of local failure from PET/CT features. (a) Model order selection using leave-one-out cross-validation. (b) Most frequent model selection using bootstrap analysis where the y-axis represents the model selection frequency on resampled bootstrapped samples. (c) Plot of local failure probability as a function of patients binned into equal-size groups showing the model prediction of treatment failure risk and the original data.

3.3.2 Predicting distant metastasis in sarcoma using PET/MR

A dataset of 51 patients with histologically proven STS was retrospectively analyzed. All patients had pre-treatment FDG-PET and MR scans. MR data comprised of: T1-weighted (T1w), T2 fat-saturated (T2FS) and T2 short tau inversion recovery (STIR) sequences as shown in Figure 3.4.

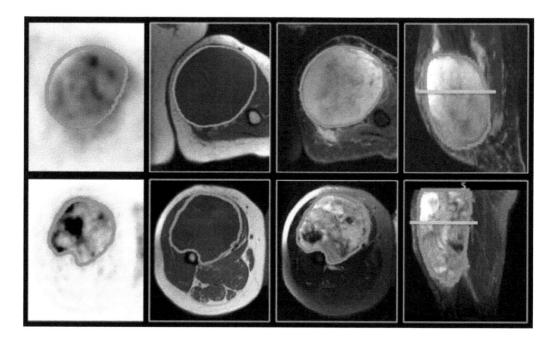

Figure 3.4 FDG-PET and MR diagnostic images of two patients with soft-tissue sarcomas of the extremities. Top row: patient that did not develop lung metastases. Bottom row: patient that eventually developed lung metastases. Fro left to right: FDG-PET, MR T1w, T2FS, and STIR (sagittal). The lines in the images of the 4th column correspond to the plane shown in the previous images were taken.

A volume fusion process was carried out to combine information from two different volumes (PET and MR) into a single composite volume that is potentially more informative for texture analysis. Fusion of the scans was performed using the discrete wavelet transform (DWT) and a band-pass frequencies enhancement technique [123]. In total, 41 different texture features were extracted out of the tumor regions of 5 different types of scans: FDG-PET, T1w and T2FS, fused FDG-PET/T1 and fused FDG-PET/T2FS scans. The texture features consisted of 3 features from first-order histograms, 7 features from GLCM, 13 features from GLRLM, 13 features from GLSZM and 5 features from NGTDM. Optimal features were found using texture optimization based on imbalance-adjusted 0.632+ bootstrap resampling method [124]. The resulting model consisted of four texture features representing variations in size and intensity of the different tumor sub-regions. It yielded a performance estimate in bootstrapping evaluations, with an area under the receiver-operating characteristic curve (AUC) of 0.984 ± 0.002, a sensitivity of 0.955 ± 0.006, a specificity of 0.926 ± 0.004 and an accuracy of 0.934 ± 0.003 as shown in Figure 3.5.

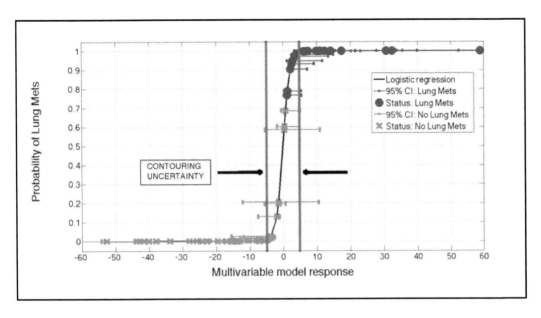

Figure 3.5 Probability of developing lung metastases as a function of the response of the radiomics model in soft-tissue sarcoma.

3.4 CONCLUSIONS

The use of imaging in outcome modeling of radiotherapy response has witnessed rapid increase in recent years adding more value to already existing use of imaging in cancer treatment in general and radiotherapy in particular. However, there are several issues that are currently limiting its rapid progression. It is well recognized that image acquisition protocols may impact the reproducibility of extracted features from image modalities, which may consequently impact the robustness and stability of these features for treatment prediction. This includes static features such as SUV/HU descriptors and texture features. Interestingly, texture-based features were shown to have a reproducibility similar to or better than that of simple SUV descriptors [125]. This demands protocols for standardized acquisition. In addition, factors that may impact the stability of these features also include signal-to-noise ratio (SNR), partial volume effect, motion artifacts, parameter settings, re-sampling size, and image quantization [101, 126]. Nevertheless, advances in hardware and software technologies will further facilitate wider application of advanced image processing techniques to medical imaging to achieve better clinical results. For instance, pre-processing methods such as denoising and deconvolution methods already help in mitigating such artifacts [127, 121, 128], however, more advanced image restoration methods based on nonlocality and sparsity may be more fruitful [129]. Outcome modeling using logistic regression has become a de facto standard, however, more advanced modeling techniques may provide further predictive power particularly when dealing with more complex and nonlinear relation-ships among features and between clinical outcomes. We believe that the synergy between image analysis and machine learning [16] could provide powerful tools to strengthen and further the utilization of image-based out-come modeling in clinical practice towards improved clinical decision making and personalized medicine in the future.

CHAPTER 4

Dosimetric data

Issam El Naqa

Randall K. Ten Haken

CONTENTS

ABSTRACT

Dosimetric data plays a major role in radiation oncology for predicting treatment response. These metrics were originally based on prescribed radiation dose fractionation and more recently are derived from 3D radiation dose distributions. In this chapter, we discuss the extraction and the application of dosimetric variables for predicting radiotherapy outcomes. We will also describe by example the different steps involved in such dosimetric analysis.

Keywords: outcome models, radiotherapy, dose-volume-histogram, dose-volume metrics.

4.1 INTRODUCTION

D OSIMETRIC modeling is based on the application and extraction of descriptive information from radiotherapy treatment plans and relating this information to radiation-
induced clinical endpoints [130]. Radiotherapy is targeted localized treatment using ablative
high-energy radiation beams to kill cancer cells. More than half all cancer patients, particularly patients with solid tumors such to the brain, lung, breast, head and neck, and
pelvic area receive radiotherapy as part of their treatment. A typical radiotherapy planning
process would involve the acquisition of patient image data (typically fully 3-D computed
tomography (CT) scans and other diagnostic imaging modalities such as positron emission
tomography (PET) or magnetic resonance imaging (MRI)). Then, the physician outlines the
tumor and important normal structures on a computer, based on the CT scan. Treatment
radiation dose distributions are simulated with prescribed doses (energy per unit mass). The
treatment itself could be delivered externally using linear accelerators (Linacs) or internally
using sealed radioisotopes (Brachytherapy) [131].

Recent years have witnessed tremendous technological advances in radiotherapy treatment planning, image-guidance, and treatment delivery [132, 133]. Moreover, clinical trials
examining treatment intensification in patients with locally advanced cancer have shown
incremental improvements in local control and overall survival [131]. However, radiation-
induced toxicities remain major dose-limiting factors [134, 135]. Therefore, there is a need
for studies directed towards predicting treatment benefit versus risk of failure. Clinically,
such predictors would allow for more individualization of radiation treatment plans. In other
words, physicians may prescribe a more or less intense radiation regimen for an individual
based on model predictions of local control benefit and toxicity risk. Such an individualized regimen would aim towards an optimized radiation treatment response while keeping
in mind that a more aggressive treatment with a promised improved tumor control will
not translate into improved survival unless severe toxicities are accounted for and limited
during treatment planning. Therefore, improved models for predicting both local control
and normal tissue toxicity should be considered in the optimal treatment planning design
process.

Treatment outcomes in radiotherapy are usually characterized by tumor control probability (TCP) and the surrounding normal tissues complications (NTCP) [133, 136]. Traditionally, these outcomes are modeled using information about the dose distribution and
the fractionation scheme [137, 138]. The main premise is to maximize TCP and minimize
or maintain NTCP below a certain maximum tolerance level as shown in Figure 4.1.

Traditionally, TCP/NTCP were modeled based on simplistic understanding of experimentally observed irradiation effects using *in vitro* assays, which constituted the basis
for developing analytical or sometimes refereed to as mechanistic models of radiotherapy
response. These models have been applied widely for predicting TCP and NTCP and designing radiotherapy clinical trials over the past century and will be the subject of Chapter
7. However, these traditional modeling methods have fallen short of providing sufficient
predictive power when applied prospectively to personalize treatment regimens. Therefore,
more recent approaches have focussed on using data-driven models utilizing advanced informatics tools in which dose-volume metrics are used alone or are mixed with other patient
or disease-based prognostic factors in order to improve outcomes prediction [138].

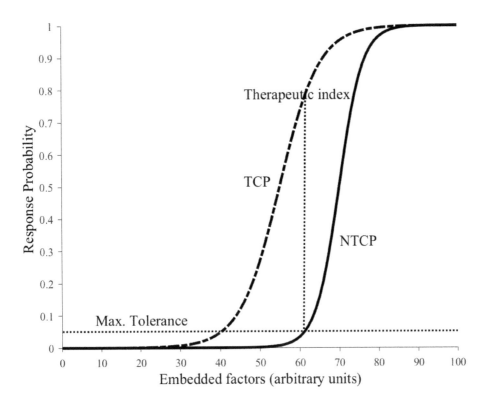

Figure 4.1 A general TCP/NTCP modeling as a function of damage variable.

4.2 DOSE VOLUME METRICS

This type of data is related to the treatment planning process in radiotherapy, which involves radiation dose simulations using computed tomography imaging; specifically dose-volume metrics derived from dose-volume histograms (DVHs) graphs [139]. The so-called direct DVH is simply a frequency distribution of the dose levels for the organ or tissue of interest (a graph of how often each dose value occurs (raw frequency, % or cm^3 of the volume) vs. the dose levels themselves. For normal tissues it is more common to display *cumulative* DVHs where the frequency or volume or each dose bin of the direct DVH is summed (starting at the high dose end) together with the volumes in each higher dose bin such that ordinate (y-axis) now represents volume receiving greater than or equal to each dose level. Although the inspection of DVHs for treatment plan evaluation is quite common and widespread, their use has certain shortcomings, most notable the loss of spatial information. An example of a prostate plan with its DVH is in shown in Figure 4.2.

That is, one can use a DVH to easily read off, for example, D3%, the minimum dose to the hottest 3% of the volume, but one cannot determine from the DVH alone whether that dose is concentrated in one continuous region or whether it comprises multiple small disconnected hot spots. Similarly, in plan comparison, a normal tissue cumulative DVH that has volume values at all dose levels lower than a DVH from a rival plan can generally be assumed to be safer. However, the relative risks of two normal tissue cumulative

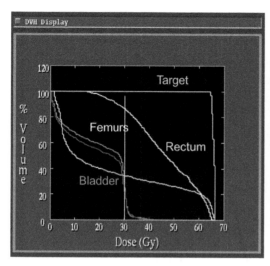
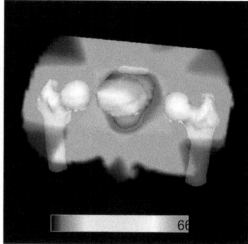

Figure 4.2 A prostate 3D plan (left) with its corresponding dose-volume histogram (DVH).

DVHs that cross each other (e.g., one showing greater volume at low dose, but less volume at high dose) cannot be determined without knowledge of the tissue type and the relative effects of high dose to small volumes versus low dose to large volumes as shown in Figure 4.3.

Dose-volume metrics have been extensively studied in the radiation oncology literature for outcomes modeling [140, 141, 142, 143, 144, 145, 146, 134]. These metrics are extracted from the DVH such as volume receiving certain dose (Vx), minimum dose to x% volume (Dx), mean, maximum and minimum dose, etc. [130]. Moreover, a dedicated software tool called "DREES" was developed for deriving these DVH metrics and modeling of radiotherapy response [122]. Some of the most common DVH metrics are summarized below.

Dose statistics: Maximum dose, minimum dose, mean dose, median dose, standard deviation, etc.

$\mathbf{V}_x$: The volume receiving $\geq$ dose "x". This could be presented absolute (cm^3) or relative to the total organ volume.

$\mathbf{D}_x$: The minimum dose to the hottest "x" % or cm^3 of the volume.

There are also in the literature other derived metrics that are less common used in planning optimization such as MOH_x and MOC_x, which are the mean doses of the hottest, and coldest x% regions [147].

4.3 EQUIVALENT UNIFORM DOSE

It is numerically unfeasible for most commonly used TCP/NTCP models to act directly on the 3-D dose distribution or the DVH. This leads to a further reduction of information (already down from the 3-D dose distribution itself to the DVH) into concepts such as effective volume (V_{eff}) or generalized equivalent uniform dose (EUD). In the DVH world this

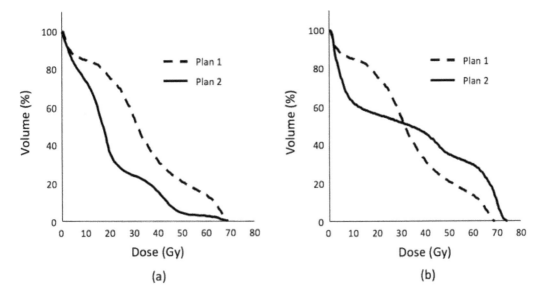

Figure 4.3 DVH analysis of an organ at risk for ranking plans (a) An easy case with plan 2 better. (b) A difficult case that would depend on the organ type and the planning objectives.

translates into transformation of the actual organ DVH (that illustrates all the effects of non-uniform irradiation) into a single-step DVH which is intended to lead to an equivalent dose effect; that is, either irradiation of a single effective volume, V_{eff}, to a single uniform dose D_0 with the rest of the volume, $1 - V_{eff}$, receiving zero dose (where D_0 is generally chosen as the prescription dose), or the uniform irradiation of the entire volume, V_0, to a single equivalent uniform dose, EUD. The mathematics of the most common DVH reduction scheme assumes a power-law relationship [148] between tolerance doses and uniformly irradiated whole or partial volumes. That is, for a direct DVH having N bins each with partial volume V_i and dose D_i, the effective volume [149] or generalized equivalent uniform dose [150, 151] can be written as:

$$V_{eff} = \sum_{i=1}^{N} V_i \left(\frac{D_i}{D_0} \right)^{1/n}, \quad \text{or} \quad EUD = \left(\sum_{i=1}^{N} \frac{V_i}{V_0} (D_i)^{1/n} \right)^n \qquad (4.1)$$

with volume effect parameter n. We note that for $n = 1$, EUD is equal to mean dose suggesting a large volume effect and large regions of the DVH matter (parallel organ); and, as n approaches $+(-)$ zero, EUD is close to the maximum (minimum) dose in the organ (serial organ). At the height of 3-D conformal therapy, the use of V_{eff} was popular [146] as first, it could be computed from a relative dose distribution (*e.g.*, doses in % of the isocenter dose) before a prescription dose was written (note the computation of V_{eff} is independent of the units of dose), and secondly because the 3-D dose distributions for irradiation of most targets in or near large volume effect tissues had large volumes receiving no dose or very low dose, such that the single step V_{eff}, D_0 DVH looked somewhat similar to the original cumulative DVH. In the current era of modern treatment planning systems and dose calculation algorithms, treatment plans are now generally computed directly as physical dose (Gy) and also, especially with IMRT, nearly all the volume of a tissue or organ of interest receives some dose and the EUD concept makes conceptual sense. The use of these concepts is beginning to appear in commercial treatment planning systems [152] and is discussed further in Chapter 12.

4.4 DOSIMETRIC MODEL VARIABLE SELECTION

Any multivariate analysis often involves a large number of variables or features [153]. The main features that characterize the observations are usually unknown. Therefore, dimensionality reduction or subset selection aims to find the *significant* set of features. Finding the best subset of features is definitely challenging, especially in the case of nonlinear models. The objective is to reduce the model complexity, decrease the computational burden, and improve the generalizability on unseen data. The straightforward approach is to make an educated guess based on experience and domain knowledge, then, apply feature transformation (e.g., principle component analysis (PCA)) [154, 155], or sensitivity analysis by using organized search such as sequential forward selection, or sequential backward selection or combination of both [155]. A recursive elimination technique that is based on machine learning has been also suggested [156]. In which the data set is initialized to contain the whole set; train the predictor (*e.g.*, SVM classifier) on the data; rank the features according to a certain criteria and keep iterating by eliminating the lowest ranked one. It should be noted that the specific definition of model order changes depending on the functional form. It could be identified by the number of parameters in logistic regression, or by the number of neurons and layers in the case of neural networks, etc. However, in any of these forms, the model order creates a balance between complexity (increased model order), and the model ability to generalize to unseen data. Finding this balance is referred to in statistical theory as the bias-variance dilemma 4.4, in which an over-simple model is expected to underfit the data (large bias and small variance), whereas a too complex model is expected to overfit data (small bias and large variance) [22]. Hence, the objective is to achieve an optimal parsimonious model, i.e., a model with the correct degree of complexity to fit the data and thus a maximum ability to generalize to new, unseen, datasets.

4.4.1 Model order based on information theory

Information theory provides two intuitive measures of model order optimality: Akaike information criteria (AIC) and the Bayesian information criteria (BIC) [2]. AIC is an estimate of predictive power of a model, which includes both the maximum likelihood principle and a model complexity term that penalizes models with an increasing number of parameters (to avoid overfitting the data). BIC is derived from Bayesian theory, which results in a penalty term that increases linearly with the number of parameters.

4.4.2 Model order based on resampling methods

Resampling techniques are used for model selection and performance comparison purposes to provide statistically sound results when the available data set is limited (which almost always the case in radiotherapy). We use two types of fit-then-validate methods: cross-validation methods and bootstrap resampling techniques. Cross-validation [155] uses some of the data to train the model and some of the data to test the model validity. The type we most often use is the leave-one-out cross-validation (LOO-CV) procedure (also known as the jackknife). In each LOO-CV iteration, all the data are used for training/fitting except for one data point left out for testing, and this is repeated so that each data point is left out exactly once. The overall success of predicting the left-out data is a quantitative estimate of model performance on new data sets. Bootstrapping [20] is an inherently computationally

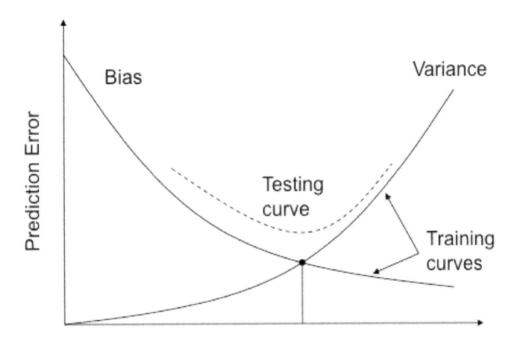

Figure 4.4 This figure illustrates a common tradeoff in modeling predictive power between prediction bias (average error) and prediction variance (square error). As model complexity increases, the average prediction error (bias) tends to decrease while the average square error tends to decrease. The point of optimal complexity tends to be near the point when average and square errors are of similar magnitude.

intensive procedure but generates more realistic results. Typically, a bootstrap pseudo-dataset is generated by making copies of original data points and randomly selected with a probability of inclusion of 63%. The bootstrap often works acceptably well even when data sets are small or unevenly distributed. To achieve valid results this process must be repeated many times, typically several hundred or thousand times. Examples of applying these methods to outcomes modeling in radiotherapy could be found in our previous work [157, 122] and are discussed in details in [130].

4.5   A DOSIMETRIC MODELING EXAMPLE

4.5.1   Data set

A set of 56 patients diagnosed with non-small cell lung cancer (NSCLC) and have discrete primary lesions, complete dosimetric archives, and follow-up information for the endpoint of local control (22 locally failed cases) is used. The patients were treated with 3D conformal radiation therapy (3D-CRT) with a median prescription dose of 70 Gy (60-84 Gy). The dose distributions were corrected for heterogeneity using Monte Carlo (MC) simulations [158]. The clinical data included age, gender, performance status, weight loss, smoking, histology, neoadjuvant and concurrent chemotherapy, stage, number of fractions, tumor elapsed time, tumor volume, and prescription dose. Treatment planning data were de-archived and potential dose-volume prognostic metrics were extracted using CERR [159]. These metrics included Vx (percentage volume receiving at least x Gy), where x was varied from 60 to 80 Gy in steps of 5 Gy, mean dose, minimum and maximum doses, center of mass location in the craniocaudal (COMSI) and lateral (COMLAT) directions. This resulted in a set of 23 candidate variables to model TCP. The modeling process using nonlinear statistical learning starts by applying PCA to visualize the data in two-dimensional space and assess the separability of low-risk from high-risk patients. Non-separable cases are modeled by nonlinear kernels. This step is preceded by a variable selection process and the generalizability of the model is evaluated using resampling techniques as discussed earlier and explained below [160].

4.5.2 Data exploration

In Figure 4.5a, we show a correlation matrix representation of these variables with clinical TCP and cross-correlations among themselves using Spearman's coefficient (rs). Note that many DVH-based dosimetric variables are highly cross-correlated, which complicate the analysis of such data. In Figure 4.5b, we summarize the PCA analysis of this data by projecting it into 2-D space for visualization purposes. It shows that two principle components are able to explain 70% of the data and reflects a relatively highly overlap between patients with and without local control; indicating potential benefit from using nonlinear kernel methods.

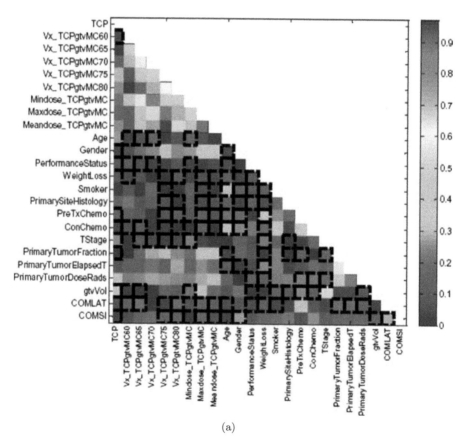

(a)

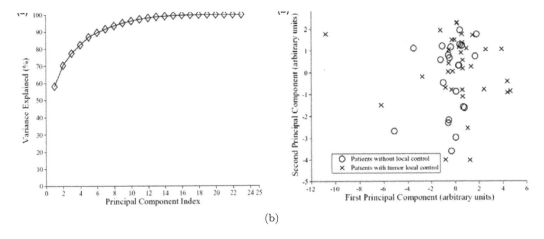

(b)

Figure 4.5 (a) Correlation matrix showing the candidate variables correlations with TCP and among the other candidate variables. (b) Visualization of higher dimensional data by principle component analysis (PCA). Left The variation explanation versus principle component (PC) index. Right The data projection into the first two principal components space. Note the cases overlap.

4.5.3 Multivariate modeling with logistic regression

The multi-metric model building using logistic regression is performed using a two-step procedure to estimate model order and parameters. In each step, a sequential forward selection strategy is used to build the model by selecting the next candidate variable from the available pool (23 variables in our case) based on increased significance using Walds statistics [157]. In Figure 4.6a, we show the model order selection using the LOO-CV procedure. It is noticed that a model order of two parameters provides the best predictive power with rs=0.4. In Figure 4.6b, we show the optimal model parameters selection frequency on bootstrap samples (280 samples were generated in this case). A model consisting of GTV volume ($\beta = -0.029, p = 0.006$) and GTV V75 ($\beta = +2.24, p = 0.016$) had the highest selection frequency (45% of the time). The model suggests that increase in tumor volume would lead to failure, as one would expect due to increase in the number of clonogens in larger tumor volumes. The V75 metric is related to dose coverage of the tumor, where it is noticed that patients who had less than 20% of their tumor covered by 75 Gy were at higher risk of failure. However, this approach does not account for possible interactions between these metrics nor accounts for higher order nonlinearities.

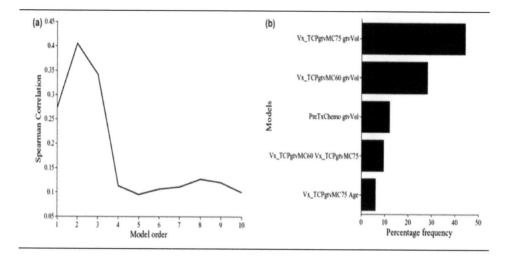

Figure 4.6 TCP model building using Logistic regression. (a) Model order selection using LOO-CV. (b) Model parameters estimation by frequency selection on bootstrap samples.

4.5.4 Multivariate modeling with machine learning

To account for potential non-linear interactions, we will apply kernel-based methods based on a support vector machine (SVM). Moreover, we will use the same variables selected by the logistic regression approach. We have demonstrated recently that such selection is more robust than other competitive techniques such as the recursive feature elimination (RFE) method used in microarray analysis. In this case, a vector of explored variables is generated by concatenation. The variables are normalized using a z-scoring approach to have a zero mean and unity variance [155]. We experimented with different kernel forms; best results are shown for the radial basis function (RBF) in Figure 4.7a. The figure shows that the optimal kernel parameters are obtained with an RBF width $\sigma = 2$ and regularization parameter $C = 10000$. This resulted in a predictive power on LOO-CV $rs = 0.68$, which represents 70% improvement over the logistic analysis results. This improvement could be further explained by examining Figure 4.7b, which shows how the RBF kernel tessellated the variable space nonlinearly into different regions of high and low risks of local failure. Four regions are shown in the figure representing high/low risks of local failure with high/low confidence levels, respectively. Note that cases falling within the classification margin have low confidence prediction power and represent intermediate risk patients, *i.e.*, patients with "border-like" characteristics that could belong to either risk group [160].

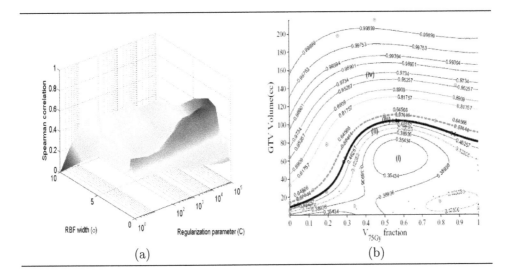

Figure 4.7 Machine learning modeling of TCP in lung cancer using the GTV volume and V75 with support vector machine (SVM) and a radial basis function (RBF) kernel. Scatter plot of patient data (black dots) being superimposed with failure cases represented with red circles. (a) Kernel parameter selection on LOO-CV with peak predictive power attained at $\sigma = 2$ and $C = 10000$. (b) Plot of the kernel-based local failure (1-TCP) nonlinear prediction model with four different risk regions: (i) area of low risk patients with high confidence prediction level; (ii) area of low risk patients with lower confidence prediction level; (iii) area of high risk patients with lower confidence prediction level; (iv) area of high risk patients with high confidence prediction level. Note that patients within the "margin" (cases ii and iii) represent intermediate risk patients, which have border characteristics that could belong to either risk group.

4.5.5 Comparison with other known models

For comparison purposes with mechanistic TCP models, we chose the Poisson-based TCP model and the cell kill equivalent uniform dose (cEUD) model. The Poisson-based TCP parameters for NSCLC were selected according to Willner *et al.* [161], in which the sensitivity to dose per fraction ($\alpha/\beta = 10$ Gy), dose for 50% control rate (D50 = 74.5 Gy), and the slope of the sigmoid-shaped dose-response at D50 ($\gamma 50 = 3.4$). The resulting correlation of this model was $rs = 0.33$. Using D50–84.5 and $\gamma 50 - 1.5$ [162, 163] yielded an $rs = 0.33$ also. For the cEUD model, we selected the survival fraction at 2 Gy (SF2=0.56) according to Brodin *et al.* [164]. The resulting correlation in this case was $rs = 0.17$. A summary plot of the different methods predictions as a function of binned patients into equal groups is shown in Figure 4.8. It is observed that the best performance was achieved by the nonlinear (SVM-RBF). This is particularly observed for predicting patients who are at high risk of local failure.

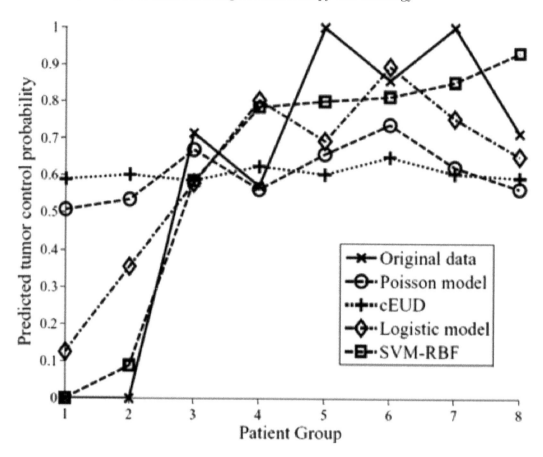

Figure 4.8 A TCP comparison plot of different models as a function of patients being binned into equal groups using the model with highest predictive power (SVM-RBF). The SVM-RBF is compared to Poisson-based TCP, cEUD, and best 2-parameter logistic model. It is noted that prediction of low-risk (high control) patients is quite similar; however, the SVM-RBF provides a significant superior performance in predicting high-risk (low control) patients.

4.6 SOFTWARE TOOLS FOR DOSIMETRIC OUTCOME MODELING

Many of the presented analytical/multi-metric methods require dedicated software tools for implementation. Two examples of such software tools in the literature are BIOPLAN and DREES. BIOPLAN uses several analytical models for evaluation of radiotherapy treatment plans [165], while DREES is an open-source software package developed by our group for dose response modeling using analytical and multi-metric methods [122] presented in Figure 4.9. It should be mention that several commercial treatment-planning systems have currently incorporated different TCP/NTCP models, mainly analytical ones that could be used for ranking and biological optimization purposes. A discussion of these models and their quality assurance guidelines is provided in TG-166 [152].

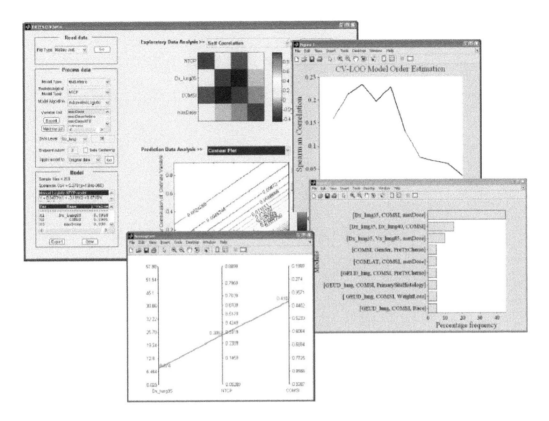

Figure 4.9 DREES allows for TCP/NTCP analytical and multivariate modeling of outcomes data. The example is for lung injury. The components shown here are: Main GUI, model order and parameter selection by resampling methods, and a nomogram of outcome as function of mean dose and location.

4.7 CONCLUSIONS

Dosimetric variables play a very important role in predicting outcomes as they represent the treatment contribution into modeling. The typical process by which these variables are extracted is by data reduction from the 3D treatment plans using DVHs, leading to loss in spatial information, which may be relevant in some instances and needs to be compensated for by other metrics (*e.g.*, tumor location). Heterogeneity in dose distributions could be corrected using EUD, and its volume effect depends on the nature of the organ and the clinical endpoint of interest.

CHAPTER 5

Pre-clinical radiobiological insights to inform modelling of radiotherapy outcome

Peter van Luijk, PhD

Robert P. Coppes, PhD

CONTENTS

ABSTRACT

Pre-clinical data can be used to better understand various patient and treatment characteristics that need to be considered when developing risk models for the prediction of treatment outcome. Here, we describe using advanced cell/organoid culture techniques and proper animal models that could help in such a process by providing means to test and generate hypotheses in a controlled environment, which is not possible in clinical trials.

Keywords: preclinical data, cell culture, animal models, outcome modeling.

5.1 VARIABILITY IN RESPONSE TO HIGHLY STANDARDIZED RADIOTHER-APY

The main challenge of radiation oncology is to find a balance between benefits of irradiating a target volume and risks of exposure of surrounding normal tissues (Figure 1, solid lines). Both the probability of cure and the risk of toxicity are dependent on radiation dose. However, even though radiotherapy treatments have been standardized to a high degree, individuals respond differently to the same treatments. This indicates that, besides dependence on dose, also individual differences between patients may influence outcome. Such differences may arise both from intrinsic differences in radiation response and differences that arise from e.g. tumor localization, that may lead to variations in the dose distribution over the normal tissues. Consequently, the optimal treatment may differ between patients (Figure 5.1, dashed lines vs. solid lines). Therefore, understanding dose-effect relations and biological mechanisms of responses of cells, organs and organisms are central themes in radiobiology providing information that may improve prediction modeling of outcome.

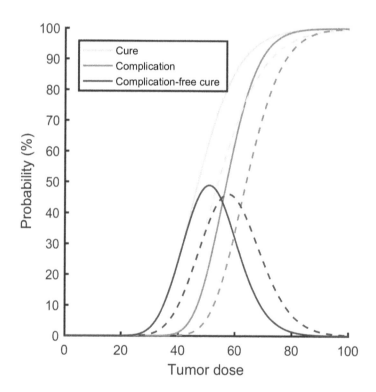

Figure 5.1 The relation between probabilities of cure, normal tissue damage and the probability of achieving complication-free cure (solid lines). The dashed lines indicate similar curves for a hypothetical patient with differing radiation sensitivity. This can be seen to lead to a different optimal treatment dose, indicating the importance of insight into individual radiation responses.

In the following sections biological and clinical evidence for individual responses will be described as well as possible routes for using such information to develop or improve predictive models.

5.2   VARIATION IN SENSITIVITY TO RADIATION

Already for a long time it has been hypothesized that individual variations in response to radiation may be predictive of treatment outcome [166]. Based on the mechanisms described in the previous section, such variation can originate from intrinsic variations in sensitivity of the target cells for organ dysfunction.

Initially obtaining information on individual radiosensitivity has been pursued by determining the surviving fraction of cells in a 2D culture after irradiating with 2 Gy. However, the radiation response of surrogate or representative normal tissue response material such as fibroblast and tumor cell lines have yielded disappointing correlations with the actual patient response [167]. This may have been, at least in part, related to non-representativeness of cell populations and culture conditions. In 2D culture , (Figure 2A) cells assume an unnatural morphology; they are flattened out and lack interaction with each other and extracellular matrix components which play a major role in radiosensitivity [168] (figure 5.2A). To overcome this, improvements have been achieved in developing organoids cultured from the patient's pre-treatment normal tissue (Figure 5.2B) and tumor biopsy samples.

Recently, 3D organoid cultures have been developed from normal and transformed tissue resembling tumor/tissue architecture, multilineage differentiation, and stem cells [169, 170, 171, 172] that allow the modeling of cancer [173, 174, 175]. In 3D cultures, the cells take more the shape they have in their natural environment and multiply to form a spheroid/organoid structure that consists of multiple cells originating from single cells, resembling the tissue of origin (Figure 5.2B). Cell/cell and cell/matrix interactions are more close to the natural situation thereby resulting in a radiation dose response relation that is closer to the actual response of the tissue [168, 176].

While whole-genome sequencing and transcriptome, epigenome, and proteome analysis reveals the complexity of cancer genomes and regulatory processes [177, 175], we recently, developed a method that allows the determination of chemoradiation response of (cancer) stem cells (CSC) cultured as organoids [176]. The ability to culture patient specific tumor and normal tissues as organoids might allow for the determination of in vitro response that can be related to the actual clinical response. Further selection of the relatively clean genetic background of organoids can be obtained by omics of CSC derived organoids that survive radiation. Moreover, organoids can also provide functional assays needed to validate resistant signatures. As such, organoid-based assessment of individual radiosensitivity and response may represent a development that may overcome the limitations of classical survival assays .

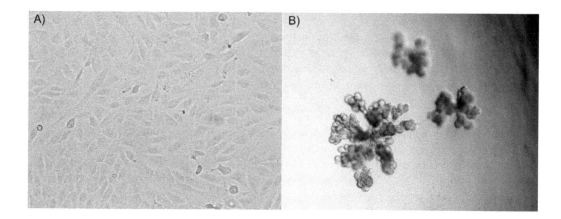

Figure 5.2 Culture conditions change cell behavior. Panel A) Cells cultured in a 2D culture system take an unnatural morphology; they are flattened out and lack interaction with each other and extracellular matrix components which play a major role in radiosensitivity. Panel B) Stem cells cultured in a 3D culture system develop into structures resembling the tissue they originated from.

5.3   UNDERSTANDING DOSE-RESPONSE OF TISSUES AND ORGANS

Though radiation can interact with cells via multiple targets, the most critical event is DNA damage. DNA damage has been identified as the primary target for cell death [178, 179]. In addition, DNA damage and uncontrolled wound healing responses in surviving cells are known to contribute to normal tissue damage [180]. Even though these local processes are the primary effect of radiation, the translation into a change in organ function and possibly a complication depend on many factors.

5.4   ANIMAL MODELS TO STUDY RADIATION RESPONSE

The ability to use organ function is a feature of animal models that facilitates investigation of processes that translate initial local radiation effects into effects on organ function and, for some toxicities, into endpoints resembling clinical complications.

Many different animal models are available. Species are usually selected based on possibilities to perform specific experiments or on anatomical or functional resemblance to patients. For example, the use of mice has the benefit of availability of genetically altered variants that lack specific mechanisms, which offers the possibility to investigate the role of these mechanisms in radiation responses. Alternatively, larger animals like rats offer better opportunities for high-precision irradiation of parts of organs. Even larger animals such as dogs, pigs and sheep have the added benefit of better resemblance to human anatomy and/or physiology. However, these large animals are rarely used due to cost of acquisition and housing.

5.5   PROCESSES GOVERNING OUTCOME

It has been long known that the amount of irradiated normal tissue is an important determinant of the risk of toxicity. Initially it has been hypothesized that this role of irradiated volume is related to the amount of function that is put at risk. To translate the effect of dose into a risk of organ failure a function organization of sub-volumes is assumed [181]). This approach relies on the assumption that the effects and consequences of dose radiation are strictly local and independent of the location where dose is deposited.

However, evidence is mounting that the consequences of damage may vary strongly with cell type and localization [182, 183, 184]. In the rat spinal cord it was demonstrated that the lateral column of white matter was more sensitive to develop necrosis than the dorsal column, leading to a regional difference in tolerance to radiation for the development of paralysis [183]). Similar observations were made in rat lung that was found to tolerate higher doses to the mediastinal as compared to the lateral half of the lungs [185]). This variation in response may be related to a relatively larger proportion of alveolar tissue in the lateral fields, which increases the impact of cell loss in the lateral region in comparison to the central region [185].

Moreover, secondary to direct radiation, damage may occur in non-irradiated regions [186, 187, 184] or even organs [187]. Tissue homeostasis is a critical process in maintaining organ function. Stem cells are indispensable for durable homeostasis, since they are the only cell type that is capable of self-renewal. As such, stem cells inactivation has been shown to lead to a critical mechanism in the development of salivary gland dysfunction [188]. Interestingly, the stem cells of the parotid gland are predominantly located in the larger ducts causing the impact of dose to the limited sub-volume containing these to be much larger than dose elsewhere [184]). In fact, in rats selective irradiation of that sub-volume was shown to lead to global degeneration of the gland [189, 184].

Another mechanism that can cause the effect of radiation to extend beyond the location where dose is deposited is on the level of organ function. Heart and lungs strongly depend on each others integrity in performing respiratory function. Indeed, it has been observed that combined irradiation leads to an increased response when compared to irradiating the lung alone [190]). This interaction can be explained by the observation that both irradiation of heart and lung lead to damage that impacts similar cardio-pulmonary physiological parameters, leading to compensatory responses and damage in both of them [187]. Specifically, heart irradiation affects both the cardiac microvasculature and myocardium. This can lead to increased end-diastolic pressure and left ventricle diastolic dysfunction, which promotes pulmonary interstitial edema [187]. Lung irradiation may impair left ventricle diastolic function indirectly by causing radiation-induced pulmonary hypertension [187], pulmonary perivascular edema, and the resultant effect on left ventricle relaxation time. As such, combined lung and heart irradiation can lead to enhanced cardiac diastolic dysfunction via both mechanisms, which may result in bi-ventricle dysfunction explaining the observed increased cardiopulmonary dysfunction [187]. This suggests that, besides dose to lungs, also dose to the heart can contribute to the development of pulmonary toxicity. By identifying organs, tissues or anatomical structures involved in the development of a complication, such mechanistic study informs modeling about prognostic factors that should be considered when developing risk models.

As such, detailed knowledge of mechanisms and target structures for the development of normal tissue damage is critical, since the identification of risk factors is a central issue in the development of risk models.

5.6 PATIENT-INDIVIDUAL FACTORS / CO-MORBIDITY

Besides individual radiosensitivity influencing the risk of cell death, also other patient individual factors influence the risk of toxicity. Various categories of factors can be distinguished. Firstly, pre-existing conditions can influence the risk of toxicity. Especially in the treatment of thoracic tumors various examples have been documented. Cardiac comorbidity was observed to be an important predictor for the development of signs of radiation pneumonitis [191], which can be understood when considering the functional connection between heart and lung [187]. Similarly, also pre-existing interstitial lung disease [192] and reduced pulmonary blood flow [193] were found to predict the occurrence of radiation pneumonitis. Especially, the latter can be understood by the observation in rats that remodeling of the pulmonary microvasculature and the consequent disturbance in pulmonary blood flow are associated with the classical signs of radiation pneumonitis. These observations suggest that reduced pre-treatment capacity reduces tolerance to further loss of capacity.

Other, less well-understood phenomena include complaints of mild xerostomia increase the risk of developing severe xerostomia post-treatment [194], increased risk for development of various gastro-intestinal complications after radiotherapy in patients with a history of abdominal surgery [195] and pre-treatment performance status as an independent predictor of radiation pneumonitis [192].

5.7 USE IN MODELS

Taken together the previous sections indicate that various patient and treatment characteristics need to be considered in the development of risk models for the prediction of treatment outcome. Preclinical investigation using advanced cell/organoid culture techniques and proper animal models could help in this process by providing means to test and generate hypotheses in controlled environment, which is not possible in clinical trials.

Various approaches to using such information are discussed in subsequent Parts II and III.

5.8 CONCLUSION

Pre-clinical models can be used as a valuable tool to test hypotheses generated from outcome models, particularly, when such model suggestions may divert drastically from clinical norms and a clinical trial may be a premature or too risky option. In addition, preclinical data can be a valuable resource to augment missing knowledge from existing clinical data when its application is plausible and supported by scientific evidence.

CHAPTER 6

Biological data: The use of -omics in outcome models

Issam El Naqa

Sarah L. Kerns, PhD

James Coates, PhD

Yi Luo, PhD

Corey Speers, MD PhD

Randall K. Ten Haken

Catharine M.L. West, PhD

Barry S. Rosenstein, PhD

CONTENTS

ABSTRACT

Advances in patient-specific biological information and biotechnology have contributed to a new era of computational biomedicine. Different -omics (genomics, transcriptomics, proteomics, metabolomics) have emerged as valuable resource of data for modeling outcomes and complementing current clinical and imaging information. Here, we provide an overview of these of -omics with specific focus on radiogenomics. We highlight the current status and potential of this data in outcome modeling.

Keywords: -omics, biomarkers, radiogenomics, outcome models.

6.1 INTRODUCTION

O MICS biotechnology has witnessed tremendous growth in recent years with the ability to generate massive amounts (high-throughput) molecular biology data for a wide range of applications from basic biology to deciphering the genetic code of cells or different tumors types by whole genome sequencing of deoxyribonucleic acid (DNA) molecule delivering the field of *Genomics*. Genomics as a term was originally coined by T.H. Roderick, a geneticist, over a beer in Bethesda, MD [196]. This notion has been since expanded to include the analysis of all messenger ribonucleic acid (mRNAs) known as *transcriptomics* , proteins (*proteomics*) and metabolites (*metabolomics*) of a specific biological sample in a non-targeted and non-biased manner using holistic data-driven rather than traditional hypothesis-driven approaches [197].

In the context of radiotherapy, the use of genomics has been coined as *radiogenomics*. This was originally used to describe the study of genetic variations associated with response to radiation therapy to distinguish it from radiomics, which involves the analysis of large amounts of quantitative features from medical images using data-characterization algorithms associated with prognosis and treatment response. However, the latter also could fall under radiogenomics when attempting to correlate imaging features with genetic markers. Radiogenomics have witnessed tremendous growth recently as highlighted by the national and international efforts underpinned by the establishment of the international Radiogenomics Consortium to coordinate and lead efforts in this area [198]. The focus of this chapter will be on the biological description of genomics or radiogenomics as a valuable resource for outcome modeling. More details on radiogenomics could be found in our recent review [17].

6.2 BIOMARKERS AND THE WORLD OF "-OMICS"

Molecular biology has witnessed rapid growth in recent years due to the extraordinary advances in biotechnology and the success of the Human Genome Project (HGP) and its offshoots. These technologies provide powerful tools to screen a large number of biological molecules and to identify biomarkers that characterize human disease or its response to treatment. These biomarkers follow from the central dogma of biology [199], in which biological information is expressed via the sequential transcription of the DNA genetic code into RNA and subsequent translation of the RNA into proteins, which further interact with intermediaries of metabolic reactions (metabolites) to determine cell function or fate as depicted in figure 6.2. Epigenetic modifications of DNA and associated histone molecules can also contribute an additional layer of regulatory influence on this process [200].

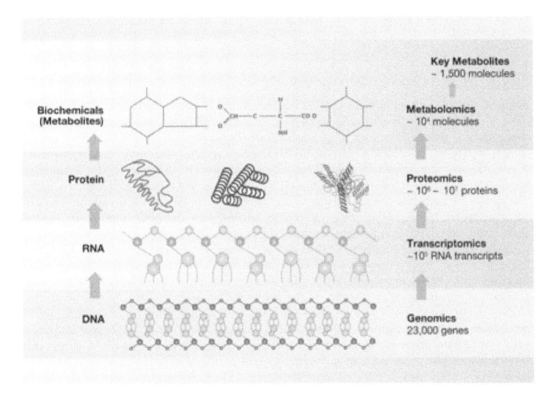

Figure 6.1 Useful biological markers (biomarkers) could be extracted from a biological specimen by whole analysis of its DNA (genomics), RNA (transcriptomics), proteins (proteomics) or metabolites (metabolomics).

A *biomarker* is formally defined as "a characteristic that is objectively measured and evaluated as an indicator of normal biological processes, pathological processes, or pharmacological responses to a therapeutic intervention" [201]. Biomarkers can be divided into prognostic and predictive biomarkers. A *prognostic biomarker* informs about a patient's likely cancer outcome (*e.g.*, disease recurrence, disease progression, death) independent of the treatment received, *e.g.*, Prostate-specific antigen (PSA) serum levels at the time of a prostate cancer diagnosis [202]. A *predictive biomarker* informs about the likely benefit from a therapeutic intervention, *e.g.*, mutations in the epidermal growth factor receptor (EGFR) gene predicting benefit from gefitinib [203]. Biomarkers can further be classified as: (1) *expression biomarkers*, measuring baseline or changes in gene expression, protein or metabolite levels; or (2) *genomics biomarkers*, based on structural variations of tumors or normal tissues, in the underlying DNA genetic code.

6.2.1 Structural variations

There are several classes of genomic variants, all of which may potentially impact response to treatment particularly side effects such as normal tissue radiosensitivity and susceptibility for development of toxicity following radiotherapy. The following will focus on the two most common ones.

6.2.1.1 *Single nucleotide polymorphisms (SNPs)*

Single nucleotide polymorphisms (SNPs) and small insertions and deletions (indels) are the most widely studied genomic variants to date. Common SNPs are single base pair variable sites that are generally present in at least 1% of the population. SNPs can affect protein coding sequences, non-coding introns, and intergenic regions that may include cis-acting and/or trans-acting regulatory sites. There are approximately 10 million SNPs in the average human genome, occurring approximately every 300 nucleotides. In contrast to common SNPs, rare variants are present in less than 1% of the population. Rare variants can also occur in protein coding sequences, non-coding introns, and intergenic regions. Because of their rarity, very large sample sizes are needed to detect associations with disease, but rare variants may be more causally related to disease than SNPs. SNPs generally tag a genomic locus, or region, that contains one or more causal variants. Advances in technology, and decreases in costs, for large-scale genotyping and sequencing projects have enabled the application of genome-wide approaches to the study of disease. In contrast to candidate-gene studies, genome-wide association studies (GWAS) have demonstrated more robust and reproducible results. For instance, SNPs in *TANC1* were found to be significantly associated with overall late urinary and rectal toxicity in prostate cancer patients in a three-stage GWAS [53]. *TANC1* plays a role in regeneration of damaged muscle tissue, representing a pathway not previously implicated in radiotherapy toxicity. A meta-analysis of four GWAS, also in prostate cancer patients, identified two additional loci: SNP rs17599026, which lies in KDM3B, was associated with increased urinary frequency; and rs7720298, which lies in DNAH5, was associated with decreased urine stream [54]. Neither of these genes was previously implicated in cellular radiation response, and they appear to be novel radiosensitivity genes. While these initial results are promising, they likely suffer from overfitting and require testing in independent studies. Evidence from polygenic models of other complex diseases suggests that many SNPs will be required to develop predictive models with high sensitivity and specificity. Indeed, results from a simulation study of radiogenomics outcomes suggests that tens to hundreds of SNPs will be required for good performance models, depending on the allele frequency and effect size of each SNP [204].

6.2.1.2 *Copy number variations (CNVs)*

Gene dosage, or copy number variations (CNVs), are a second genetic variable that may be relevant to radiotherapy toxicities [205, 206]. CNVs capture a much larger part of the genome, spanning thousands to millions of base pairs, whereas SNPs each describe changes specific to only one nucleotide. Thus, CNVs are indicative of more macro-scale changes in the genome that may play more significant roles when using a datamining approach. The objective of CNV analysis is to identify the chromosomal regions at which the number of copies of a gene deviates from two. These could be gains (CNV > 2) or losses (CNV < 2) [207]. CNVs have been identified in several complex diseases such Crohn's disease [208], psoriasis [209]), autism [210, 211], and susceptibility to cancer [212, 213]. CNVs are less numerous than SNPs but they affect up to 10% of the genome, disrupting coding sequences or interrupting long range gene regulation, therefore, accounting for more differences among individuals [214]. In addition, there is evidence to suggest that there is limited overlap between SNPs and CNVs indicating potential complementary effect [215]. As an example application, the *XRCC1* CNV has been shown as a potential risk factor of radiation-induced toxicities in prostate cancer [216].

6.2.2 Gene expression: mRNA, miRNA, lncRNA

Gene expression profiling entails the use of techniques to investigate the relative rate of expression of many mRNA transcripts at a single time with respect to some baseline genes (called housing keeping genes) or across different time points. Microarray techniques using fluorophores and unbiased sequencing approaches (RNA-seq) are the most commonly used methods for determining such expression levels. The advent of high-throughput expression profiling techniques has meant that mRNA levels can be readily quantified and integrated into outcome models. For example, several mRNA signatures have been developed to predict the efficacy of neoadjuvant chemotherapy in breast cancer patients. Several signatures have been derived that reflect tumor hypoxia and predict benefit from combining hypoxia-modifying treatment with radiotherapy [217, 218].

Micro RNAs (miRNA) have also become of interest in oncology in recent times. They have been identified as playing oncogenic as well as tumor suppressor roles and may play a role in disease progression [219]. As an example, miRNA has been implicated to play a role in resistance to radiotherapy [220, 221, 222].

Another class of RNA transcripts that are gaining interest are long non-coding RNAs (lncRNAs), which are abundant in the genomes of higher organisms and seem to be part of the genome regulatory machinery. They are defined as transcripts of 200 up to 100,000 bp lacking an open reading frame [223]. Recent studies suggest that lncRNA plays a role in intercellular communication (gene-silencing) in response to chemoradiation [224, 225].

6.2.3 Protein expression

Protein expressions, in particularly, cytokines, which is a family of signaling polypeptides, mainly proteins that are secreted by different immune cells that mediate inflammatory and immune reactions [226]. Circulating cytokines such as transforming growth factor-β1 (TGF-β1) and the interleukins have been shown to play an important role in radiation-related inflammatory responses. For instance, TGF-β1 has been widely related to play an important role in radiation pneumonitis (RP) and fibrosis [227, 228, 229, 230]. However, reports contradict each other concerning its role in radiation pneumonitis, possibly due to its dual role of as a pro- and anti-inflammatory factor [231] could be the reason for the conflicting reports about its involvement in radiation pneumonitis in breast and lung cancers. Another related set of markers is comprised of variations in adhesion molecules (ICAM-1), which have been reported to occur during irradiation of the lung causing increase in the arrest of inflammatory cells (macrophages and neutrophils) in the capillaries [232].

Cytokines have been demonstrated in several recent animal studies to act by mediating the activation and translocation of nuclear transcription factor kappa B (NF-κB), which has been identified as a good candidate for therapeutic intervention by inhibiting NF-κB activation via CAPE (Caffeic acid phenethyl ester) [233, 234, 235]. In addition, humoral factors including chemokines and other adhesion molecules such as (MCP-1, MIP, and selectins) [236, 237] can further modulate the expression of fibrotic cytokines (bFGF and TGF-β [238, 228, 229, 234], IL-1α [239, 240], and IL-6 [241, 239].

To identify robust biomarkers for predicting a treatment limiting side effect of thoracic irradiation such radiation pneumonitis (RP) in lung cancer patients who received radiotherapy as part of their treatment, mass spectrometry (MS) was performed on peripheral blood samples from a longitudinal 3 × 3 matched-control cohort. To compensate for the large number variables to samples, a graph-based scoring function was developed to rank and identify the most robust biomarkers [242]. The proposed method measured the proximity between candidate proteins identified by MS analysis utilizing prior reported knowledge in the literature of known markers of RP as shown in Figure 6.2a. The α-2-macroglobulin (α2M) protein was ranked as the top candidate biomarker. As an independent validation of this candidate protein, an enzyme-linked immunosorbent assay (ELISA) was performed on independent cohort of 20 patients' samples resulting in early significant discrimination between RP and non-RP patients (p = 0.002) as shown in Figure 6.2b.

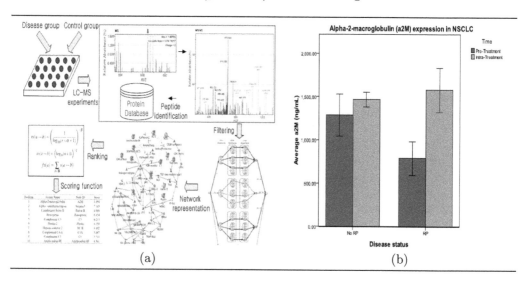

(a) (b)

Figure 6.2 Incorporation of prior knowledge in proteomics analysis. (a) Graph-based proteomics analysis to incorporate prior knowledge in which mass spectrometry data is analyzed in conjunction with known biomarkers of RP using filtering and network analysis. The approach identified α2M as the top ranked candidate. (b) Independent validation using ELISA analysis. Interestingly, it is noted that α2M acts as a radioprotector (higher expression leads to less incidences of RP) and as a biomarker (Patients likely to develop RP experience large increase during therapy) [242, 17].

6.2.4 Metabolites

Metabolites are products of metabolism that drive essential cellular functions, such as energy production and storage, signal transduction and apoptosis. Moreover, metabolites can regulate epigenetic changes and embryonic stem cell behavior. *Metabolomics* is the profiling of such metabolites in biofluids, cells and tissues, is routinely applied as a tool for biomarker discovery [243]. The best studied feature of cancer metabolism is central carbon metabolism and the relationship between glycolysis, the tricarboxylic acid (TCA) cycle and oxidative phosphorylation [244]. Recent metabolomics studies have investigated the role of mitochondrial enzyme serine hydroxymethyltransferase (SHMT2) in human glioblastoma cells, as a regulator of their survival [245]. An interesting application of metabolomics is characterizing hepatocellular carcinoma and its progression from liver cirrhosis by comparing metabolite levels in serum between hepatocellular carcinoma patients and cirrhosis controls revealing peaks of significant interest between spectra from each group [246]. In terms of treatment response, metabolomics analysis from metformin-treated breast cancer patients revealed disruptions to glucose and insulin metabolism [247] and identified platinum resistance in ovarian cancer [248]. In case of radiotherapy, Wibom *et al.* observed systematic metabolic changes induced by radiotherapy treatment in glioblastoma patients indicating the possibility of detecting metabolic marker patterns associated to early treatment response [249].

6.3 RESOURCES FOR BIOLOGICAL DATA

There are no dedicated web resources for outcome modeling studies in oncology per se. Nevertheless, oncology biological markers studies can still benefit from existing bioinformatics resources for pharmacogenomic studies that contain databases and tools for genomic, proteomic, and functional analysis as reviewed by Yan [250]. For example, the National Center for Biotechnology Information (NCBI) site hosts databases such as GenBank, dbSNP, Online Mendelian Inheritance in Man (OMIM), and genetic search tools such as BLAST. In addition, the Protein Data Bank (PDB) and the program CPHmodels are useful for protein structure three-dimensional modeling. The Human Genome Variation Database (HGVbase) contains information on physical and functional relationships between sequence variations and neighboring genes. Pattern analysis using PROSITE and Pfam databases can help correlate sequence structures to functional motifs such as phosphorylation [250]. Biological pathways construction and analysis is an emerging field in computational biology that aims to bridge the gap between biomarkers findings in clinical studies with underlying biological processes. Several public databases and tools are being established for annotating and storing known pathways such as KEGG and Reactome projects or commercial ones such as the IPA or MetaCore [251]. Statistical tools are used to properly map data from gene/protein differential experiments into the different pathways such as mixed effect models [252] or enrichment analysis [253].

In the case of radiogenomics specifically, there has been substantial progress in recent years to identify the genetic/genomic factors associated with the development of normal tissue toxicities following radiotherapy. One of the major factors responsible for the progress that has been made in this field of research is due to formation of the Radiogenomics Consortium (RGC) in 2009 [198]. It was recognized that in order to conduct definitive studies, it would be necessary to both markedly increase the size of cohorts being examined and to include multiple cohorts for meta-analyses and replication studies. To achieve this aim, the RGC was established and became a National Cancer Institute/NIH-supported Cancer Epidemiology Consortium (http://epi.grants.cancer.gov/Consortia/single/rgc.html.). The

RGC currently consists of 217 investigators at 123 institutions in 30 countries. The common aims of the RGC investigators are to identify SNPs associated with radiotherapy adverse effects and to develop predictive assays ready for clinical implementation. The goal of the RGC is to bring together collaborators to pool samples and data for increased statistical power of radiogenomic studies. An important function of the RGC is that it has facilitated cross-center validation studies, which are essential for a predictive instrument to achieve widespread clinical implementation [8]. Initiatives for validation are equally important. The REQUITE Project sponsored by the European Union aims to develop a centralized database for validation of biomarker data [97].

6.4 EXAMPLES OF RADIOGENOMIC MODELING

6.4.1 Prostate cancer

Different modeling schemes could be used to model response to therapy using genetic variants. For instance, a logistic regression model for rectal bleeding using LASSO, a shrinkage approach for variable selection is shown in Figure 6.3 [204]. The figure shows comparison between the predicted incidence of grade 2+ rectal bleeding and the actual incidence of grade 2+ rectal bleeding. The predicted outcomes were produced after applying the logistic regression to outputs of the LASSO model using 2 principal components on the validation data set with 484 SNPs that entered the LASSO. Based on the sorted predicted outcomes, the patients were binned into 6 groups, with the first being the lowest toxicity group and the sixth being the highest. The ratio above each group represents the observed number of patients who experienced grade 2+ rectal bleeding and the total number of patients in the group [204].

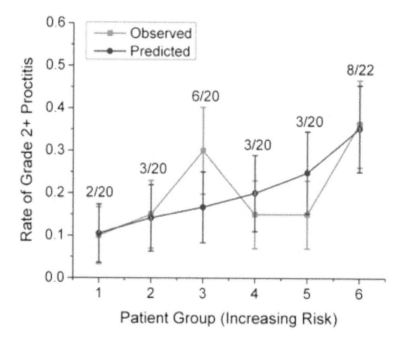

Figure 6.3 Radiogenomic model with SNPs. Prediction of rectal bleeding post-radiotherapy from SNPs using LASSO [204].

6.4.2 Breast cancer

Another modeling example using an ensemble machine learning technique is in presented in the case of breast cancer toxicity. A panel of 147 gene signature was recently correlated with radiosensitivity using random forest (RF) machine learning, which is an ensemble of decision trees. The signature was further refined to 51 genes that were enriched for concepts involving cell-cycle arrest and DNA damage response. It was validated in an independent dataset and shown to be the most significant factor in predicting local recurrence on multivariate analysis outperforming all clinically used clinicopathologic features as shown in Figure 6.4 [254].

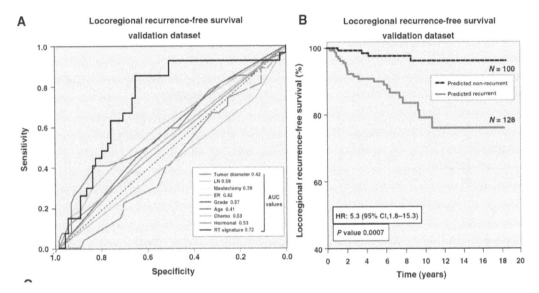

Figure 6.4 Radiogenomic model of breast cancer. Ten-year receiver operating characteristic (ROC) curves (A) and Kaplan-Meier survival estimate (B) analysis in validation dataset of radiosensitivity in breast cancer using random forest machine learning [254].

6.4.3 Lung cancer

Radiation-induced lung inflammation known as radiation pneumonitis (RP), is a limiting toxicity to promising dose escalation radiotherapy techniques of lung cancer patients. By employing advanced bioinformatics tools, robust biomarkers for RP were identified from large-scale proteomic studies. These findings were further validated in an independent dataset using an enzyme-linked immunosorbent assay (ELISA) [242]. These biomarkers identified from proteomics analysis along with other candidate cytokines and dosimetric variables (*i.e.*, mean lung dose, mean heart dose, $V_2 0$, etc.) obtained from published reports were used for constructing a Bayesian network (BN) for RP, which provided better prediction than known individual biomarkers or a combination of variables using conventional regression modeling counterparts [255]. In a recent analysis of lung cancer patients following radiotherapy, BN models of RP were generated that used genetic markers including SNP and miRNA in addition to clinical, dosimetric, and cytokine variables before and during the course of the radiotherapy [256] as shown in figure 6.5. It was noted that the performance of the network improved by incorporating during treatment cytokine changes achieving an area under the receiver-operating characteristic (ROC) curve of 0.87.

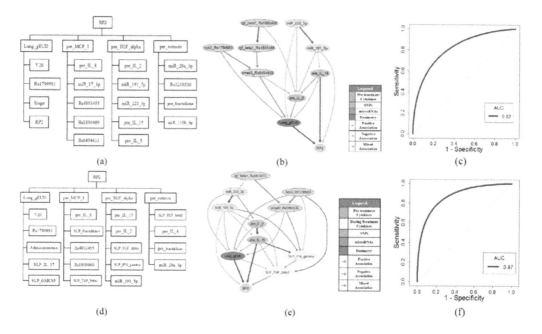

Figure 6.5 Radiogenomic model of lung cancer.An outcome model using systems biology techniques based on Bayesian networks is used for modeling radiation pneumonitis (RP) in lung cancer. First row shows pre-treatment BN modeling of RP. (a) Markov blanket, (b) BN structure, and (c) ROC analysis on cross-validation. The second row shows during-treatment BN modeling of RP. (d) Markov blanket, (e) BN structure, and (f) ROC analysis on cross-validation [256].

6.5 CONCLUSIONS

Due to advances in biotechnology, and complex biologic analysis the amount of potential data available for association with outcomes has grown exponentially in the past decade. This -omics revolution has provided new opportunities for re-shaping our understanding of treatment outcome response. The addition of relevant biomarkers to clinical or image-based outcomes models is likely to improve prediction performance by capturing more of the underlying manifestation of the pathophysiology of disease in question. However, with this large amount of patient-specific biological information, in addition to other resources of data, precautions should be taken to avoid pitfalls of under- or over-fitting during the building process of outcome models.

II

Top-down Modeling Approaches

CHAPTER 7

Analytical and mechanistic modeling

Vitali Moiseenko, PhD

Jimm Grimm, PhD

James D. Murphy, PhD

David J. Carlson, PhD

Issam El Naqa, PhD

CONTENTS

ABSTRACT

We provide a comprehensive review of mathematical models of cell survival following exposure to ionizing radiation, and their connection to clinical outcomes. We emphasize mechanistic models which aim to account for mechanisms governing cell response to radiation. We discuss approaches to describe the process of translating primary radiation-induced DNA damage into lethal lesions, and role of repair and binary misrepair in this process. We describe reaction-rate models and models accounting for track structure and DNA organization in the cell nucleus, as well as the connection between these models and the popular linear-quadratic model. We discuss the importance of accounting for dose fractionation or protraction by different models. We present different approaches to incorporate biological data revealed by molecular and functional imaging, for example the effect of hypoxia, to model cell survival and tumor response. Finally, we present models specifically developed to predict cell survival and tumor response following large doses per fraction, as in intracranial stereotactic radiosurgery/radiotherapy and stereotactic body radiotherapy.

Keywords: radiation, cell survival, dose-response, linear-quadratic model, tumor control probability.

7.1 INTRODUCTION

MECHANISTIC models must account for our understanding of how the physical, chemical, and biologic processes lead to the endpoint under investigation. Our knowledge and interpretation of how radiation-induced biological effects develop, from the atomic and cellular levels up to tumor and normal tissue response have evolved over the years. However, detailed understanding of certain steps involved in the development of radiation-induced damage, and factors mediating this process, remains limited. Despite this, mechanistic models have been efficiently used to probe the relationships of physics, chemistry and biology over different spatial and temporal scales on integral endpoints, such as cell survival, normal tissue and tumor response.

While tumor and normal tissue effects are of primary interest to the radiation oncology community, these tissue-level effects are mechanistically connected to underlying radiation-induced DNA damage and cell death, even though the target cell population may not be well defined. This relationship permeates multi-scale models and provides a framework for describing the radiosensitivity of tumor and normal tissue in response to modifiers, such as dose fractionation/protraction, hypoxia, radiation quality, etc. While statistical models, also known as empirical or phenomenological models, are merely designed to quantitatively describe observed data, mechanistic models allow us to explore the underlying biological causes of observed experimental phenomena and assess the significance of specific factors that alter the response to radiation.

Ongoing mechanistic modeling efforts to address the complex relationship between factors affecting radiation-induced cell death closely follow our understanding of underlying physical and biological mechanisms. Basic components of the process leading to cell kill, specifically, induction of primary lesions, repair, misrepair, dependence of radiation sensitivity on the phase of cell cycle, dose rate/fractionation effects, have been established since the early days of radiation biology. Attempts to combine these components into models have been a central research theme for radiobiologists and radiation physicists. Niederer and Cunningham [257] proposed a sophisticated model to describe cell survival following fractionated irradiation in 1976. The authors considered cell proliferation and kinetics, in particular mitotic delay, as they mediate biological response. Notably, this early attempt to produce a mechanistic model was made before detailed atomistic representation of DNA, Monte Carlo-generated particle tracks, binary effects in DNA misrepair, effect of chromosome domains and other important biological phenomena which have moved to the forefront of modeling biological effects on the DNA and cellular levels.

Many mechanistic models of cell survival aim at accounting for these various factors, including the track structure of particles, DNA organization, the complexity and spatial distribution of DNA damage, and the relationship between these factors to DNA repair and misrepair. Applicability of these models in the radiation therapy world, however, needs to be rigorously tested for robustness using actual clinical data. Most models currently implemented in biologically-guided radiation therapy (BGRT) remain empirical. This chapter describes models which are not strictly mechanistic but driven by our understanding of underlying biophysical processes. These models will be connected to the linear-quadratic (LQ) model, which is by far the most commonly used model in the radiation oncology community to describe the biological effect of radiation dose, dose per fraction, and dose rate for radiotherapy treatments. This chapter focuses mainly on reproductive cell death (also known as mitotic or proliferative), *i.e.*, when cells die while attempting to divide. This is the dominant pathway of cell kill as it relates to modeling radiation therapy outcomes and the narrative below applies to the typical doses and fractionations used in radiation therapy. Other pathways, for example apoptosis, are not considered in this chapter.

7.2   TRACK STRUCTURE AND DNA DAMAGE

Radiation-induced cell death in mammalian cells has been conclusively linked to radiation-induced DNA damage [258, 259, 260]. Cell survival models are based on concepts of primary lesions and lethal lesions, the latter resulting from primary lesions following processes described below. DNA double-strand breaks (DSB), or at least a subset of DSB, have been identified as the precursors which ultimately lead to formation of lethal lesions [261]. Exploring the mechanisms of DNA damage induction by radiation, damage complexity as it relates to DNA repair, and the spatial distribution of DNA lesions in the cell nucleus and among chromosome domains are vital steps required to understand radiation-induced cell kill. Understanding these relevant mechanisms requires detailed knowledge of energy deposition pattern and chemical evolution of a particle track on the DNA scale.

The sequence of events leading to radiation-induced cell kill has been established, although our knowledge of mechanisms governing each step varies greatly between steps. The overall sequence and timing are outlined in Figure 7.1. Note that time scale for stages following the pre-chemical stage are strongly dependent on cell type and repair capacity. In a typical cellular environment, chemical evolution is complete by $10^8 - 10^7$s [262]. Energy depositions by electrons and heavy charged particles are presently well understood, at least for energy depositions in liquid water. Monte Carlo codes to simulate energy depositions in matter have been developed, reported, and reviewed [263, 264]. Further, track structure codes have been applied to model radiation-induced damage, *e.g.*, PARTRAC [265], NOREC [266], and GEANT4-DNA [267].

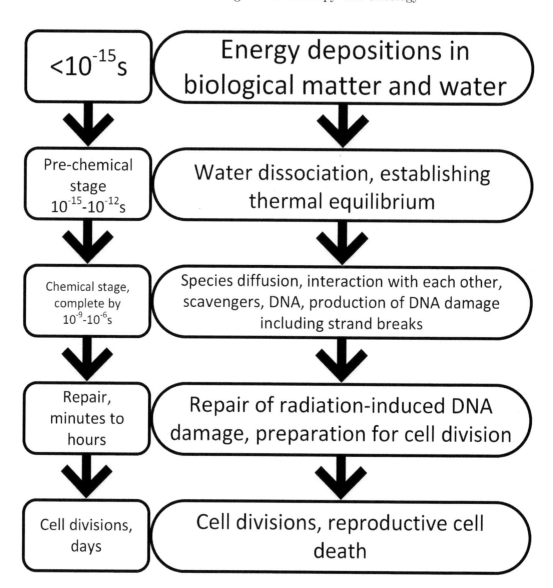

Figure 7.1 Sequence of events governing mammalian cell response to ionizing radiation.

Monte Carlo track structure codes model energy deposition in water whereas the target of interest is DNA. While modeling indirect effects requires consideration of producing ionizations and excitations in water molecules, modeling direct effects has to rely on a representative model of DNA. This representation has progressed dramatically through the years starting with simple geometric-volumetric approach, with DNA modeled as a cylinder structure with segments assigned as bases and sugar-phosphate backbone [268]. This simplified representation has been replaced by detailed atomistic models including hydration shells. Figure 7.2 shows the basic components required to model cellular DNA linear DNA and a nucleosome including linker DNA [269, 270]. Higher levels of DNA organization have been successfully modeled [265, 271] and are described in details in chapter 10. These models, combined with detailed track structure consideration, bridge the gap between DNA damage and chromosome aberrations, some of which are known to be lethal.

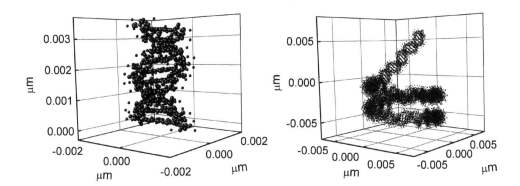

Figure 7.2 Left panel: ten-base pair segment of linear DNA (large size symbols) including the first hydration shell (small size symbols); right panel: one nucleosome including the linker DNA.

These models are capable of probing for impact of damage complexity, geometrical distribution of DNA damage within and between chromosome domains, and the relation of these characteristics to radiation quality. Therefore, accounting for damage complexity when considering repair, or accounting for geometrical distribution of DSB when considering formation of exchanges, can be achieved.

7.3   LINEAR-QUADRATIC MODEL

The linear-quadratic (LQ) model is the most commonly used model in radiation therapy to account for dose fractionation/protraction, to propose and compare fractionation schedules, and to compensate for missed fractions. Broad use of this simple but versatile model owes in large part to efforts by Dr. J.F. Fowler, who pioneered the use of radiobiological modeling in radiation therapy [272]. Commonly used concepts of biologically effective dose (BED), distinguishing between early and late radiation response and how they connect to radiobiological parameters have been developed by Dr. Fowler and will remain in the forefront of radiation therapy [14].

The LQ equation, in a form accounting for dose fractionation/protraction, has been originally applied to describe the yield of chromosome aberrations, Y, as a function of dose, D [273]:

$$Y = \alpha D + G \beta D^2 \tag{7.1}$$

Coefficients α and β describe two pathways of producing chromosome aberrations , one-track and two-track action, respectively. The Lea-Catcheside factor , G, accounts for dose fractionation/protraction, and, if we assume mono-exponential repair as a function of time:

$$G = \frac{2}{D^2} \int_{-\infty}^{\infty} \dot{D}(t)dt \int_{-\infty}^{t} e^{-\lambda(t-t')} \dot{D}(t')dt' \tag{7.2}$$

where is $\dot{D}$ is dose rate as a function of time, t, and λ is a repair constant. The link between the number of chromosome aberrations and cell survival was established in the seminal report by Cornforth and Bedford [274], although cell survival is not solely dependent on the presence of certain types of aberrations [275]. The authors showed that the proportion of cells surviving and the proportion of cells free from certain types of chromosome aberrations closely match. This implies that certain types of chromosome aberrations are indeed lethal. These chromosome aberrations are called unstable because their frequency decreases through cell divisions as cells containing these aberrations are incapable of progressing through mitosis. This is in contrast to stable chromosome aberrations, for example symmetric translocations, which may corrupt the genetic material but do not interfere with cell division.

Studies of dose dependence for chromosome aberrations also reveal that while certain aberrations, for example terminal deletions, show linear dose-response, other aberrations such as dicentrics show a pronounced quadratic term for low-LET radiation. This connects to the binary exchange concept which is the cornerstone of the LQ model [276]. Table 7.1 shows the general pathway of inducing stable and unstable chromosome aberrations. This table also demonstrates how the induction of chromosome aberrations, some of which are lethal, connects to the one-track/two-track foundation of the LQ model. Chromosome aberrations shown in the fourth column of the bottom three rows require two DSBs. If each DSB is produced by a separate track (two-track action), the frequency of these aberrations will depend on dose squared. This also illustrates how the quadratic term in the Equation 7.1 is dependent on dose rate as both DSBs have to be in the same cell at the same time, and capable of interaction, for an exchange to form.

The debate whether LQ is a mechanistic or purely empirical model has been on-going and proponents of the model point to one-track/two-track pathways, and accounting for dose protraction/fractionation as mechanistically-based basics of the LQ formalism [275]. It was pointed out that the basis of the model does not apply to large doses per fraction. Specifically, the LQ model over-predict cell lethality for doses relevant to stereotactic radiosurgery (SRS) and body radiotherapy (SBRT) [277]. Counter-arguments question the very foundation of the model from both microdosimetric and statistical point of view [278]. Despite limitations on the model, it remains the current standard model used in radiation oncology, in particular when it comes to isoeffect calculations [14]. The basic equation connecting cell survival and tumor control probability (TCP) based on Poisson statistics is [279]:

$$S = e^{-\alpha D - G\beta D^2} \tag{7.3}$$

$$TCP = e^{-N_S} = e^{-N_0 S} = e^{-N_o exp(-\alpha D - G\beta D^2)} \tag{7.4}$$

where N_S is the number of surviving tumor cells capable of regrowing the tumor (clonogens) , N_0 is the initial number of clonogens, and S is the surviving fraction. For a specific case of fractionated treatment with dose D delivered in n fractions we can assume that time to deliver a fraction is too short for repair to impact the quadratic term, $i.e.$, when accounting for intra-fraction dose-rate effects $G = 1$. Time between fractions is typically

TABLE 7.1   Example of outcomes following induction and repair/misrepair of DNA DSBs. Faithful restitution means that the cell is viable. Failure to repair makes the cell non-viable. Binary events may leave the cell viable (third column) or result in chromosome aberrations which lead to cell death, although not necessarily in the first division (fourth column).

Number of DSB and distribution among chromosomes/arms	DSB depicted on chromosome level	Faithful repair or viable binary misrepair outcomes (stable chromosome aberrations)	Failure to repair or lethal binary misrepair (unstable chromosome aberrations)
One chromosome, one DSB		Faithful restitution	Terminal deletion
One chromosome, two DSB, different arms		Pericentric inversion	Centric ring
One chromosome, two DSB, same arm		Paracentric inversion	Acentric ring
Two chromosomes, one DSB each		Symmetric translocation	Dicentric

sufficient for full repair to preclude interactions between lesions produced by different fractions, i.e., for inter-fraction effects $G = 0$, overall value of the Lea-Catcheside factor is well approximated by $G = 1/n$. To further account for tumor cell proliferation after a lag period of T_K:

$$TCP = e^{-Ns} = e^{-N_0 S} = e^{-N_o exp(-\alpha D - \beta d D + \gamma(T - T_K))} \tag{7.5}$$

where d is dose per fraction D/n, T is overall treatment time, and γ is the repopulation term $\gamma = ln2/T_p$, where T_p is the doubling time for tumor clonogens. Lag period, T_K, was estimated at four weeks for head and neck and lung cancer [163, 280]. The repopulation term only apples for $T > T_K$. This formulation is obviously simplified. At minimum these simplifications assume that there is no inter-patient variability in cell radiosensitivity or initial number of clonogens in a tumor, all cells within a tumor are equally sensitive, and dose is uniform through the tumor volume.

Accounting for different sources of heterogeneity will require additional model parameters, which is highly undesirable. Moreover, there is reciprocity in heterogeneity, for example mathematically increasing the initial number of clonogens can be compensated by assuming radioresistance. This is clearly demonstrated by the inter-patient heterogeneity model proposed by Webb and Nahum [281]. The authors suggested ignoring the quadratic term in the LQ equation, then, accounting for inter-patient heterogeneity with the following equation:

$$TCP_{IPH} = \int \psi(\alpha, \bar{\alpha}, \sigma_\alpha) e^{-N_0 exp(-\alpha D)} d\alpha \tag{7.6}$$

where ψ describes the distribution of α among patients, this distribution is often assumed Gaussian, with the mean value $\bar{\alpha}$ and standard deviation σ_α. This equation clearly shows how variation in the number of clonogens and variation in α can be re-cast mathematically. For any arbitrary dose distribution and clonogen density distribution we can divide the volume into voxels small enough so that dose and clonogen density are uniform. Then the overall TCP is dependent on the product of voxel control probabilities, VCP. If voxel i receives dose D_i, there are N_i clonogens in the voxel i, and there are M voxels the above equation can be generalized:

$$TCP_{IPH} = \int \psi(\alpha, \bar{\alpha}, \sigma_\alpha) \prod_{i=1}^{M} VCP_i d\alpha = \int \psi(\alpha, \bar{\alpha}, \sigma_\alpha) \prod_{i=1}^{M} e^{-N_i exp(-\alpha D_i)} d\alpha \tag{7.7}$$

This equation is generic and for practical applications has to be connected to realistic distributions. Moreover, use of functional imaging to probe for properties affecting radiotherapy outcome, has been growing. Accounting for these properties is shown in Section 7.5.

The LQ expression can be derived using multiple methods including detailed reaction-rate models as shown in Section 7.4, or a premise which bypasses mechanistic considerations. Specifically, the theory of dual radiation action [282] connects biological response of a cell as a second order reaction of sublesions, and the number of radiation-induced sublesions in a sensitive volume proportional to a microdosimetric quantity, specific energy, z. Assumption of binary interactions is again at the core of the formulation. For a dose D, the number of lesions, $E(D)$, which result from sublesions, is:

$$E(D) = K(\bar{z}_D + D)D \tag{7.8}$$

where $\bar{z}_D$ is dose-mean specific energy and K is a constant characteristic of a cell. At low doses when the quadratic term can be ignored this model gives a simple microdosimetrically driven method to calculate RBE, which is simply a ratio of dose-mean specific energy values. This formalism also implies that the quadratic term is cell-line specific, and is not connected to microdosimetric quantities.

7.4   KINETIC REACTION RATE MODELS

The LQ model, while providing a versatile formalism to account for time-dose-fractionation, in particular for isoeffect calculations, does not explicitly describe the production of primary lesions (DSB), their repair and misrepair, and the production of lethal lesions. The kinetic reaction rate model describes the processes of the production of primary lesions and their processing using equations of chemical kinetics, hence the name kinetic. The repair-misrepair (RMR) and lethal-potentially-lethal (LPL) models [283, 284] formulated in the mid 1980s serve as a benchmark. These two models contained binary exchanges as a pathway of lethal lesion production which has become an integral feature of refined models. Overall, reaction rate models follow the pathways outlined in table 7.1, at least as far as binary exchanges are concerned [285]. Notably, interpretation of the events which downstream lead to the linear term in dose-response remains uncertain. Although our knowledge of DNA damage, repair mechanisms and pathways of cell lethality are much more advanced compared to the 1980s, we still know more about behavior of the linear term, α, as a function of radiation quality, and how it varies between cell lines, rather than how it comes about [286].

It is well established that as dose rate decreases the contribution of the quadratic term to cell death also decreases, while the α term remains unchanged; that is, unless dose rate becomes ultra-low and other factors like repopulation or reassortment start playing a role. An interpretation of events leading to a terminal deletion, Table 7.1, varies from model to model, including in the RMR and LPL. While some models assume conversion of sublethal lesions into lethal lesions via lethal restitution [284], others assume that a proportion of sublethal lesions fail to go through either repair or misrepair pathway and therefore remain as residual by the time a cell reaches mitosis [287]. Other alternative models have been proposed and extensively reviewed [285]. The LPL model makes a considerably different assumption in that it proposes a pathway of direct production of a lethal lesion by a single track. This difference, while appearing minor, leads to different formalism for lethal lesion production. In RMR, a certain proportion of sublethal lesions is converted into lethal lesions via lethal restitution, and at any time is dependent on the number of sublethal lesions present and rate of lethal restitution [284]. Conversely, in the LPL model the rate of producing single-track lesions will be proportional to dose rate [283]. Most notably, under specific circumstances the solution to the reaction rate models reduces to the LQ equation [285].

7.4.1   Repair-misrepair and lethal-potentially-lethal models

These models have been extensively reviewed in the literature though only a brief description is presented in this chapter. The RMR model is based on the pathways outlined in Table 7.1 and assumes existence of uncommitted lesions. The productions of these lesions (*e.g.*, a DSB) is linearly correlated with dose, their repair is assumed mono-exponential, and binary misrepair pathways of lethal lesion production is an important feature of this model [284].

The equations describing the rate of producing primary lesions, U, their repair, misrepair and production of lethal lesions, L [285]:

$$\frac{dU}{dt} = \delta \dot{D} - \lambda U - kU^2 \tag{7.9}$$

$$\frac{dL}{dt} = (1 - \varphi)\, \lambda U + (1 - \psi)kU^2 \tag{7.10}$$

Primary lesions are produced with a rate δ, repaired with a time constant λ, and binary misrepair occurs at a rate k. A certain proportion of repair $(1-\varphi)$ leads to lethal restitution and a proportion of binary exchanges $(1-\psi)$ are lethal. Assuming that the Poisson statistics applies the surviving fraction is simply the exponent $exp(-L(t))$. For certain simple initial conditions the solution is possible, in particular for a situation relevant to radiation therapy, when full dose is delivered in a period of time too short for repair to take place, and full repair is allowed.

Sachs *et al.* [285] argued that $(1-\psi)=1/4$, which, as illustrated in Table 7.1 means that for one dicentric we need four DSBs because a dicentric results from 2 DSBs half of the time, the other half results in a symmetric translocation. Then, following the narrative of Sachs et al, if repair dominates misrepair, $kU^2 << \lambda U$, and repair/misrepair are allowed to run full course, which is to say we can justify setting time to infinity, the solution reduced to LQ. Equation 7.3 can therefore be obtained under these assumptions and:

$$\alpha = \delta\,(1 - \varphi) \tag{7.11}$$

$$\beta = \frac{(\varphi - \psi)\, k\delta^2}{2\lambda} \tag{7.12}$$

The LPL model assumes presence of potentially lethal lesions, n_{PL}, with their production and repair/binary misrepair identical to shown above for RMR. However, this model is distinctly different from RMR in describing production of lethal lesions, L. As discussed above it assumes a direct one-track production of lethal lesions [283], consequently:

$$\frac{dL}{dt} = \alpha \dot{D} + ckn_{PL}^2 \tag{7.13}$$

While these models are different in their formulation, notably, the solutions can-be recast from one model to another [285].

7.4.2 Refined models

As any other model, RMR and LPL models are based on assumptions and as our knowledge of events leading to radiation-induced cell kill progresses these assumptions change. Two aspects of the RMR and LPL models are in particular simplifications and these have been addressed in refined models. First, both models disregard DNA organization in the cell nucleus. While DSB yield is known to linearly depend on dose at least in the dose region of interest for radiation therapy, yield of lethal lesions and frequency of chromosome aberrations are non-linear functions of dose. This conversion of linear to non-linear has been conclusively attributed to repair/misrepair, and in particular binary events [288]. Ideally, modeling of events leading to cell lethality should to account for track structure and how this track structure interplays with DNA organization. The latter concept has received plenty of attention in the literature with chromatin fiber, loops, and chromosome domain organization in interphase modeled in fine detail [289, 290, 291]. Second, both models assume exponential repair, which is a simplification [292]. In addition to the presence of repair pathways characterized by different time constants [286], *e.g.*, slow and fast repair [293], other aspects of repair behavior as a function of time and dose have been addressed, for example repair saturation [294].

Attention to track structure and its interplay with DNA organization was in part prompted by increased interest in proton therapy, and investments in carbon ion radiotherapy. It has been long accepted that not all DSBs are created equal and accompanying strand breaks or base damage contribute to their complexity, prompting researchers to designate complex DSB as DSB+ or DSB++ [295]. Another hypothesis is that these DSB may present a stronger challenge to repair enzymes [296] and may be prone to misrepair [297]. The question of radiation quality is then not only if high-LET radiation produces more DSB per unit dose compared to X- or γ-rays, but also if these DSB are functionally different [298, 299]. This, combined with track structure-driven proximity effects, i.e., spatial distribution between DSB and their distribution in fiber, loops and domains, are pre-requisite to mechanistically address effects of radiation quality on cell kill [300, 289, 301]. Some of these refined models are briefly described below.

7.4.3 The Giant LOop Binary LEsion (GLOBE)

The Giant LOop Binary LEsion (GLOBE) model [291] accounts for DNA organization and is motivated by the concept of chromatin loops approximately 2 Megabasepair in size. The model distinguishes between isolated DSB within a loop iDSB and clustered damage producing multiple DSB in one giant loop, cDSB. Cell survival is assumed to obey the Poisson distribution, that is, the exponent to the expectation value for the number of lethal lesions is equal to survival probability. Spatial-temporal production of DSB can then be converted into survival:

$$S = exp(-\varepsilon_i n_i - \varepsilon_c n_c) \tag{7.14}$$

where n_i and n_c are expectation values for iDSB and cDSB, and $\varepsilon_i n_i + \varepsilon_c n_c$ is the expectation value for the number of lethal lesions. The model allows for binary exchanges between DSB produced in different domains, however does not account for correlation of the number of DSB in proximal loops. This model has been applied to describe cell survival data for photons including ultra-soft photons, and comparisons with RMR and LPL models have been presented [291, 302, 286]. Notably, the authors demonstrated that until dose is large (tentatively >10 Gy, however exact value is cell line-dependent), LQ and GLOBE models agree in their predictions and LQ parameters can be re-cast from the GLOBE model parameters [303].

7.4.4 Local Effect Model (LEM)

To specifically address modeling of cell survival to serve as a radiobiological foundation for ion beam therapy a model tailored to account for the energy deposition pattern characteristic of ion beam was proposed. The Local Effect Model (LEM), which went through a number of updates (LEMI-LEMIV) in an attempt to account for energy deposition pattern through the concept of local dose. The premise of the LEM is that microscopic local dose distribution pattern determines biological effect, and equal local doses lead to equal local effects, irrespective of radiation quality. This model has been put forward to account for spatially correlated distribution of DSB following traversal of a cell nucleus by a heavy charged particle. This pattern of producing correlated DSB within a single loop may further lead to release of DNA segments not attached to the nuclear matrix [291, 304]. The mean number of DSB in a subvolume is derived from photon data, showing a yield of approximately 30 DSB per cell per Gy. From a local average number of DSB, the spatial distribution of DSB is calculated using Monte Carlo simulations amorphous track structure pattern in place of the probability density function of DSB. Indirect effects, *i.e.*, DNA damage by radicals following water radiolysis, are accounted for in LEM by smearing radial dose profile. The authors propose use of a cluster-index:

$$C = \frac{N_{cDSB}}{N_{cDSB} + N_{iDSB}} \tag{7.15}$$

where N_{cDSB} and N_{iDSB} are numbers of loops containing clustered and isolated DSB. This index quantitatively captures the effect of clustering as heavy ions will be likely to deposit energy locally and produce correlated breaks within a loop. The model has been demonstrated to accurately describe the RBE dependence on LET for V79 cells [291].

7.4.5 Microdosimetric-kinetic model (MKM)

The microdosimetric-kinetic model (MKM) builds on microdosimetric concepts which served as a foundation of the dual radiation action theory [282]. Similar to LPL and RMR, MKM describes production of potentially lethal lesions and their processing into lethal lesions as differential equations [305, 306, 307, 308]. The basic premise of the MKM is that the cell nucleus is divided into equal size compartments (or domains) which fill the volume of a nucleus like a jigsaw puzzle and their number can be from a few hundred to a few thousand. If reshaped into spheres these domains will have a diameter of 0.5-1 μm. Equal size does not imply equal DNA mass, which is allowed to vary between domains. A radiation-induced DNA lesion is confined to a domain and interactions between lesions in different domains is not allowed. Dose variation from domain-to-domain is accounted for to mimic a traversal of the cell nucleus by a charged particle. Similar to RMR or LPL, potentially lethal lesions can undergo faithful restitution, conversion to a lethal lesion and binary exchanges confined to a domain. The MKM model has been used to predict RBE as a function of LET and a connection to LQ parameters, suitable for RBE analysis has been reported [307].

7.4.6 The Repair-misrepair-fixation model

As DSB are the most critical form of initial DNA damage formed by radiation, it is conceptually appealing to link trends in the α and β parameters of the LQ model with particle type and LET to DSB induction. The Repair-Misrepair-Fixation (RMF) model directly links DSB induction and processing to clonogenic cell death [309]. In the RMF, a coupled system of non-linear ordinary differential equations is used to model the time-dependent kinetics of DSB induction, rejoining, and pairwise DSB interaction to form lethal (and non-lethal) chromosomal damage. The RMF model assumes that it is the formation of lethal point mutations and chromosome damage rather than the initial DSB themselves that are the dominant mechanism underlying the effects of particle LET on cell death. Earlier kinetic reaction-rate models such as the RMR and LPL models also assume both linear and quadratic terms for the repair of sub-lethal and potentially lethal damage. However, in the RMF model, the initial formation of DSB is assumed to follow a compound Poisson process instead of a simple Poisson process as in the RMR and LPL models. Reproductive cell death by mitotic catastrophe, apoptosis, or other cell death modes are explicitly linked to DSB induction and processing in the RMF model. Table 7.1 shows all potential DSB processing pathways. The effect of radiation quality q on the α and β radiosensitivity parameters of the LQ model can be effectively modeled using [309, 310, 311]:

$$\alpha\left(q\right) = \theta\Sigma(q) + k\bar{z}_F(q)\Sigma^2(q), \tag{7.16}$$

and

$$\beta\left(q\right) = (k/2)\,\Sigma^2(q) \tag{7.17}$$

Here, Σ is the initial number of DSB per Gray per giga base pair ($Gy^{-1}\ Gbp^{-1}$), z_F is the frequency-mean specific energy (Gy), and θ and κ are tumor- or tissue-specific parameters related to the biological processing of initial DSB into more lethal forms of damage (*e.g.*, chromosome aberrations). θ is the fraction of DSB that undergo lethal first-order misrepair and damage fixation, whereas κ is the fraction of initial DSB that undergo pairwise damage interaction. An attractive feature of Equations 7.16 and 7.17 is that estimates of α and β for any particle type, including electrons and photons, are determined by a well-defined, purely physical parameter that is the same in vitro and in vivo (z_F), two biological parameters (θ and κ) that are tissue-specific but independent or a weak function of radiation quality, and a biological parameter (Σ) that is a strong function of radiation quality (track structure) but is approximately the same in vitro and in vivo and among all mammalian cells. Sophisticated stochastic models, such as the Monte Carlo Damage Simulation (MCDS), are also available to predict trends in Σ with radiation quality (*e.g.*, [312, 313], and references therein).

For a reference radiation x with known radiation sensitivity parameters α_x and β_x, the radiation quality-dependent LQ parameters can also be expressed by:

$$\alpha\left(q\right) = \frac{\Sigma(q)}{\Sigma_x}\left(\alpha_x + 2\frac{\beta_x}{\Sigma_x}(\Sigma\left(q\right)\bar{z}_F\left(q\right) - \Sigma_x\bar{z}_{F,x})\right), \tag{7.18}$$

and

$$\beta\left(q\right) = \left(\frac{\Sigma(q)}{\Sigma_x}\right)^2\beta_x \tag{7.19}$$

Equations 7.18 and 7.19 provide a convenient formalism to relate published radiosensitivity parameters for photons and other radiations with low-LET to radiosensitivity parameters for other particle types, including protons and other heavier ions. It has been demonstrated [310, 311, 314] how mechanistic models, such as the RMF, can be used in particle therapy to predict the relative biological effectiveness (RBE) of heavy ions and optimize radiotherapy treatments using protons, helium, and carbon ions. This is discussed in Chapter 14.

7.5 MECHANISTIC MODELING OF STEREOTACTIC RADIOSURGERY (SRS) AND STEREOTACTIC BODY RADIOTHERAPY (SBRT)

SRS and Stereotactic radiotherapy (SRT) are special cases of extreme hypofractionation and refer to stereotactic radiosurgery in single-fraction or multiple fractions of irradiation for treating intracranial lesions with a very high degree of dose conformity. This approach has been recently extended to extracranial radiosurgery and is often referred to as stereotactic body radiation therapy (SBRT) or stereotactic ablative radiotherapy (SABR) in the literature. It should be noted that available in vitro data mainly support application of the standard LQ model in the range of 1-5 Gy [315], and above this range is currently a subject of intense debate in the literature. This includes application to SRS and

SBRT/SABR, which typically apply higher dose per fraction than 5 Gy as discussed below [316, 317, 318, 319, 320, 321].

There are currently two schools of thought on the current clinical success of SRS/SBRT treatments. One argument is that classic radiobiology principles (5Rs) and the LQ model are adequate to explain the effects of large dose per fractions and the main observed advantages are mainly attributed to the "Geometrical Window of Opportunity" that is expanded by advancements in image-guidance and treatment delivery techniques [322, 316, 320, 321]. Others highlight potential limitations of the LQ model for accurately describing high dose per fraction data and laboratory evidence suggesting the possible existence of radiobiology that may involve new mechanisms related to stem cells, vascular damage, and immune-mediated effects or combination of these [317, 319].

Classical radiobiology principles suggest that cancers with low α/β (e.g., melanoma, liposarcoma, prostate and some breast cancers) will benefit from hypofractionation regimens while typical cancers with high α/β (i.e., around 10 Gy or higher), will benefit from conventional or hyperfractionated regimens in terms of sparing late-reacting normal tissue. These effects could be adequately discerned from the LQ model. However, application of these concepts to higher doses per fraction have been questioned in the light of recent SBRT results such as RTOG-0236, which is a phase II multi-institutional trial of early stage lung cancer patients with a prescription dose of 18 Gy per fraction x 3 (54 Gy total), which yielded an estimated 3-year primary tumor control rate of 97.6% (95% CI: 84.3%-99.7%), disease-free survival of 48.3% (95% CI, 34.4%-60.8%), and overall survival of 55.8% (95% CI, 41.6%-67.9%) [323]. These results represent remarkable improvement over conventional fractionation, which in theory according to the LQ model would be the fractionation of choice for lung cancer given its high α/β.

Proponents of a unique high-dose biology cite that under certain SRS and SBRT regimens classical radiobiology fails to explain outcomes. For instance, SRS or SBRT delivery for each session is relatively longer than conventional fractionation allowing for higher rates of sub-lethal damage SLD repair, where a fraction delivery in excess of 30 minutes may lead to loss of BED [324]. Moreover, at large doses (e.g., > 10 Gy/fraction), enhanced vascular damage may occur [325], leading to tumor microenvironment hypoxia and acidification, which prevent reoxygenation. High doses prevent cell cycle progression and cells undergo interphase death in the cell cycle phases where they are irradiated rather than being redistributed. Repopulation during the shortened 1-2 weeks period may be as well negligible. Furthermore, they assert that the LQ model may not be appropriate at the high doses per fraction encountered in SRS/SBRT because it was derived largely from in vitro, rather than in vivo data, and thus does not consider the impact of ionizing radiation on the supporting tissues nor does it consider the impact of the subpopulation of radioresistant clonogens (i.e., cancer stem-cell) or vascular damage [317, 319, 325]. On the other hand, the clinical relevance of these unique high-dose biological mechanisms remains to be determined. The other school of thought argue that the 5Rs and the LQ model are supported by available preclinical and clinical data and there is little evidence to suggest otherwise [322, 316, 320, 321]. For instance, in the case of lung cancer, Brown et al. suggest that dose escalation (BED > 100 Gy), rather than new biology may be adequate to explain the excellent clinical efficacy of SBRT in early stage non-small cell lung cancer (NSCLC) patients as shown in Figure 7.3 [320]. This issue is still a subject of ongoing debate given the uncertainties in the existing published data.

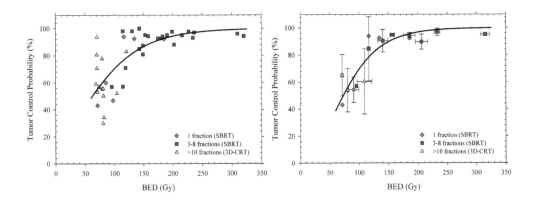

Figure 7.3 TCP as a function of BED [320].

7.5.1 LQ limitations and alternative models

It is recognized that the LQ is an approximation to more sophisticated kinetic reaction models [283, 284], see Section 7.4. Therefore, several modifications have been introduced to the LQ model to allow better fit to high doses per fraction. These modifications to the LQ model effectively aim to straighten the survival curve at high doses, where it fails. This could be achieved by simply having higher α/β values in cases of rapidly proliferating and hypoxic tumors [316, 326] or by developing alternate models to the standard LQ model such as the linear-quadratic-linear (LQL) [327, 328], linear-quadratic-cubic (LQC) [329], the universal survival curve (USC) [330], and the generalized LQ (gLQ) (Wang et al. 2010) among others. Below, we briefly discuss some of these alternate LQ models intended for hypofractionation regimens.

The phenomenological Modified LQ or its equivalent kinetic LQL model modify the standard LQ by including an additional parameter (δ), which characterizes the bending of the isoeffect curve at higher doses replacing the standard protraction factor of Equation 7.2 as follows [327, 328]:

$$S = e^{-\alpha D - \beta G(\delta D)D^2} \tag{7.20}$$

$$G = \frac{2}{(\delta D)^2}(e^{-\delta D} + \delta D - 1), \tag{7.21}$$

where δ can vary from 0.04 to 0.4 Gy^{-1}. In this case, G(δD) term increases survival for an acute dose fraction and supposedly accounts for the reduction in repairable lesions due to interactions with forming lesions [327].

The LQC accounts for the linear response at higher doses by adding a term proportional to the cube of the dose:

$$S = e^{-\alpha D - \beta D^2 + gD^3} \tag{7.22}$$

where $g = \beta/(3D_L)$ and $D_L = 18$ Gy at which the survival curve is presumed to become a straight line.

The USC model creates a hybrid between the LQ model and the classical multi-target model [330]:

$$lnS = \begin{pmatrix} -n\left(\alpha d + \beta d^2\right), \ d \leq D_T \\ -n\left(\frac{d}{D_0} - \frac{D_q}{D_0}\right), \ d \geq D_T \end{pmatrix} \tag{7.23}$$

where the transition dose (D_T), is given by $D_T = 2Dq/(1 - \alpha D_0)$. In this case, it is assumed that the LQ model gives a good description for conventional fractionation (i.e., the shoulder effect), while for ablative doses beyond the shoulder, the survival curve is better described as a straight line as predicted by the multi-target model.

The gLQ aims to continuously encompass the entire range of possible dose delivery patterns [331]:

$$S = e^{-\alpha D - G\beta D^2} \tag{7.24}$$

$$G = \frac{2}{D^2} \sum_{i=1}^{n} \{\frac{R_i^2}{\varepsilon_i^2}(\varepsilon_i T_i - 1 + exp(\varepsilon_i T_i)) +$$

$$\sum_{j=1}^{i-1} \frac{R_i R_j exp(-\beta D_{ij})}{\varepsilon_i \varepsilon_j} [exp(\varepsilon_j t_j) - exp(\varepsilon_j t_{j-1})][exp(-\varepsilon_i t_{i-1}) - exp(\varepsilon_i t_i)]\} \tag{7.25}$$

where β_2 is the second hit component of $\beta = \beta_1 \beta_2/2$, and the dose rate pattern is given by $R(t) = R_i, t_{i-1} < t \leq t_i, 1 = 1, \ldots, N$.

It should be noted that an abundance of caution should be used when interpreting and applying these models to clinical data and comparing their performances in terms of the assumed parameters, dose prescription, isodose lines, and the size of irradiated volume. Empirical or semi-empirical modifications to the LQ model are attempts to straighten cell survival response (on a log scale) at high doses [327, 328, 330, 332]. However, the underlying biological mechanisms implied in these modifications are often unclear and the question remains whether the addition of unsubstantiated parameters into the modeling process could help improve the predictions of clinical outcomes than the standard LQ without further validation. For fractionated stereotactic body radiation therapy (SBRT), several investigators found that the standard LQ model suffices when modeling local tumor control [333, 334, 316, 321] and pulmonary toxicity [335, 336]. Alternative high-dose models did not provide better fits of the clinical data than the two-parameter LQ model. Mehta *et al.* [337] found that when analyzing the available clinical data for early-stage non-small cell lung cancer (NSCLC), the clinical significance between the LQ and USC models at therapeutic doses in an SBRT regimen is minimal. A recent analysis of the available local control data for 3,000 early-stage NSCLC and brain metastases patients undergoing single- and multi-fraction SBRT found that the LQ model with heterogeneous radiosensitivity provided a significantly better fit over the entire range of treatment doses than did any of the high-dose alternative models [321].

The analysis of published local control data for lung tumors and brain metastases from patients undergoing stereotactic radiotherapy suggests that unique tumoricidal mechanisms do not determine tumor control at high doses per fraction[320, 316, 321]. However, more homogeneous clinical data sets are needed to further test this hypothesis. In addition, the negative impact of tumor hypoxia on biological effectiveness may be more significant in the context of hypofractionation [316, 338], particularly for single-fraction SBRT [333, 334, 321].

7.6 INCORPORATING BIOLOGICAL DATA TO DESCRIBE AND PREDICT BIOLOGICAL RESPONSE

Biological imaging, specifically molecular and functional imaging, provides a means to probe tumor and normal tissue properties which may impact response to radiation. The most commonly used modalities including positron emission tomography (PET), single photon emission tomography (SPECT) and functional magnetic resonance imaging (fMRI) provide quantitative tools to assess spatial distribution of a property of choice for an individual patient, a field currently known as radiomics (see Chapter 3). These properties include glucose metabolism, hypoxia, perfusion, diffusion, cell proliferation, membrane synthesis and amino acid metabolism. Emphasis thus far has been on probing tumor properties and designing trials to account for tumor biology. There is convincing evidence that select properties such as hypoxia or metabolic activity assessed by biological imaging before or during the course of radiotherapy relate to outcomes. In particular, a relationship between tumor control and tumor biological properties was demonstrated for non-small lung cancer [339], rectal cancer [340], and cervical cancer [341]. Boosting the dose to regions which show properties associated with higher risk recurrence, e.g., large concentration of tumor cells, hypoxia or rapid proliferation, is an intuitive yet clinically unproven approach to individualize radiotherapy treatments for specific subgroups of patients. Designing these boost targets ideally should rely on dose-response and a mechanistic understanding of how a particular property affects this response.

Segmenting the regions for boosting and designing discrete levels for dose boosts represents an inviting method to enhance radiotherapy planning because tools readily available in commercial treatment planning systems can be used [342]. This is a practical take on the dose painting-by-numbers approach [343]. The biggest obstacles in utilizing this approach are that the desired dose must be connected to the specific property, and this relationship, as well as the quantitative relationship between biological imaging and actual property, remain uncertain. The simplest approach to design boost doses connected to a property, which is described by a voxel intensity, I, is to assume a linear relationship between dose and intensity, under constraints that there is a minimum baseline dose required, D_{min}, and there is maximum dose to ensure safe radiotherapy, D_{max}. This leads to a simple relationship between dose and voxel intensity:

$$D\left(I\right) = D_{min} + \frac{I - I_{min}}{I_{max} - I_{min}}\left(D_{max} - D_{min}\right). \tag{7.26}$$

This formalism is void of any relationship between biological imaging and modification of response. The LQ equation based on Poisson statistics, Equation 7.3, is versatile and can readily accommodate factors affecting outcomes. Depending on the nature of the factor this can be incorporated either through modifying dose-response characteristics, or weight of a voxel. In the simplest case of non-uniform spatial distribution of tumor cells the number of surviving cells can be rewritten as:

$$N_s = \sum v_i \rho_i S_i \tag{7.27}$$

where ρ_i is cell density and v_i is fractional volume. Factors which modify dose-response have to cast through LQ model parameters. Hypoxia in particular received attention as use of PET and MRI to probe for hypoxia has been expanding [344, 345]. Modeling efforts have concentrated on accounting for hypoxia through well-established modifying factors, *e.g.*, oxygen enhancement ratio (OER), focusing on deriving OER estimated from in vitro and in vivo data [346], including the effects of intermediate levels of hypoxia on treatment response for different dose fractionations [338, 347], and separating the effects of chronic and acute hypoxia. Preclinical studies have demonstrated promise, for example, a model was developed and tested on a rat sarcoma data set [348]. The premise of the approach is to modify LQ parameters, α and β, established for oxygenated conditions to account for presence of hypoxia, for example for chronic hypoxia LQ parameters α_{ch} and β_{ch}:

$$OER_ch = \frac{\alpha_0}{\alpha_{ch}} = \left(\frac{\beta_0}{\beta_{ch}} \right)^{(1/2)} \tag{7.28}$$

Surviving fraction for fractional dose d then:

$$S = e^{\frac{-\alpha d}{OER_{ch}} - \frac{\beta d^2}{OER_{ch}^2}} \tag{7.29}$$

Accounting for acute hypoxia can be done in a similar manner except acute hypoxia, by definition is present only a portion of the time. The model has been further translated from applying to preclinical studies to predicting TCP for NSCLC following SBRT. Presence of hypoxia may be an important factor in hypofractionated treatment [349] because of reduced potential for reoxygenation compared to conventional fractionation [338]. Owing to the versatility of the Poisson-based TCP model tumor characteristics revealed by biological imaging, for example the presence of a quickly proliferating pools of cells can be incorporated [348, 350]. As a result modeling-based optimal fractionation can be derived [351].

A compartment approach has also been tested to attempt a more sophisticated modeling approach [352]. The authors accounted for the interplay between proliferation and the presence of hypoxia. Three compartments were assumed to be present: proliferative, intermediate, and hypoxic. This formalism is sophisticated and includes both radiation-induced effects, and effects caused by environmental factors, for example cell loss due to lack of access to nutrients. The model allows for cells migrating between compartments, and accounts for hypoxia and proliferation. The authors connect the model to biological imaging, in particular FDG-PET and investigate response modulation by different FDG profiles [353, 352].

7.7 CONCLUSIONS

Mathematical models of cell survival, tumor and normal tissue response to radiation are meant to describe observed data and make predictions for doses and fractionation schedules considered for clinical practice and clinical trials. These models help assessing changes in biological response, particularly in the absence of clinical data. The impact of these predictive models is particularly important in the modern radiation era which allows conformal delivery of radiation to the tumor and boosting doses to subvolumes showing higher risk of recurrence. Mechanistic models are most appropriate for this purpose compared to empirical models as they rely on quantitative interpretation of mechanisms governing cell, tumor and normal tissue response to radiation. The models presented in this chapter can be used to guide selection of new or alternate radiotherapy regimens. It is important to remember that uncertainties in model predictions relate to uncertainties in radiosensitivity parameter estimates. These uncertainties will have a large impact on calculated dose prescriptions when the alternate treatment differs greatly from the established therapy regimens such as when converting standard fractionated regimens to the high dose per fraction used with stereotactic radiation. It is therefore prudent to make small incremental changes away from conventional regimens instead of drastic changes to reduce the possibility of mis-estimation. Ultimately, clinical data from prospective randomized trials are the gold standard in medicine, and mathematical models, in particular mechanistic models, can only assist guiding clinical radiation oncology.

CHAPTER 8

Data driven approaches I: conventional statistical inference methods, including linear and logistic regression

Tiziana Rancati

Claudio Fiorino

CONTENTS

ABSTRACT

Multivariable methods of statistical analysis are commonly used in oncology. This type of analysis considers multiple variables to predict a single outcome, with multivariable methods exploring the relation between two or more predictors (independent variables) and one outcome (dependent variable). These models serve two purposes: (1) they can predict the value of the dependent variable for new values of the independent variables and (2) they can help describe the relative contribution of each independent variable to the dependent variable, controlling for the influences of the other independent variables. Linear regression and logistic regression are among the most popular multivariable models. They have some mathematical similarities but differ in the expression and format of the outcome variable. In linear regression, the outcome variable is a continuous quantity, such as blood pressure or plasma level of a biomarker. In logistic regression, the outcome variable is usually a binary event, such as alive versus dead, or case versus control. In this chapter, linear and logistic regression will be presented from basic concepts to interpretation.

Keywords: outcome models, statistical modeling, linear regression, logistic regression.

8.1   WHAT IS A REGRESSION

Ehind a model, there is always the attempt to simplify a complex phenomenon starting from the analysis of a limited set of data, supposed or estimated to be sufficiently large to give a correct representation of it. Then, the need for creating a model starting from measured data mainly depends on the intrinsic limitation due to the impossibility of measuring "the whole Universe" when referring to a certain event. In oncology, an event may typically be, for instance, the relapse of a tumour, the death of a patient or the onset of a side-effect. In other words, a model is a way to build conjectures that may synthetically and efficiently represent our data, with, moreover, the implicit and important aim to construct a tool able to predict, within certain limitations, the risk of experiencing the event in the future [354].

Modelling real data always denotes finding those variables with an optimal combination and weighting to allow for better customizing prediction to both represent discovering dataset patterns and possibly describe a large number of independent datasets, including future data. All these models, grounded on the observation of association in systematic and extended collections of data, are called parametric (statistical) models.

Regression is a statistical method that involves prediction of a continuous or quantitative output (e.g., the level of a biomarker expression), this includes prediction of the probability of occurrence of a defined event (*e.g.*, death or insurgence of a side-effect). A regression assesses whether the values of predictor features (the independent variables) resolve variability in the outcome measure (the dependent variable).

A regression model comprises Equations (the underlying mathematical formalism of the relationship between independent and dependent variables) and parameters (the weighting coefficients to be applied to the different predictors to calculate the predicted outcome). Equations are established by the type of regression shape (*e.g.*, linear, logistic, ecologic, logic, Bayesian, or quantile), while parameters are estimated by fitting the observed data to the regression formulae. In mathematical language, given a set of N independent variables $\vec{X} = (X_1, X_2, \cdots X_N)$ and a dependent outcome variable Y, we assume that the values of Y are related to $\vec{X}$ according to the following relationship:

$$Y = f(\vec{X}, \vec{\beta}) \tag{8.1}$$

where f is the regression formula and $\vec{\beta}$ is the vector of regression parameters $\vec{\beta} = (\beta_1, \beta_2, \ldots, \beta_N)$ to be associated with the vector of independent variables $\vec{X} = (X_1, X_2, \cdots X_N)$.

In this chapter, we will focus on widely used linear and logistic regression. These topics are the subjects of several statistics books; we suggest [355, 356, 357, 358] to expand the subject matter. Linear and logistic regression formalisms are simple but nevertheless often adequate to first approach the data and common relationships often encountered in the oncology field. Their results are easily interpreted and allow insights into the possible meaning of the influence of predictors over the outcome, thus also providing a basis for designing further "biological" experiments or clinical trials devoted to deepening the understanding of phenomena and searching for causal relationships explaining the statistical relationships highlighted from regression.

8.2 LINEAR REGRESSION

Continuous, quantitative outcomes are quite common in medical oncology studies, even if they are not often considered for clinical prediction models. Multiple linear regression is a useful tool for predicting this type of quantitative response.

8.2.1 Mathematical formalism

The multiple linear regression model can be written as follows:

$$Y = \beta_0 + \sum_i \beta_i \times X_i + \varepsilon \tag{8.2}$$

where Y is the quantitative outcome, β_0 is the intercept, β_i denotes the regression coefficients that relate the value of the predictor features X_i to the value of the outcome Y and ε is the error. The error is defined as the difference between the observed outcome Y and the predicted outcome $\widehat{Y}$. This entity is also known as the residual of the prediction of Y. In linear regression, we assume that the distribution of the residuals has a normal shape (normality) and does not depend on X_i (homoscedasticity) [358].

8.2.2 Estimation of regression coefficients

In practice, the intercept β_0 and the regression coefficients β_i are not known and have to be estimated from data. Let's consider a dataset consisting of M cases, each described by its outcome Y_j and by its N observed independent features $X_{j,i}$ (with the index j running on the M cases and i running on the N recorded features). Our goal is to find an intercept $\widehat{\beta}_0$ and N regression coefficients $\widehat{\beta}_i$ so that the line resulting from Equation 8.2 is as close as possible to the M observation of Y_j.

This is usually accomplished by minimizing the sum of squared residuals. Given

$$\widehat{Y}_j = \widehat{\beta}_0 + \sum_i \widehat{\beta}_i \times X_{i,j} \tag{8.3}$$

i.e., the prediction of Y_j based on the jth values of X_i, $\varepsilon_j = Y_j - \widehat{Y}_j$ is the j_{th} residual. The residual sum of squared (RSS) is defined as follows:

$$RSS = \sum_j \varepsilon_j^2 = \sum_j \left(Y_j - \widehat{Y}_j \right)^2 = \sum_j \left(Y_j - \widehat{\beta}_0 - \sum_i \widehat{\beta}_i \times X_{i,j} \right)^2 \tag{8.4}$$

The values of $\widehat{\beta}_0$ and $\widehat{\beta}_i$ are calculated as those that minimize RSS in Equation 8.4.

In the particular case of linear regression involving only one predictor (i.e., $i = 1$), $\widehat{\beta}_0$ and $\widehat{\beta}_1$ can be calculated by using the method of least squares; calculations lead to the following expressions:

$$\widehat{\beta}_1 = \frac{\sum\limits_j (X_j - \bar{X})(Y_j - \bar{Y})}{\sum\limits_j (X_j - \bar{X})^2}, \quad \widehat{\beta}_0 = -Y - \widehat{\beta}_1 - X \tag{8.5}$$

where $\bar{Y}$ and $\bar{X}$ are the sample means, defined as $\bar{Y} \equiv \frac{1}{M} \sum\limits_{j=1}^{M} Y_j$ and $\bar{X} \equiv \frac{1}{M} \sum\limits_{j=1}^{M} X_j$.

When more predictors are involved, multiple regression coefficient estimates may have more complicated forms; their explicit analytical expressions are discussed in Chapter 12. Any statistical software package includes libraries that can compute these coefficient estimates. Later, at the end of this section, we will give an example with calculations in the freely available package R [359].

As a general point, any multiple linear regression involving N predictors can be graphically represented in a $N+1$ dimensional space. The predicted values $\widehat{Y}_j$ lie in the hyperplane defined by Equation 8.3, the residuals correspond to the vertical distance between the data points and the plane, and the multiple regression coefficients $\widehat{\beta}_i$ are those minimizing the global deviation of all data points from the plane. In the example at the end of this section, we explicitly consider the case of a linear regression involving two variables, with a graphical example for this particular case.

8.2.3 Accuracy of coefficient estimates

In linear regression modeling, we start from the assumption that the true relationship between Y and X_i is given by Equation 8.2, involving β_0 and β_i, and we use a given set of data to estimate $\widehat{\beta}_0$ and $\widehat{\beta}_i$. Equation 8.3 is thus considered the best approximation of the true relationship between Y and X_i.

A different dataset could generate different estimates for β_0 and β_i, and if the same is pursued for a large number of datasets, a distribution of $\widehat{\beta}_0$ and $\widehat{\beta}_i$ would be obtained. If we use the means of these distributions to estimate β_0 and β_i, these estimates are unbiased. In other words, on the basis of one dataset we could overestimate the coefficients, while based on another dataset we could underestimate them. However, if a large number of observations is available, the average estimated values would exactly equal to the regression coefficients of Equation 8.2 (the real relationship between Y and X_i).

In clinical trial, we almost always have the possibility to fit data to one single dataset; in this case, it is of interest to quantify how far the single estimates of $\widehat{\beta}_0$ and $\widehat{\beta}_i$ are from β_0 and β_i. This is accomplished by computation of the coefficient intervals. Calculation of standard errors (SE) and, consequently, of coefficient intervals are beyond the scope of this chapter; some details can be found in [357]. Common statistical softwares always provide estimates of regression coefficients together with their SE; thus, the 95% coefficient intervals for β_0 and β_i are $\beta_0 = \widehat{\beta}_0 \pm 1.96 \cdot SE(\widehat{\beta}_0)$ and $\beta_i = \widehat{\beta}_i \pm 1.96 \cdot SE(\widehat{\beta}_i)$.

8.2.4 Rejecting the null hypothesis

After having determined regression coefficients and their SE, it could be of interest to perform hypothesis testing on the coefficients to assess their significance. The most common test involves testing the null hypothesis $H_0 : \beta_1 = \beta_2 = \ldots = \beta_N = 0$ (i.e., there is no relationship between Y and X_i, and all coefficients equal 0) vs the alternative hypothesis $H_\alpha : \beta_i \neq 0$ for at least one β_i. This test is conducted by computing F-statistics [357]:

$$F = \frac{\frac{TSS - RSS}{N}}{\frac{RSS}{M - N - 1}} \tag{8.6}$$

where N is the number of predictors included in the model, M is the number of observations in the dataset, RSS is the residual sum of squared (as defined in Equation 8.4) and TSS is the total sum of squares, $TSS = \sum_j (Y_j - \bar{Y})^2$, with $\bar{Y} \equiv \frac{1}{M} \sum_{j=1}^{M} Y_j$ being the outcome sample mean.

If there is no relationship between the outcome and the predictors, F is close to 1. On the other hand, if F is greater than 1, there is evidence against the null hypothesis, with at least one predictor with $\beta_i \neq 0$. The exact value of F highly depends on M and N. For a large M, low values for F can be obtained, still supporting evidence against the null hypothesis. When H_0 is true and the errors ϵ_i are normally distributed, F follows an F-distribution. Of note, F follows an F-distribution even if the errors are not normally distributed, as long as M is large. All statistical packages provide p-values for F-statistics for every combination of M and N, thus allowing us to evaluate how confidently the null hypothesis could be rejected [357].

8.2.5 Accuracy of the model

Once H_0 has been rejected, we assess how well the model fits the observed outcome. The accuracy of a linear regression is usually assessed using the residual standard error (RSE) and R^2 statistics. RSE is related to the average gap between model predicted outcomes and observed outcomes; it is defined as follows:

$$RSE = \sqrt{\frac{1}{M - N - 1}\sum_{j}(Y_j - \widehat{Y}_j)^2} = \sqrt{\frac{1}{M - N - 1}RSS} \qquad (8.7)$$

RSE is considered an absolute measure of lack of fit of the model to the data. It is a very intuitive measure; nevertheless, it is given in units of Y. For this reason, it is not always clear which value defines a good RSE. Of note, from Equation 8.7, RSE could increase with increasing the number of predictors (N) included in the model. This can happen when a small decrease in RSS due to the addition of a variable is counterbalanced by the effect of an increase N in the denominator of Equation 8.7.

R^2 statistics overcomes this limitation and can provide a widely used alternative: it has the form of a proportion, the proportion of the variance explained, thus always taking a value between 0 and 1, which is independent from the scale of Y. R^2 statistics is defined as:

$$R^2 = \frac{TSS - RSS}{TSS} \qquad (8.8)$$

TSS is the total variance in the outcome Y, while RSS is a measure of the variability that is still unexplained after application of the model. TSS-RSS is the variability that was removed by application of the model, i.e., the variability explained by the model. R^2 close to 1 implies that RSS is close to 0, i.e., predicted outcome values are all very close to observed outcomes. R^2 getting closer to 0 denotes a large RSS, with large differences between predicted and observed values.

An additional tool for model accuracy is the plotting of data. Graphical visualization can help in stressing issues that are not captured by standard statistical measures, such as non-linearity of relationships or presence of interaction effects (synergy between variables). Examples of how a graphical representation could help in highlighting the presence of a non linear relationship or of an effect of interaction between two variables are presented in Figures 8.1 and 8.2.

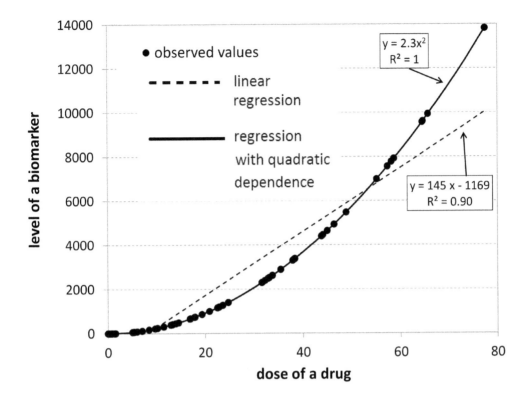

Figure 8.1 Example using a simulated dataset consisting of measurement of plasma levels of a biomarker as a function of dose of a drug. Solid circles represent observed data. Linear regression apparently fits the experimental points, $y = 145x - 1169$, with $R^2 = 0.90$, which could indicate an accurate model. On the other hand, a graphical representation of data stresses the existence of a non linear relationship, with increasing distance between predicted and observed values for increasing doses. The graphical representation reveals that a quadratic regression is more appropriate.

8.2.6 Qualitative predictors

Multiple linear regression can be performed even in the presence of qualitative predictors, also known as factors. If a factor only has two possible values (levels), it can be introduced into the model in a straightforward way by defining a dummy variable that takes only two possible numerical values. The exact values assigned to the dummy variable (*e.g.*, 0/1, 1/0 or -1/1) have no influence on the magnitude of predicted outcomes, *i.e.*, the final predictions will be identical, regardless of the selected coding rule. The only difference is in the resulting regression coefficients to be applied to the factor and in the interpretation of their meaning. When the binary coding 0/1 is used, the category with the value set to 0 is considered as the reference category (and prediction in this category is not influenced by the coefficient of the dummy variable); meanwhile, the other one is the subgroup adding a difference to the prediction, and the coefficient of the dummy variable can be interpreted as the average difference in outcome between the two categories. This choice (0-1 coding) is often applied

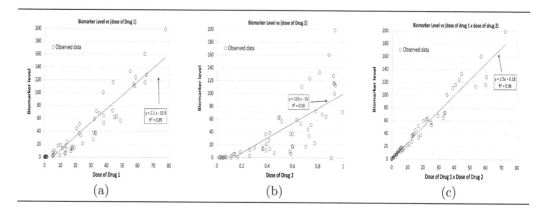

Figure 8.2 Example using a simulated dataset consisting of measurement of plasma levels of a biomarker as a function of doses of two drugs. Empty circles represent observed data. Panel (a) presents biomarker levels as a function of dose of drug 1, together with linear regression (which is apparently sufficiently accurate, $R^2 = 0.89$); panel (b) reports biomarker levels as a function of dose of drug 2, together with linear regression (which result is not accurate, $R^2=0.56$); panel (c) presents biomarker levels as a function of the interaction between dose of drug 1 and dose of drug 2 (*i.e.* a new variable defined as the product between dose of drug 1 and dose of drug 2 was considered), together with linear regression with the interaction variable (which result is more accurate than regression only considering linear relationships, $R^2=0.96$). The graphical representation helped in revealing more appropriate relationships.

when there is a "reason" for indicating a reference category, *e.g.*, not having a comorbidity can be considered a baseline/reference situation with respect to having the comorbidity. When there is no plausible choice for a reference category, the coding -1/1 is often selected, *e.g.*, when considering gender. In this case, the regression coefficient can be interpreted as the average in outcome of each category with respect to the average outcome of all observations, regardless of their category.

When a factor has more than two levels, a single dummy variable cannot be used to account for all possible values. A set of variables should be considered; *e.g.*, for a three-level factor Z with possible values $A/B/C$, a first dummy variable X_1 can be set to 1 if the factor$=A$ and 0 if the factor$\neq A$; a second variable X_2 can be set to 1 if the factor$=B$ and 0 if the factor$\neq B$. In this way if $Z = A \rightarrow X_1 = 1$ and $X_2=0$, $Z = B \rightarrow X_1 = 0$ and $X_2 = 1$ and $Z = C \rightarrow X_1 = 0$ and $X_2 = 0$. X_1 and X_2 can be inserted into a linear regression with the possibility of describing all levels of variable Z.

8.2.7 Including interactions between variables

In some situations, the simple linear additive model may be incorrect. In Equation 8.2, increases in Y due to an increase in any selected X_i are completely independent from what occurs with all the other variables included in the model.

A possible way to extend this type of model is to allow insertion of a further predictor that considers the interaction between variables. For simplicity, let us consider a two variable model [357]:

$$Y = \beta_0 + \beta_1 X_1 + \beta_2 X_2 + \varepsilon \tag{8.9}$$

We could consider the possibility of including an interaction term between X_1 and X_2:

$$Y = \beta_0 + \beta_1 X_1 + \beta_2 X_2 + \beta_{1,2} X_1 X_2 + \varepsilon \tag{8.10}$$

This can be re-written as:

$$Y = \beta_0 + \beta_1 X_1 + \beta_2 X_2 + \varepsilon \tag{8.11}$$

where $\tilde{\beta}_1 = \beta_1 + \beta_2 X_2$, thus indicating that the coefficient for X_1 is now dependent on the value of X_2.

The presence of synergy between variables is of great importance in the oncology field, e.g., the effect of the dose of one drug can have an (increasing or decreasing) interaction effect on the response outcome due to the dose of another drug.

Importantly, when including interaction terms, the hierarchical principle has to be considered. This principle establishes that when interaction between two variables is considered, both variables have to be included also as main effects, even if the coefficients associated with their main effects are not significantly different from 0.

8.2.8   Linear regression: example

As a practical example of what was described in the above sections, we report the analysis of a simulated dataset. The dataset includes simulated data on 60 patients created starting from a set of patients who underwent radical radiotherapy for prostate cancer and who participated in a study on the association of plasma level of inflammatory molecules with acute radio-induced toxicity [360]. In particular, this dataset contains data on baseline and end of radiotherapy levels for the chemokine (C-C motif) ligand 2 (CCL2) in pg/mL, variables **baselineCCL2** and **endRTCCL2**, age (**age**), body mass index (BMI), presence of diabetes (**diabetes**), hypertension (**hypertension**), previous abdominal surgery (**surgery**), presence of neoadjuvant androgen deprivation (**neoadjAD**) and use of anticoagulants (**anticoagulants**). The aim of the multiple linear regression analysis is to determine the relationship between variables measured before radiotherapy and CLL2 levels at the end of radiotherapy, which are supposed to be associated with acute toxicity.

Analyses are performed in R, The R Foundation for Statistical Computing Platform [361] (https://cran.r-project.org/bin/windows/base/ for free download of the package). Basic functions to perform linear regression are included in the base distributions; the **library()** function can be used to load specific libraries that allow the use of more sophisticated functions. In the following, we need the libraries **rms** for Regression Modelling Strategies and **scatterplot3d**, which includes functions for 3-dimensional plotting.

```
> install.packages("rms")
> install.packages("scatterplot3d")
> library(rms)
> library(scatterplot3d)
```

Data are included in the `dataex` object, a matrix with cases in lines and variables in columns. The following commands allow reporting of variables included in `dataex` and some basic descriptive statistics.

```
> names(dataex)
[1] "age"           "BMI"          "diabetes"       "hypertension"
[5] "surgery"       "neoadjAD"      "anticoagulants" "baselineCCL2"
[9] "endRTCCL2"
#
> datadist(dataex)
                 age   BMI diabetes hypertension surgery neoadjAD
Low:effect     62.25 24.00        0            0       0        0
Adjust to      68.00 25.00        0            0       0        0
High:effect    73.00 26.00        1            1       1        1
Low:prediction 57.75 21.85        0            0       0        0
High:prediction 78.05 31.05       1            1       1        1
Low            53.00 19.00        0            0       0        0
High           79.00 32.00        1            1       1        1
                anticoagulants baselineCCL2 endRTCCL2
Low:effect                   0      122.125  142.0725
Adjust to                    0      137.330  163.8550
High:effect                  1      151.270  183.3400
Low:prediction               0       97.777  109.6250
High:prediction              1      209.308  223.7960
Low                          0       66.700  102.6300
High                         1      215.560  248.8200

Values:

BMI : 19 22 24 25 26 27 31 32
diabetes : 0 1
hypertension : 0 1
surgery : 0 1
neoadjAD : 0 1
anticoagulants : 0 1
```

The following lines have the goal of telling R that the variables to be considered are in `dataex` and that some variables are qualitative (factors).

```
> attach(dataex)
> diab.f=factor(diabetes)
> hyper.f=factor(hypertension)
> surgery.f=factor(surgery)
> neoadjAD.f=factor(neoadjAD)
> antico.f=factor(anticoagulants)
```

A matrix of scatterplots can help in visualizing what types of relationships are present between data. In the following, to preserve visualization, we only used a sub-group of variables, but the command line can be extended with the inclusion of all features in the dataset.

```
> pairs(~age+baselineCCL2+BMI+surgery.f+antico.f+endRTCCL2,
data=dataex, main="Scatterplot matrix")
```

The plot in Figure 8.3 is generated; each panel in the figure is a scatterplot of a pair of variables, which are identified based on the corresponding row and column labels.

We can now start with simple linear regression including the variable that seems more associated with endRTCCL2; in this particular situation, it is baselineCCL2. The lm() function performs linear regression; the summary() function provides details on regression results. For convenience, results from the lm() function are stored in the lm.fit object (which can be later included in the arguments of R functions). The results for simple linear regression can be visualized in a scatterplot together with the regression line.

```
> lm.fit=lm(endRTCCL2~baselineCCL2,data=dataex)
#
> summary(lm.fit)

Call:
lm(formula = endRTCCL2 ~ baselineCCL2, data = dataex)

Residuals:
    Min      1Q  Median      3Q     Max
-34.251 -21.341  -6.084  10.524  79.392

Coefficients:
             Estimate Std. Error t value Pr(>|t|)
(Intercept)   83.8072    16.3770   5.117 3.67e-06 ***
baselineCCL2   0.5691     0.1138   5.001 5.61e-06 ***
---
Signif. codes:  0 '***' 0.001 '**' 0.01 '*' 0.05 '.' 0.1 ' ' 1

Residual standard error: 28.67 on 58 degrees of freedom
Multiple R-squared:  0.3013,    Adjusted R-squared:  0.2892
F-statistic: 25.01 on 1 and 58 DF,  p-value: 5.606e-06
#
> confint(lm.fit, level=0.95)
                   2.5 \%      97.5 \%
(Intercept)   51.0251291 116.5893244
baselineCCL2   0.3412738   0.7968626
#
> plot(baselineCCL2, endRTCCL2, main="Scatterplot CCL2 level end RT
vs baseline CCL2", xlab="baseline CCL2", ylab="end RT CCL2", pch=19)
> abline(lm.fit, col="blue")
```

Summary() provides information on estimated regression coefficients and their standard errors, the distribution of residuals, and R^2 and F-statistics. Adjusted R^2 is also reported, which is a modification of R^2 that adjusts for the number of explanatory variables in a model and for the sample size. It is an attempt to consider the phenomenon of R^2 automatically and artificially increasing when extra predictors are added to the model. The difference between R^2 and adjusted R^2 becomes small for large datasets and decreasing numbers of predictors included in the model. `confint()` allows for determination of confidence intervals for regression coefficients at any given level (95% confidence intervals in the example). `plot()` and `abline()` functions help produce a graphical representation of results. The plot in Figure 8.4a is generated (scatter plot of `endRTCCL2` versus `baselineCCL2`, together with the model regression line).

Scatterplot matrix

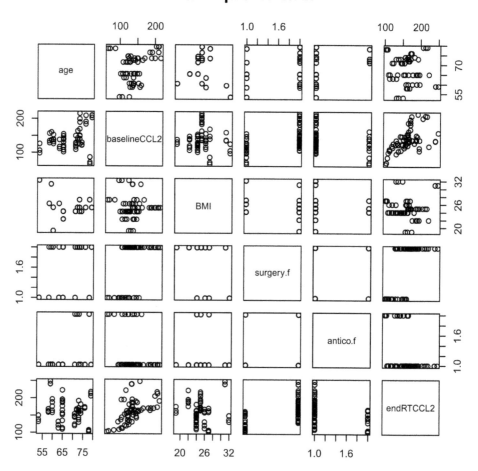

Figure 8.3 The dataex dataset: plot of information about baseline and end of radiotherapy levels for the chemokine (C-C motif) ligand 2 (CCL2) in pg/mL, age, body mass index, previous abdominal surgery), and use of anticoagulants (anticoagulants).

Going from one to two predictors, we can represent each observation inside a three-dimensional space. The values predicted by the multiple linear regression lie in the plane $\widehat{Y} = \widehat{\beta}_0 + \widehat{\beta}_1 X_1 + \widehat{\beta}_2 X_2$.

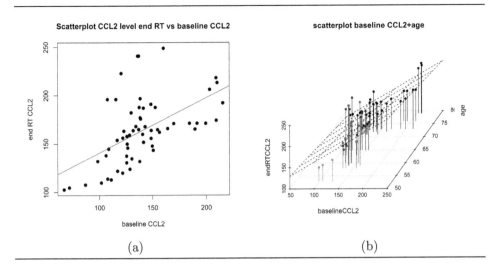

Figure 8.4 (a) scatter plot of endRTCCL2 vs baselineCCL2, together with the model regression line. (b) three-dimensional scatterplot of endRTCCL2 vs baselineCCL2 and age, together with the plane representing the regression model.

```
> lm2.fit=lm(endRTCCL2~baselineCCL2+age,data=dataex)
> summary(lm2.fit)

Call:
lm(formula = endRTCCL2 ~ baselineCCL2 + age, data = dataex)

Residuals:
    Min      1Q  Median      3Q     Max
-38.285 -20.294  -5.407  14.018  68.378

Coefficients:
              Estimate Std. Error t value Pr(>|t|)
(Intercept)   169.6625    33.6449   5.043 4.97e-06 ***
baselineCCL2    0.6654     0.1124   5.919 1.95e-07 ***
age            -1.4687     0.5114  -2.872  0.00572 **
---
Signif. codes:  0 '***' 0.001 '**' 0.01 '*' 0.05 '.' 0.1 ' ' 1

Residual standard error: 27.03 on 57 degrees of freedom
Multiple R-squared:  0.3896,    Adjusted R-squared:  0.3682
F-statistic: 18.19 on 2 and 57 DF,  p-value: 7.764e-07
#
> scatterplot3d(baselineCCL2,age,endRTCCL2, pch=20, highlight.3d=TRUE,
type="h", main="scatterplot baseline CCL2+age", angle=40, box=F)
> s3d=scatterplot3d(baselineCCL2,age,endRTCCL2, pch=20, highlight.3d=TRUE,
type="h", main="scatterplot baseline CCL2+age", angle=55, box=F)
> s3d$plane3d(lm2.fit)
```

The three-dimensional plot generated by scatterplot3d is shown in figure 8.4b. endRTCCL2 vs baselineCCL2 and age are reported, together with the plane representing the regression model.

We can now try a multiple linear regression, including all variables that appear to contribute to the end radiotherapy levels of CCL2. The function predict() computes predicted values using the fitted model. To visualize the results, we can plot predicted values vs observed end radiotherapy levels of CCL2 (Figure 8.5).

```
> lm4.fit=lm(endRTCCL2~baselineCCL2+age+BMI+surgery.f)
> summary(lm4.fit)

Call:
lm(formula = endRTCCL2 ~ baselineCCL2 + age + BMI + surgery.f)

Residuals:
    Min      1Q  Median      3Q     Max
-51.828 -14.527  -4.267  12.239  57.969

Coefficients:
             Estimate Std. Error t value Pr(>|t|)
(Intercept)  127.0070    43.0650   2.949 0.004671 **
baselineCCL2   0.3950     0.1041   3.795 0.000370 ***
age           -1.5546     0.4279  -3.633 0.000617 ***
BMI            2.1785     1.1107   1.961 0.054913 .
surgery.f1    43.3599     7.8644   5.513 9.69e-07 ***
---
Signif. codes:  0 '***' 0.001 '**' 0.01 '*' 0.05 '.' 0.1 ' ' 1

Residual standard error: 22.08 on 55 degrees of freedom
Multiple R-squared:  0.607,      Adjusted R-squared:  0.5784
F-statistic: 21.24 on 4 and 55 DF,  p-value: 1.238e-10

> pred=predict(lm4.fit)
> plot(endRTCCL2,pred)
```

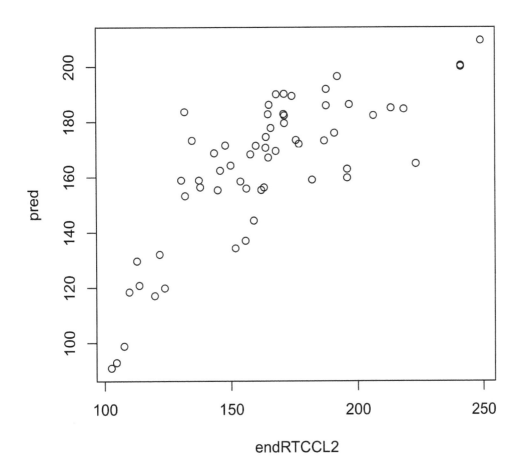

Figure 8.5 Plot of predicted values for end of radiotherapy level of CCL2 (4 variable linear model including baseline level of CCL2, age, BMI and previous abdominal surgery) vs observed end of radiotherapy level of CCL2.

The selection of variables to be included in a model is an important topic. If the number of variables is not too high, we can perform all-subsets regression using the `leaps()` function from the leaps package. In the following code, `nbest` indicates the number of subsets of each size to be reported. Here, the five best models will be reported for each subset size (1 predictor, 2 predictors, and so on).

```
> leaps=regsubsets(endRTCCL2~baselineCCL2+age+BMI+surgery.f,
data=dataex, nbest=5)
> summary(leaps)
Subset selection object
Call: regsubsets.formula(endRTCCL2 ~ baselineCCL2 + age + BMI + surgery.f,
```

```
      data = dataex, nbest = 5)
4 Variables  (and intercept)
              Forced in Forced out
baselineCCL2     FALSE       FALSE
age              FALSE       FALSE
BMI              FALSE       FALSE
surgery.f1       FALSE       FALSE
5 subsets of each size up to 4
Selection Algorithm: exhaustive
          baselineCCL2 age BMI surgery.f1
1  ( 1 ) " "              " " " " " " "*"
1  ( 2 ) "*"              " " " " " " " "
1  ( 3 ) " "              "*" " " " " " "
1  ( 4 ) " "              " " "*" " " " "
2  ( 1 ) "*"              " " " " " " "*"
2  ( 2 ) " "              "*" " " " " "*"
2  ( 3 ) " "              " " "*" " " "*"
2  ( 4 ) "*"              "*" " " " " " "
2  ( 5 ) "*"              " " "*" " " " "
3  ( 1 ) "*"              "*" " " " " "*"
3  ( 2 ) "*"              " " "*" " " "*"
3  ( 3 ) " "              "*" "*" " " "*"
3  ( 4 ) "*"              "*" "*" " " " "
4  ( 1 ) "*"              "*" "*" " " "*"
> plot(leaps,scale="r2")
```

The `plot()` function plots a table of models (Figure 8.6) showing variables in each model and with models ordered based on the selected statistics. In the reported example, R^2 statistics were selected. Another possibility is to select adjusted R^2 statistics.

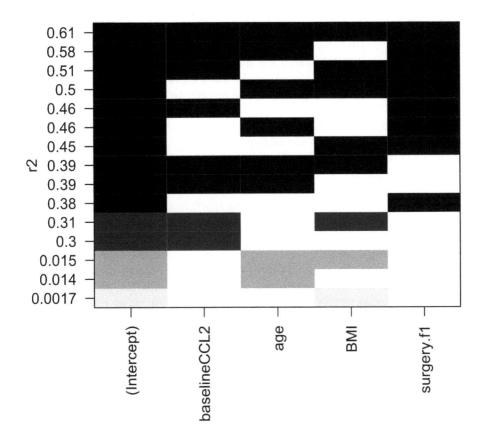

Figure 8.6 Table showing the result of all-subsets regression using the `leaps()` function from the leaps package. The table displays the variables in each model and with models ordered by the selected statistics. In the reported example R^2 statistics was selected. The model with four variables has the highest R^2.

8.3 LOGISTIC REGRESSION

8.3.1 Modelling of qualitative (binary) response

It is certain that most of the events of interest in oncology are discrete or qualitative in nature, rather than quantitative. An event occurs or does not occur, thus exhibiting a binary discrete nature. For example, in the case of the outcome of a therapy, the success may be expressed as "yes" or "no" when considering if the patient is alive or dead, or if the patient's disease is under control or not. When considering a treatment side effect, similarly, we are generally interested in the presence or absence of a symptom, or in the quality of life/patient reported scores. Although the reality may be much more complex (this is especially true when looking at the spectrum of possible treatment-induced reactions and still more when considering quality of life scores), the binary framework is very attractive because it permits us to simplify the complexity of the reality into a "success/failure" scheme. This approach is also highly useful for easily assessing and comparing the outcomes of patients treated in different ways, times and countries.

Binary phenomena usually take the form of a dichotomous factor, or a dummy variable. Although it would be possible to employ any number, using a variable with 0 and 1 values has some advantages. Particularly, the mean of such a defined dummy variable equals the proportion of cases with the value=1; this can be interpreted as a probability.

At first glance, multiple linear regression could seem suitable to handle binary responses; in these cases, regression coefficients can be interpreted as the increase/decrease of the probability of exhibiting the binary response as a function of the values of the independent variables. Despite this simple and straightforward interpretation of the regression coefficients for use of a binary outcome, the linear regression estimates confront two types of problems [358]. The first one is conceptual; it is related to the fact that a probability is bound to belong to the [0, 1] interval, while the linear regression predictions can extend above and below the [0, 1] interval; see, for example, the situation described in Figure 8.7. The second problem is statistical in nature; it is related to the fact that linear regression with a binary dependent variable violates the assumptions of normality and homoscedasticity. Both these problems are due to the fact that the outcome variable Y has only two possible values. Linear regression assumes that in the population a normal distribution of error values around the predicted Y is associated with each X_i value, and the dispersion of errors value for each X_i value is the same. Yet, with only two possible values for Y ($0 =$ no event and $1 =$ event), only two possible residuals exist for any single X_i value. For example, in case of linear regression involving only one variable, for any value of $X_{1,j}$, the predicted probability is $\beta_0 + \beta_1 X_{1,j}$, which determines only two possible values for residuals: $1 - (\beta_0 + \beta_1 X_{1,j})$, when $Y_j = 1$ and $0 - (\beta_0 + \beta_1 X_{1,j})$, when $Y_j = 0$. This determines violation of homoscedasticity (distribution of residuals depend on X) and of normality (the distribution of residuals cannot be normal as it has only two possible values).

For these reasons, it is recommended to approach the issue of modeling binary outcomes in the framework of logistic regression. This type of regression belongs to the class of generalized linear models; it is flexible, as it can incorporate both continuous and qualitative predictors, and it can easily manage non-linear transformations and the introduction of interaction terms. Correspondingly to the linear regression, logistic regression links the probability of the outcome Y to a linear combination of a set of predictors X_i and of regression coefficients β_i. The logistic link function is used to restrict predicted values to the [0, 1] interval, thus allowing modelling of the probability of occurrence of the outcome Y, rather than of its value.

In the previous few lines, there is probably the major motivation for why logistic regression is so popular in modelling (radiation) oncology and, more generally, medical data. Several statistical textbooks face the issue of detailed considerations of logistic regression modelling [355, 356, 357, 358]; in this Chapter, we are going to discuss the main ideas related to modelling medical problems.

Logistic regression is recognized as a member of the binomial family of regression models, with the underlying probability distribution being the Bernoulli function. Among its relevant characteristics, there is the "sigmoid-like" S shape of the resulting probability curve: this characteristic is particularly interesting in radiation dose-effect studies. In this field, radiobiological and clinical experiments support the belief that the occurrence of toxicities is guided by the absorbed dose in a probabilistic manner. At each dose value, it is only possible to associate a risk of damage and it is not possible to predict damage in a deterministic way. In the literature, the Normal Tissue Complication Probability (NTCP) is usually described by a sigmoid curve that is a function of a given dose parameter with a shape that is determined by the radio-sensitivity of the considered tissue [362]. Logistic regression can thus be included also in the framework of radiobiological models, assessing dose-effect curves with the typical and well-known characteristic of a monotone sigmoid dose-response shape.

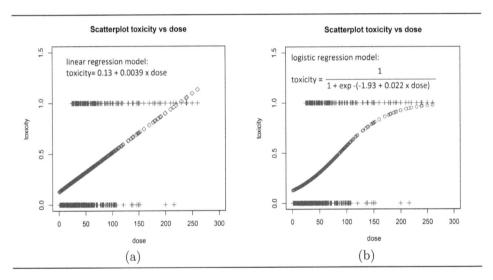

Figure 8.7 Example using a simulated dataset with data on dose and on onset of toxicity. Panel (a), toxicity observation as a function of dose (0=no toxicity, 1=toxicity) and prediction of toxicity probability using a linear regression model. Estimated probabilities > 1 are present. Panel (b), toxicity observation as a function of dose and prediction of toxicity probability using a logistic regression model. Estimated probabilities lie in the [0,1] interval.

8.3.2   Mathematical formalism

The probability that the dependent outcome Y is 1, given a set of N independent variables $\vec{X} = (X_1, X_2, \cdots X_N)$, can be written as

$$\Pr\left(Y = 1 | \vec{X}\right) = p(\vec{X}) \tag{8.12}$$

In the logistic regression framework, $p(\vec{X})$ is modeled using the logistic function [358]:

$$p(\vec{X}) = \frac{e^{\beta_0 + \sum_i \beta_i \cdot X_i}}{1 + e^{\beta_0 + \sum_i \beta_i \cdot X_i}} = \frac{1}{1 + e^{-(\beta_0 + \sum_i \beta_i \cdot X_i)}} \tag{8.13}$$

The quantity

$$\frac{p(\vec{X})}{1 - p(\vec{X})} = e^{\beta_0 + \sum_i \beta_i \cdot X_i} \tag{8.14}$$

is called the odds and can take any value between 0 and ∞. The larger the odds value, the higher the probability of the outcome Y being equal to 1. The log-odds or logit is defined as the natural logarithm of the odds:

$$\ln\left(\frac{p(\vec{X})}{1 - p(\vec{X})}\right) = \beta_0 + \sum_i \beta_i \cdot X_i \tag{8.15}$$

From the above equation, we argue that the logistic model has a logit transformation, which is a linear combination of the dependent variable values X_i. When β_i is positive, $p(\vec{X})$ increases with increasing X_i (if all other variables are fixed, and in this case X_i is a risk factor), while a negative β_i indicates that $p(\vec{X})$ decreases with increasing X_i (X_i is a protective factor). The rate of change in $p(\vec{X})$ per unit change in X_i is not a constant; it depends on the value of X_i.

An important quantity that is defined in logistic regression is the odds ratio (OR), which is defined for each variable X_i as

$$OR_i = \frac{p(X_i + 1)}{p(X_i)} = e^{\beta_i} \tag{8.16}$$

This gives an assessment of the effect size for the variable X_i, *i.e.*, OR represents the proportional increase/decrease of the probability of the outcome for a unit change in X_i. Thus, it gives an estimate, in a single number, of the effect of the specific variable on the total probability.

In the particular case of a binary variable X_i (a factor) taking only two possible values, 0=absence of the factor and 1=presence of the factor, the OR is:

$$OR_i = \frac{p(X_i = 1)}{p(X_i = 0)} = e^{\beta_i} \tag{8.17}$$

which gives a direct measure of the proportional increase/decrease of the probability of the outcome in presence of the factor with respect to absence of the factor.

Note again that ORs represent proportional increases, as they are defined as probability ratios. ORs do not depend on the value of X_i; they are a constant in the model. ORs are always positive: ORs > 1 indicate that the probability increases when the value of the variable increases, ORs < 1 indicate a decreasing probability of the outcome when the value of the variable increases. An OR=1 (corresponding to a $\beta = 0$) is related to a predictor whose value has no influence on the outcome probability.

8.3.3 Estimation of regression coefficients

The regression coefficients β_0 and β_i are not known and have to be estimated from the data. Let's assume we have a dataset consisting of M cases, each described by its outcome Y_j (with Y_j taking only two possible values: 0=no or 1=yes) and by its N observed independent features $X_{j,i}$ (with the index j running on the M cases and i running on the N recorded features). Estimation of the regression coefficients is usually accomplished using maximum likelihood (ML) techniques. The idea behind ML is that given a set of M cases, with P observations of $Y_j = 1$ and $M - P$ observed $Y_j = 0$, the probability of observing that particular composition of the outcome is given by:

$$L\left(\beta_0, \beta_i\right) = \prod_{j:Y_j=1} p(\vec{X}) \cdot \prod_{j:Y_j=0} (1 - p(\vec{X})) \tag{8.18}$$

with $p(\vec{X})$ defined by Equation 8.13. $L\left(\beta_0, \beta_i\right)$ is called the likelihood function. It is a function of the regression coefficients, and the best estimate for these coefficients $\left(\widehat{\beta}_0, \widehat{\beta}_i\right)$ is the one which maximizes $L\left(\beta_0, \beta_i\right)$, i.e., the best estimate of coefficients is the one maximizing the probability of observing the dataset that was observed experimentally. For convenience in numerical estimation, the natural logarithm of $L\left(\beta_0, \beta_i\right)$, i.e., the log-likelihood $LL\left(\beta_0, \beta_i\right)$, is often considered. It is calculated by taking the logarithm of Equation 8.18:

$$LL\left(\beta_0, \beta_i\right) = \sum_{j:Y_j=1} ln\left(p(\vec{X})\right) + \sum_{j:Y_j=0} ln\left(1 - p(\vec{X})\right) \tag{8.19}$$

Due to the monotonic relation between Equation 8.18 and Equation 8.19, coefficients maximizing Equation 8.19 also maximize Equation 8.18. Of note, as probabilities $p(\vec{X}) \in (0, 1)$, $L\left(\beta_0, \beta_i\right) \in (0, 1)$, while $LL\left(\beta_0, \beta_i\right)$ ranges from $-\infty$ to 0; thus, 0 is the maximum possible value for $LL\left(\beta_0, \beta_i\right)$. A perfect fitting model would have $LL\left(\beta_0, \beta_i\right) = 0$. In medical problems, perfect predictions cannot be made (unless we consider a completely deterministic model) and $LL\left(\beta_0, \beta_i\right)$ is a negative, with better fitting models having $LL\left(\beta_0, \beta_i\right)$ closer to 0.

Providing details of ML computation is beyond the scope of this chapter. Any statistical software package includes libraries that can compute these coefficient estimates. Later in this section, we will give an example with calculations in R.

The estimated coefficients $\left(\widehat{\beta}_0, \widehat{\beta}_i\right)$ can be inserted in Equation 8.13 to estimate the probability of $Y = 1$ given a vector of independent variables $\vec{X}$. $\widehat{P}\left(\vec{X}\right)$ is the predicted probability:

$$\widehat{P}\left(\vec{X}\right) = \widehat{Pr}\left(Y = 1, \vec{X}\right) = \frac{1}{1 + e^{-\left(\widehat{\beta}_0 + \sum_i \widehat{\beta}_i X_i\right)}} \tag{8.20}$$

Estimated ORs can also be computed: $\widehat{OR}_i = e^{\widehat{\beta}_i}$.

8.3.4 Accuracy of coefficient estimates

Similar to the linear regression, even in the case of logistic regression, a different dataset could generate different estimates for β_0 and β_i, and if the same is pursued for a large number of datasets, a distribution of $\widehat{\beta}_0$ and $\widehat{\beta}_i$ will be obtained. Calculations of SEs and of confidence intervals give us the possibility to quantify how far the single estimates of $\widehat{\beta}_0$ and $\widehat{\beta}_i$ are representative of the true β_0 and β_i. This is accomplished by computation of the coefficient intervals. Statistical software always provides estimates of regression coefficients together with their SE; thus, the 95% coefficient intervals for β_0 and β_i are $\beta_0 = \widehat{\beta}_0 \pm 1.96 \cdot SE(\widehat{\beta}_0)$ and $\beta_i = \widehat{\beta}_i \pm 1.96 \cdot SE(\widehat{\beta}_i)$.

The 95% coefficient intervals for ORs can also be determined from $SE(\widehat{\beta}_i)$: $OR_i \in \left(e^{\widehat{\beta}_i - 1.96 \cdot SE(\widehat{\beta}_i)}, e^{\widehat{\beta}_i + 1.96 \cdot SE(\widehat{\beta}_i)} \right)$.

8.3.5 Rejecting the null hypothesis, testing the significance of a model

The log-likelihood of a model can be thought of as the deviance from a perfect model (with $LL(\beta_0, \beta_i) = 0$). Although LL increases with increasing efficacy of the regression parameters, it also depends on the sample size (M) and on the number of parameters (N) included in the model. For these reasons, it has a poor intuitive meaning. To overcome these limitations and to have a measure to test the significance of the model, the likelihood ratio test is introduced.

The first step is the definition of the likelihood of the null model (L_0), *i.e.*, the baseline model, where the dependent outcome probability is not related to any independent predictor, in the null model $\beta_i = 0 \, \forall \, i = 1, .., N$. Only $\beta_0 \neq 0$, thus defining a constant probability. This is equivalent to using the mean probability (*i.e.*, the ratio between the number of observations with $Y = 1$ and the sample size M) as the predicted value for all cases. Intuitively, the greater the difference between L_0 and $L(\beta_0, \beta_i)$ resulting from a model including some $\beta_i \neq 0$, the more significant the model is with respect to the baseline model.

The ratio $\frac{L(\beta_0, \beta_i)}{L_0}$ can be used in hypothesis testing, *i.e.*, to evaluate null hypothesis $H_0 : \beta_1 = \beta_2 = ... = \beta_N = 0$. Test statistics are built, starting from the likelihood ratio and taking its logarithm [358]:

$$LR = -2(LL_0 - LL(\beta_0, \beta_i)) \tag{8.21}$$

LR statistics follows the chi-squared (χ^2) distributions, with degree of freedom equal to the number of independent variables (not including the constant); in this way, a p-value can be associated with LR and used to accept/reject the null hypothesis. This test reveals whether improvement in likelihood due to all independent predictors happened by chance beyond a pre-determined significance level.

The same test can be used to evaluate nested models (*i.e.*, models differing for inclusion of only one variable); in this case, $LR(N \text{ vs } N+1) = -2(LL_N - LL_{N+1})$, which follows the χ^2 distributions with one degree of freedom. This test is used to evaluate the appropriateness of adding one more variable to the model.

8.3.6 Accuracy of the model

The next step is trying to assess how well the model fits the observed outcome. The central idea in quantifying the accuracy of a (logistic) model is that the distance between the predicted probability and the observed probability has to be low.

These distances between observed and predicted probabilities are tightly related to the concept of goodness-of-fit, with better models having smaller distances between predicted and observed outcomes. Goodness-of-fit measures are evaluated in the same dataset used to develop the model. When the distance between observed and predicted probabilities is calculated in a new dataset, we speak of predictive performance.

For goodness-of-fit of models considering binary outcomes (which is the case of logistic regression), the Hosmer-Lemeshow (HL) [363] test is often used. The HL test assesses whether the observed event rates match predicted event rates in subgroups of the model population. The HL test specifically considers subgroups as the deciles of predicted probabilities (*i.e.*, any of the nine values that divide the sorted probabilities into ten equal sized groups such that each group represents 1/10 of the observed population). The HL test statistic is defined by:

$$H = \sum_{g=1}^{n} \frac{(O_g - E_g)^2}{N_g p_g (1 - \widehat{P}(\overrightarrow{X})_g)} \tag{8.22}$$

If decile grouping is used, $n = 10$, g is the gth risk decile group, O_g denotes observed events in the gth risk decile group, E_g are expected events, N_g denotes total observations in the gth group ($N_g = M/n$) and $\widehat{P}(\overrightarrow{X})_g$ is the predicted probability.

The HL test statistic asymptotically follows a χ^2 distribution with $n-2$ degrees of freedom. This test should not be statistically significant; here, the null hypothesis H_0 is that there is no difference between expected and observed values. The number of risk groups (n) may be adjusted depending on how many events are included in the dataset. This helps to avoid singular groups (*i.e.*, groups without events).

Calibration is another possible measure of the agreement between the observed probability of the outcome and the probability predicted by the model. It is usually presented graphically in a calibration plot, where groups of observed (average) probabilities are plotted against predicted (average) probabilities. A line of identity helps with orientation: perfect predictions should be on the 45° line (=the calibration line).

Particularly, observed outcome (average) probabilities based on the decile of predictions can be plotted, which makes the calibration plot a graphical illustration of the HL goodness-of-fit test. We note, however, that such grouping (reflecting the HL test) is arbitrary and other choices can be performed.

The calibration plot is characterized by the intercept and slope. For perfectly calibrated models, the intercept should be 0; a value$\neq$0 indicates that the extent predictions are systematically too low or too high (so-called calibration-in-the-large) [355, 356]. The calibration slope should be 1.

The Brier score (BS) [355] is used to indicate overall model performance. It is a quadratic scoring rule based on residuals, where the squared differences between actual binary outcomes Y and predictions $\widehat{P}(\overrightarrow{X})$ are calculated. For a sample including M subjects, it is defined as follows:

$$BS = \frac{1}{M} \sum_{j} \left(\widehat{P}(\overrightarrow{X})_j - Y_j \right)^2 \tag{8.23}$$

The Brier score ranges from 0 for a perfect model to 0.25 for a non-informative model, with a 0.5 incidence of the outcome. When the outcome incidence is lower, the maximum score for a non-informative model is lower. As a general rule, for an outcome incidence k ($k \in (0,1)$), a non-informative model has a BS value ($BS_{MAX,k}$), which can be computed as follows:

$$BS_{MAX,k} = k \cdot (1-k)^2 + (1-k) \cdot k^2 \tag{8.24}$$

To have an absolute evaluation of the BS, which does not depend on the incidence of the outcome, a scaled BS can be defined (BS_{SCALED}), which ranges between 0 and 1 [355]:

$$BS_{SCALED} = \frac{1-BS}{BS_{MAX,k}} \tag{8.25}$$

8.3.7 Qualitative predictors

Logistic regression can be performed, even in the presence of qualitative predictors, by introducing dummy variables, in the same way as was described in the section devoted to the introduction of qualitative factors in multiple linear regression.

As already pointed out, when a factor is considered, the OR (defined by Equation 8.17) directly expresses the effects size of the factor: it gives the estimation of proportional probability increase (decrease) for patients harbouring the risk (protective) factor with respect to patients not having the considered characteristic, *i.e.*,

$$OR = \frac{\text{Probability of outcome for patients with the factor}}{\text{Probability of outcome for patients without the factor}} \cdot$$

8.3.8 Including interaction between variables

Logistic models can be extended to allow for the insertion of interaction between variables, following the same procedure described in the section devoted to this issue for multiple linear regression.

In the simplest case of a two-variable model, the logit (defined by Equation 8.15) introducing interaction between the two predictors has the following form:

$$logit(X_1, X_2, X_1X_2) = \beta_0 + \beta_1 X_1 + \beta_2 X_2 + \beta_{1,2} X_1 X_2 \tag{8.26}$$

8.3.9 Statistical power for reliable predictions

In the case of logistic regression, the statistical power calculation should be aimed at guaranteeing the reliability of predictions. Overfitting is the main issue involved with this topic; this is largely related to the number of predictors included in a model, *i.e.*, the more variables that are included, the more likelihood is enhanced when calculated with the sample used to develop the model. If a large number of predictors is kept in a model, there is a higher risk that the model is "too much" tuned on the developing sample data, thus fitting the noise of the dataset together with the true signal, making the model less generalizable to external data.

The number of predictors that can be safely inserted into a model is related to the number of events (*i.e.*, the number of observations with $Y = 1$) or to the number of non-events, whichever is lower.

A largely applied rule of thumb is the events-per-variable rule, which requires at least ten events per each predictor included in the model [364, 365]. Models with less than 10 events per variable have a high probability of being overfitted. When the number of predictors largely increases (*i.e.*, studies considering genetic variables coming from genome-wide or microarray experiments), a higher number of events per variable is requested, possibly at least 50 events per predictor.

8.3.10 Time consideration

Note that logistic regression does not take time-to-event into consideration. This fact has to be considered when modelling outcomes that can arise at any given time point after beginning observation, *e.g.*, tumour control failures or insurgence of late toxicities.

The correct procedure in this case is to choose a follow-up time point that is long enough to reasonably assume that the large majority of considered outcome events occurred during this period. Observations with a maximum follow-up smaller than the fixed time point must be excluded from the sample (whether or not they registered an event); observations with an event at a time $\leq$ the selected time point are included as "exhibiting the event", while observations with an event at a time $>$ the analysis time point are included as "not exhibiting the event."

The closer the analysis time point is to the maximum time interval in which the observed outcome can occur, the less the resulting model is biased by time-to-event considerations.

8.4 MODEL VALIDATION

The topic of model validation is of top importance. It is very possible to develop a model that apparently provides satisfactory predictions but fails when applied to out-sample-data. Three classes of validation procedures can be generally defined [355]:

1. *apparent validation*, when accuracy and performance are evaluated within the sample used to develop the model (fitting accuracy)

2. *internal validation (reproducibility)*, when accuracy and performance are evaluated within the population distribution underlying the sample used to develop the model

3. *external validation (transportability and generalizability)*, when accuracy and performance are evaluated with a related but slightly different population

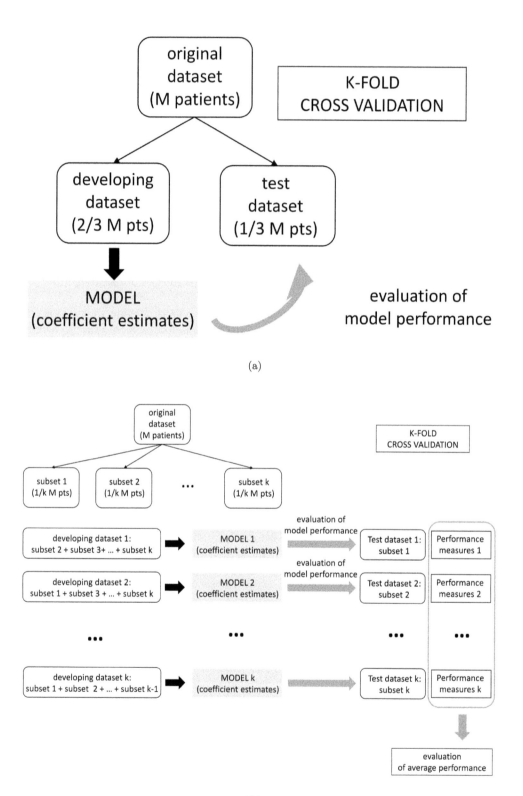

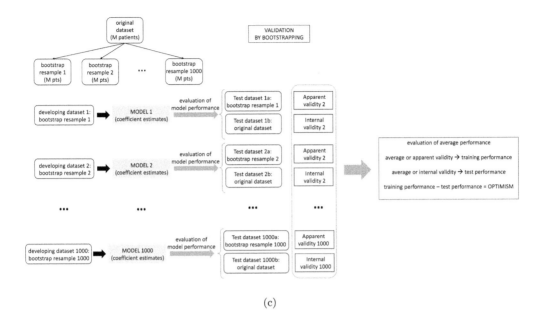

(c)

Figure 8.8 Schematic description of the concepts of (a) split sample validation, (b) k-fold cross validation and (c) internal validation based on bootstrap resampling.

8.4.1 Apparent validation

Apparent validation (fitting) is easy to perform and deals with measures of accuracy and performance, which were described in the sections devoted to linear and logistic regression. However, apparent validation generally results in optimistic estimates because the same sample is used both for development and testing. Thus, apparent model accuracy may be satisfactory as a result of a fine tuning to the development data. Application of the model to new patients may produce unsatisfactory results.

8.4.2 Internal validation

Internal validation is based on the key characteristics, *i.e.*, that model development and evaluation are based on random samples from the same original population. Internal validation is important to obtain an honest estimate of model accuracy for subjects that are similar (but not the same) to those in the development sample, *i.e.*, share the same probability distribution.

Different approaches may be followed, we here considered three internal validation techniques based on statistical resampling. A schematic description of the three main techniques is in Figure 8.8:

1. Split-sample

2. Cross-validation

3. Bootstrap

In *split-sample*, the sample is randomly divided into two groups; typical splits are 50%:50% or 2/3:1/3. This is a very classical approach and is inspired by the design of external validation studies. However, note that the split in developing and test set is random and involves subjects from the same original population. The model is created based on one group (*e.g.*, 50% of the data), and then accuracy and performance are evaluated based on the remaining sample (*e.g.*, the other 50% of the data). *Split-sample validation* has some important draw-backs: (a) if samples are split fully at random, substantial imbalances may occur with respect to the distribution of predictors and outcome. This may have negative consequences especially in case the outcome of interest is a relatively rare event. (b) only a limited part of the data is used for model fitting, thus leading to less-stable model results; and (c) the validation part is relatively small, thus possibly leading to unreliable evaluation of the model. Split-sample validation is a classical approach, which appears to be quite inefficient and rough, dating back to a time when informatics was less advanced and CPU power was available at high cost. Presently, more efficient, computer-intensive methods should be considered.

Cross-validation is an expansion of split-sample validation, leading to more stable results. The original sample is split into k random groups (we speak of k-fold cross-validation). The model is fitted to a sample consisting of $(k - 1)$ groups and then tested against the group that was left out from the sample. This process is repeated k times, each time using a different group for testing. Overall accuracy and performance are defined as the average of all k assessments. In this case, every subject participates both in model development and in model testing.

Bootstrap validation is based on bootstrap re-sampling. A bootstrap sample usually means that a random sample is drawn from the initial sample with replacement and usually yields a new sample with the same size as the original one. In this way, the bootstrap sample includes the same number of subjects (M, but some are excluded and others are included once, twice, etc.). Bootstrap validation considers bootstrapping of a very large number (k, usually at least 1000) of random samples, with development of a new model in each of them and tests of the model both with the bootstrap sample (apparent validity) and the original one (validation in a new set). The difference between the two is a measure of optimism. Even in this case, overall accuracy and performance are defined as the average of all k assessments. This method is efficient because M subjects are used every time both for development and validation.

8.4.3   External validation

The fundamental idea of external validation is that the purpose of a predictive model is to provide adequate outcome predictions for new subjects. Actually, the dataset used to develop a model is not of interest other than to learn for the future. External validation provides a measure of "generalizability" and "transportability" of the prediction model to cohorts that could be somewhat different from the one used for model development, *i.e.*, patients treated at different hospitals, at different dose levels, with different techniques, in different countries, or in different years. Note that a single external validation is not enough to label a model as "validated". The more often a model is externally validated and the more diverse these settings are, the more assurance we can gain regarding its generalizability. The process of external validation explicitly includes validation of the predictors in the new sample, performed through a comparison of odds ratios in the two populations and replications of the model in the independent populations with evaluation of performance (*e.g.*, HL test establishing predictive performance, calibration, Brier score, and discrimination slope). External validity is easy to calculate if appropriate data are available, i.e., data containing predictors and outcome variables required for the prediction model, measured in the same way and with the same timing. This type of validation may show different results from the internal validation because many aspects may be different across settings, *e.g.*, selection of patients, definition of variables, diagnostic or therapeutic procedures.

8.5 EVALUATION OF AN EXTENDED MODEL

A key interest in current medical research is whether a marker (*e.g.*, genetic, molecular, imaging) adds incremental information to an already existing model. Validation in a fully independent external dataset is the best way to compare the performance of a model with and without the new marker.

Performance measures specific to this situation are [366]: (a) Reclassification Table, *i.e.*, calculation of the percentage of individuals that change risk category when applying the new biomarker-extended model and (b) Net Reclassification Improvement, i.e., any upward movement in risk classes for subjects with outcome implies prediction improvement, while any downward movement (for subjects with outcome) indicates worse reclassification. The reverse is true for subjects without outcomes. Net reclassification is then defined as the difference between the proportion improvement and the proportion worsening.

8.6 FEATURE SELECTION

Modern prospective observational studies devoted to modeling of medical problems are often accumulating a large amount of treatment and patient-related information, which are candidates to be incorporated in a model because of their possible relationship with the investigated outcome. This situation requires attention when probability modelling is approached. A core issue is related to the selection of features, directly influencing overfitting, discrimination, personalization and generalizability.

Analysis of clinical research datasets is particularly demanding: these datasets are often characterized by low cardinality - i.e. the number of cases is overall low - and are often strongly imbalanced in the endpoint categories, *i.e.*, the number of positive cases (e.g. toxicity events or loss of disease control) is small, or even very small, with respect to the negative ones. This is mainly due to limited follow-up, low event rate, and restricted inclusion of patients. Other critical relevant issues are related to the circumstance that candidate features are often strongly correlated with each other, which hinders the reliable estimation of regression coefficients, and to the fact that probability distributions of possible predictors are often skewed, due to small subpopulations with values differing substantially from the majority, thereby strongly influencing resulting models.

A well known general rule in statistical modeling is *parsimony*, *i.e.*, the model with the fewest variables has to be chosen if it works in an adequate manner. Parsimonious regression models are certainly characterized by lower probability of overfitting and higher possibility to be generalized to external populations. Nevertheless, they may result in lower discrimination power and low perchance of personalization. This makes parsimonious models not very attractive for use in clinical practice. From these considerations, methods for selection of variables (number and type of predictors) become of great importance in the field of medical modeling.

When dealing with feature selection, a first important (not statistical) issue is related to the possible clinical/biological meaning of predictors for the selected outcome. Predictors already mentioned in known literature should be included into the investigation, and variables with no possible (at least speculative) explanation should be avoided.

Univariate analyses are then commonly used to prioritize the features, *i.e.*, testing each variable individually and ranking them on their strength of association with outcome. In this phase, a critical approach to evaluation of odds ratios should be recommended: if protective ORs are found to be associated to predictor with no chance of being protective (i.e. increasing dose of a drug or of radiotherapy, an impaired functional status, presence of co-morbidities), these predictors should be removed. Similar considerations should be applied to features selected as risk factors due to their ORs. After a first exploratory univariate analysis, methods for multivariable feature selection can be followed. These methods can be divided into three main classes: (a) classical approaches (e.g. forward selection and backward selection); (c) shrinking and regularization methods (*e.g.*, LASSO); (d) bootstrap methods.

8.6.1 Classical approaches

Forward selection [357] is based on the idea of starting with no variables in the model and then testing the addition of each variable using a predefined model comparison criterion. The variable (if any) that improves the model the most is added and the whole process is repeated until none of the remaining variables improves the model.

Backward selection works in the opposite way with respect to forward selection. It starts with all candidate features included and tests the deletion of each variable using a predefined model comparison criterion. The variable that improves the model the less is deleted, and the process is repeated until no further improvement is possible.

These two methods produce a ranking of variables that can be included in the predictive model and are grounded in evaluation of a model comparison criterion. Common criteria are F-test and the information theoretic approach of Akaike information criterion (AIC).

For two models ($m1$ and $m2$) fitted to a sample of M observations and including p_1 and p_2 predictors, respectively, where $p_2 > p_1$, the F-test statistics is defined by:

$$F = \frac{\frac{(RSS_{m1}-RSS_{m2})}{(p_2-p_1)}}{\frac{RSS_{m2}}{M-p_2}} \tag{8.27}$$

where RSS_{m1} and RSS_{m2} are the residual sum of squared of models $m1$ and $m2$, defined by Equation 8.4. F follows an F-statistics distribution with $(p_2 - p_1, M - p_2)$ degrees of freedom. This allows calculation of a p-value determining at what degree of significance the model $m2$ better fits observed outcomes with respect to model $m1$.

AIC criterion is based on information theory related to the log-likelihood function as follows [367]:

$$AIC = 2k - 2LL \qquad (8.28)$$

where LL is log-likelihood (as defined by Equation 8.19) and k the number of estimated parameters in the model. The model to be preferred is the one with minimum AIC: AIC is related to goodness of fit (it is based on likelihood), but includes a penalty for increasing number of predictors, so discouraging overfitting (remember that increasing the number of predictors always improves goodness of fit).

8.6.2 Shrinking and regularization methods: LASSO

The Least Absolute Shrinkage and Selection Operator (LASSO) [368] is a regression method that involves penalizing the absolute size of the regression coefficients. It can be applied both to linear and logistic regression. LASSO ends up in a situation where some of the parameter estimates are forced to be exactly zero, thus encompassing a reduction of the number of predictors to be included in the model. The larger the penalty applied, the further estimates are shrunk towards zero.

When logistic regression is considered, LASSO estimates regression coefficients by minimizing:

$$\sum_j \left(Y_j - \frac{1}{1 + e^{-(\beta_0 + \sum_i \beta_i X_i)}} \right) + \le \gamma \sum_i |\beta_i| \qquad (8.29)$$

where index j is running over the M subjects in a dataset and i is running over N predictors in the model. γ is a tuning parameter, to be determined separately, which serves to control the impact of the shrinking term $\gamma \sum_i |\beta_i|$ on parameter estimates.

When $\gamma = 0$, no shrinkage is performed; as γ increases, fewer variables will be included to build the model, due to the fact that an increasing number of low-sized coefficients will be constrained to 0, thus being deleted from the model. In this way, LASSO performs selection of variables. The LASSO method usually requires initial normalization of the variables, so that the penalization scheme is fair to all predictions.

Application of LASSO involves selection of the value of the tuning parameter γ. This is usually performed by considering cross-validation. For example with k-fold cross validation, k models are considered, each estimating coefficients on $k - 1$ subgroups, by using log-likelihood approach in Equation 8.19. Evaluation of the cross validation error is accomplished by considering application of every single model to the kth group which was not considered in parameter estimation. Final cross validation error (CV) is then defined by:

$$CV = \frac{1}{k} \sum_{z=1}^{k} RSS_z \qquad (8.30)$$

where RSS_z is the residual sum of squared for each of the k models.

This process is repeated over a grid of γ values, thus obtaining cross validation errors as a function of γ. The γ value corresponding to the smallest cross validation error is chosen as appropriate tuning parameter and the LASSO logistic regression model is re-fit following Equation 8.29 with inclusion of the appropriate γ.

8.6.3 Bootstrap methods

Variable selection carried out through simulations coupled to bootstrap resampling procedure can also be considered: this has the final goal of detecting the leading robust variables and minimizing the noise due to the particular dataset, thus trying to avoid both under- and overfitting.

We here propose a method [369] that follows, with adjustments, a procedure firstly introduced by El Naqa [15]. A high number of bootstrap resamplings of the original dataset are created and backward feature selection is performed on each bootstrap sample. In this way a high number of possible models (each possibly including different features) is obtained. The rate of occurrences and the placement of each variable (selected and ranked by the backward feature selection) in the bootstrapped datasets are used to classify the most robust predictors. A synthetic index, called normalized area (NArea), can be defined for ranking each predictor: it corresponds to the area under the histogram representing the number of occurrences of each variable (x-axis) at a given importance level in each re-sampled dataset. In that way, the most robust predictors are chosen among those that appeared more frequently (highest number of occurrences) and had the best placement (best ranking) in the backward feature selections performed on each of the bootstrapped datasets. NArea corresponds to a weighted occurrence sum (highest weights for the leading placements) divided by the highest achievable score (the number of bootstrapped datasets times the number of variables extracted in each backward selection) and is given by the following Equation:

$$\text{NArea}\,(X) = \sum_{i=1}^{n_a} \frac{O_i(n_a - i + 1)}{n_a \cdot N_{bootstrap}} \tag{8.31}$$

where O_i corresponds to the number of occurrences of the variable X in the ith position, n_a is the number of the allowed variables in each backward feature extraction and $N_{bootstrap}$ is the number of bootstrapped datasets considered. In that way, e.g., the first position has the highest weight na, while the last one has weight equal to 1. An example is given in Figure 8.9. This example refers to modelling of acute urinary symptoms after radiotherapy for prostate cancer, presented in [369]. Two endpoints were defined: (a) increase in the International Prostate Symptom Score (IPSS) of at least 10 points (named "delta IPSS $\geq$10", 77 events out of 280 patients) and (b) increase in the IPSS of at least 15 points (named "delta IPSS $\geq$15", 28 events out of 280 patients). Following the rule of thumb of $\approx$1 variable every 10 events, 8 variables were selected to be included in the score for delta IPSS $\geq$10 and 3 in the score for delta IPSS $\geq$15. 1000 bootstrap re-sampled datasets were created and backward logistic regression performed on each re-sampled dataset. The rate of occurrences and the placement of each variable (selected by the backward feature selection) in the 1000 bootstrapped datasets were recorded and used to calculate the normalised area for each variable (following Equation 8.31). NArea index was then used to classify the most robust predictors, which were later included in a logistic model. In Figure 8.9a, the normalized areas for the top 8 ranking variables for delta IPSS $\geq$10 are presented, while Figure 8.9b reports the top 4 ranking variables for delta IPSS $\geq$ 15. Neoadjuvant hormone therapy was ranked as the more important predictor for delta IPSS $\geq$10, while the interaction factor between dose (as measured by the absolute bladder surface receiving more than 12 Gy/week) and use of cholesterol lowering drugs was ranked as the more important predictor for delta IPSS $\geq$15.

A further step can be performed by applying a basket analysis of the total sets of predictors: this helps in identifying the predictors that appear together with higher probability. Basket analysis is a common technique used in business data mining. Each model is seen as a customer, buying a set of items (predictors), in order to maximize its "satisfaction" (accuracy and performance). The technique is called Association Analysis and it is based on the Apriori algorithm. Apriori uses a "bottom up" approach, where frequent subsets (starting from 1-item sets) are extended one item at a time using a breadth-first-search (a step known as candidate generation), and groups of candidates are tested against the data. The algorithm terminates when no further successful extensions are found. The technique identifies the "frequent item sets", in our case the sets of predictors that are most commonly "brought together" by the models. Key concepts in basket analysis are support and confidence. Support is, in this case, the fraction of models that chose a certain subset of predictors. Confidence is the probability of choosing an additional predictor, given the subset already selected. Bootstrap resampling, NArea ranking and considerations coming from basket analysis (*i.e.*, predictors often appearing together or rarely appearing together) can lead to determining the most robust set of features to be considered for a model, and the model can be fitted on these selected variables.

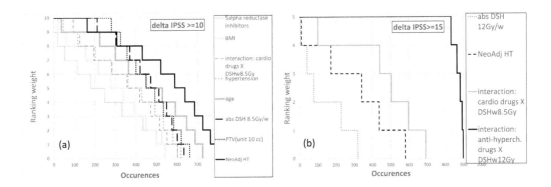

Figure 8.9 Examples for normalized areas as resulting from analysis presented in [369]. Panel (a): ranking of variables associated to acute urinary symptoms as measured by increase in the International Prostate Symptom Score of 10 points. With 77 toxicity events, 8 variables were selected to be included in the score. Neoadjuvant hormone therapy was ranked as the more important predictor. Panel (b): ranking of variables associated to acute urinary symptoms as measured by increase in the International Prostate Symptom Score of 15 points. With 28 toxicity events, 3 variables were selected to be included in the score. The normalized areas for the top 4 ranking variables are presented. The interaction factor between dose (as measured by the absolute bladder surface receiving more than 12 Gy/week) and use of cholesterol lowering drugs was ranked as the more important predictor. BMI=body mass index; abs DSH 8.5Gy/w=absolute bladder surface receiving more than 8.5 Gy/week; cardio drugs= use of drugs for cardiovascular diseases;PTV=planning target volume; NeoAdj HT=neoadjuvant hormone therapy; abs DSH 12 Gy/w=absolute bladder surface receiving more than 12 Gy/week; andi-hyperch drugs=use of cholesterol lowering drugs.

8.6.4 Logistic regression: example

As a practical example of what was described in the above sections, we report a possible analysis of a dataset including sample data of 380 patients who underwent radical radiotherapy for prostate cancer and who participated in a prospective trial devoted to modelling of urinary toxicity. The here considered endpoint is acute urinary toxicity as measured by increase in the International Prostate Symptom Score (IPSS) at the end of radiotherapy (with respect to baseline score). Full analyses were published in [369]. We here consider a reduced dataset in the number of available variables. In particular, this dataset contains data on age (age), hypertension (hypertension), use of drugs for cardiovascular diseases (cardiodrug), dose (DSH, defined as the absolute bladder surface receiving more than 8.5 Gy/week) and urinary toxicity (delta10). The aim of the logistic regression analysis is to determine the relationship between variables measured before radiotherapy and acute urinary toxicity.

Analyses are performed in R. Basic functions to perform logistic regression are included in the base distributions. We also need the libraries **rms** for Regression Modelling Strategies and **PredictABEL**, which includes functions for evaluation of performance of a model. Data are included in the **dataex** object, a matrix with cases in lines and variables in columns. The following commands allow reporting of variables included in **dataex** and some basic descriptive statistics.

```
{> names(dataex)
[1] "age"          "hypertension" "cardiodrug"   "DSH"          "delta10"
> datadist(dataex)
> library(rms)
> datadist(dataex)
                 age hypertension cardiodrug    DSH        delta10
Low:effect        67           0           0   40.96831         0
Adjust to         71           0           0   56.12607         0
High:effect       75           1           1   80.70781         1
Low:prediction    56           0           0   23.53247         0
High:prediction   79           1           1  166.75862         1
Low               46           0           0   11.56000         0
High              83           1           1  313.32404         1

Values:

hypertension : 0 1
cardiodrug : 0 1
delta10 : 0 1
```

The following lines have the goal of telling R that the variables to be considered are in dataex and that some variables are qualitative (factors).

```
> attach(dataex)
> hypertension.f=factor(hypertension)
> cardiodrug.f=factor(cardiodrug)
> delta10.f=factor(delta10)
```

The `glm()` function allows analysis in the framework of generalized linear models, the argument `family=binomial()` indicates that logistic regression is chosen. For convenience results from the `glm()` function are stored in the `glm.fit` object (which can be later included in the arguments of R functions). The `summary()` function gives details on regression results.

```
> ddist<-datadist(hypertension.f,cardiodrug.f,delta10,age,DSH)
> options(datadist="ddist")
> glm.fit=glm(delta10.f~age+hypertension.f+cardiodrug.f+DSH,
data=dataex,family=binomial())
> summary(glm.fit)

Call:
glm(formula = delta10.f ~ age + hypertension.f + cardiodrug.f +
    DSH, family = binomial(), data = dataex)

Deviance Residuals:
    Min       1Q   Median       3Q      Max
-1.0717  -0.6918  -0.5940  -0.4772   2.1492

Coefficients:
                 Estimate Std. Error z value Pr(>|z|)
(Intercept)      0.606632   1.507445   0.402  0.68737
age             -0.042061   0.021701  -1.938  0.05260 .
hypertension.f1  0.436340   0.295783   1.475  0.14016
cardiodrug.f1    0.238317   0.294535   0.809  0.41844
DSH              0.009032   0.003052   2.959  0.00308 **
---
Signif. codes:  0 '***' 0.001 '**' 0.01 '*' 0.05 '.' 0.1 ' ' 1

(Dispersion parameter for binomial family taken to be 1)

Null deviance: 383.06  on 379  degrees of freedom
Residual deviance: 368.66  on 375  degrees of freedom
AIC: 378.66

Number of Fisher Scoring iterations: 4
```

Estimated regression coefficients together with their standard errors are reported. The value of AIC is also presented.

95% confidence intervals for coefficients can be calculated, together with odds ratios and their confidence intervals.

```
> confint(glm.fit) # 95% CI for the coefficients
Waiting for profiling to be done...
                      2.5 %        97.5 %
(Intercept)     -2.408597668 3.5276955251
age             -0.084516103 0.0009071703
hypertension.f1 -0.138576032 1.0245763562
cardiodrug.f1   -0.343089443 0.8148905872
```

```
DSH                0.003064361 0.0151184929
> exp(coef(glm.fit)) # exponentiated coefficients
   (Intercept)  age hypertension.f1  cardiodrug.f1      DSH
      1.8342432  0.9588115  1.5470351 1.2691114  1.0090730
> exp(confint(glm.fit)) # 95% CI for exponentiated coefficients
Waiting for profiling to be done...
                     2.5 %     97.5 %
(Intercept)        0.08994133 34.045420
age                0.91895686  1.000908
hypertension.f1    0.87059706  2.785915
cardiodrug.f1      0.70957474  2.258929
DSH                1.00306906  1.015233
```

In the following predicted probabilities are computed and plotted, together with observed toxicity events (Figure 8.10).

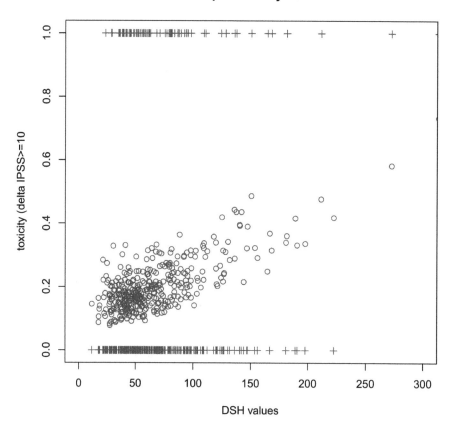

Figure 8.10 Scatterplot of observed toxicity rates vs predicted probabilities. Distribution of events (=1) and not-events (=0) are also plotted as a function of model predicted probabilities.

```
> pred=predict(glm.fit, type="response") # predicted values

> plot(DSH, delta10, main="Scatterplot toxicity vs dose",xlab="DSH values",
ylab="toxicity (delta IPSS>=10", pch=3,xlim=c(0,300),ylim=c(0,1.0),col=4)
> par(new=T)
> plot(DSH,pred,main="Scatterplot toxicity vs dose",xlab="DSH values",
ylab="toxicity (delta IPSS>=10", pch=1,xlim=c(0,300),ylim=c(0,1.0),col=4)
>
```

We had a previous guess that an interaction between dose and use of drugs for cardio-vascular diseases could be of importance for insurgence of urinary toxicity. For this reason, a second model including an interaction factor is fitted.

```
> glm.fit2=glm(delta10.f~age+hypertension.f+cardiodrug.f+DSH+
cardiodrug.f*DSH, data=dataex,family=binomial())
> summary(glm.fit2)

Call:
glm(formula = delta10.f ~ age + hypertension.f + cardiodrug.f +
    DSH + cardiodrug.f * DSH, family = binomial(), data = dataex)

Deviance Residuals:
    Min      1Q   Median       3Q      Max
-1.3514  -0.6746  -0.5863  -0.4843   2.1060

Coefficients:
                   Estimate Std. Error z value Pr(>|z|)
(Intercept)        1.003252   1.531357   0.655   0.5124
age               -0.044298   0.021844  -2.028   0.0426 *
hypertension.f1    0.390616   0.296711   1.316   0.1880
cardiodrug.f1     -0.552389   0.580435  -0.952   0.3413
DSH                0.006122   0.003521   1.739   0.0820 .
cardiodrug.f1:DSH  0.011667   0.007280   1.603   0.1090
---
Signif. codes:  0 '***' 0.001 '**' 0.01 '*' 0.05 '.' 0.1 ' ' 1

(Dispersion parameter for binomial family taken to be 1)

Null deviance: 383.06  on 379  degrees of freedom
Residual deviance: 366.00  on 374  degrees of freedom
AIC: 378

Number of Fisher Scoring iterations: 4
```

We fit a third (final) model with exclusion of hypertension.

```
> glm.fit3=glm(delta10.f~age+cardiodrug.f+DSH+cardiodrug.f*DSH,
data=dataex,family=binomial())
> summary(glm.fit3) # display results
```

```
Call:
glm(formula = delta10.f ~ age + cardiodrug.f + DSH + cardiodrug.f *
    DSH, family = binomial(), data = dataex)

Deviance Residuals:
    Min      1Q   Median      3Q      Max
-1.3259  -0.6727  -0.5845  -0.5148   2.0898

Coefficients:
                     Estimate Std. Error z value Pr(>|z|)
(Intercept)          0.886375   1.527372   0.580   0.5617
age                 -0.039716   0.021539  -1.844   0.0652 .
cardiodrug.f1       -0.459738   0.575818  -0.798   0.4246
DSH                  0.005618   0.003470   1.619   0.1055
cardiodrug.f1:DSH    0.012524   0.007237   1.731   0.0835 .
---
Signif. codes:   0 '***' 0.001 '**' 0.01 '*' 0.05 '.' 0.1 ' ' 1

(Dispersion parameter for binomial family taken to be 1)

Null deviance: 383.06  on 379  degrees of freedom
Residual deviance: 367.75  on 375  degrees of freedom
AIC: 377.75

Number of Fisher Scoring iterations: 4

> confint(glm.fit3) # 95% CI for the coefficients
Waiting for profiling to be done...
                      2.5%        97.5%
(Intercept)      -2.165081389 3.849991211
age              -0.081836258 0.002962186
cardiodrug.f1    -1.629543998 0.641071176
DSH              -0.001406660 0.012429742
cardiodrug.f1:DSH -0.001355562 0.027243947
> exp(coef(glm.fit3)) # exponentiated coefficients
        (Intercept)              age      cardiodrug.f1              DSH
          2.4263187        0.9610628          0.6314491        1.0056339
cardiodrug.f1:DSH
          1.0126028
> exp(confint(glm.fit3)) # 95% CI for exponentiated coefficients
Waiting for profiling to be done...
                   2.5%      97.5%
(Intercept)      0.1147406 46.992650
age              0.9214228  1.002967
cardiodrug.f1    0.1960189  1.898513
DSH              0.9985943  1.012507
cardiodrug.f1:DSH 0.9986454  1.027618
> pred3=predict(glm.fit3, type="response") # predicted values
```

The function lrm() also allows fitting of logistic regression. It is included in the rms package, which has some additional functions which can help in evaluating goodness of fit and internal validation.

```
> lrm.fit3<-lrm(delta10.f~age+cardiodrug.f+DSH+cardiodrug.f*DSH,data=dataex,

method="lrm.fit",model=T,x=T,y=T,se.fit=T)
> lrm.fit3


Logistic Regression Model

lrm(formula = delta10.f ~ age + cardiodrug.f + DSH + cardiodrug.f *
    DSH, data = dataex, method = "lrm.fit", model = T, x = T,
    y = T, se.fit = T)
```

		Model Likelihood Ratio Test		Discrimination Indexes		Rank Discrim. Indexes	
Obs	380	LR chi2	15.32	R^2	0.062	C	0.633
0	303	d.f.	4	g	0.486	Dxy	0.266
1	77	Pr(> chi2)	0.0041	gr	1.625	gamma	0.268
max |deriv|	2e-08			gp	0.082	tau-a	0.086
				Brier	0.154		

	Coef	S.E.	Wald Z	Pr(>|Z|)
Intercept	0.8864	1.5274	0.58	0.5617
age	-0.0397	0.0215	-1.84	0.0652
cardiodrug.f=1	-0.4597	0.5758	-0.80	0.4246
DSH	0.0056	0.0035	1.62	0.1055
cardiodrug.f=1 * DSH	0.0125	0.0072	1.73	0.0835

In the output, results for the likelihood ratio test and Brier score (Brier) are reported.

The validate() function performs internal validation. k-fold cross validation is pursued when method="cross-validation" (the argument B determines the number of subsets created from the original dataset). In the present example 5-fold cross validation was chosen.

```
> val.cross<-validate(lrm.fit3,method="crossvalidation", B=5, bw=FALSE,
rule="aic", type="residual", sls=0.05, aic=0, pr=FALSE)
> val.cross
```

	index.orig	training	test	optimism	index.corrected	n
Dxy	0.2654	0.2731	0.1908	0.0822	0.1831	5
R2	0.0622	0.0670	0.0431	0.0239	0.0383	5
Intercept	0.0000	0.0000	-0.2881	0.2881	-0.2881	5
Slope	1.0000	1.0000	0.7991	0.2009	0.7991	5
Emax	0.0000	0.0000	0.1069	0.1069	0.1069	5
D	0.0377	0.0402	0.0145	0.0257	0.0120	5
U	-0.0053	-0.0066	-0.0008	-0.0058	0.0005	5
Q	0.0429	0.0468	0.0153	0.0315	0.0114	5
B	0.1545	0.1540	0.1602	-0.0063	0.1607	5
g	0.4857	0.5090	0.3797	0.1294	0.3563	5
gp	0.0821	0.0848	0.0612	0.0236	0.0585	5

```
>
```

Output results include details on average performance in training and test datasets. Optimism is also evaluated. B is the Brier score, Intercept and Slope are the intercept and slope of the calibration plot.

With the `validate()` function, internal validation with bootstrap resampling is pursued when `method= "boot"` (the argument B determines the number of bootstrap resamplings, 1000 in the present example).

```
> val.boot<-validate(lrm.fit3,method="boot", B=1000, bw=FALSE,
rule="aic", type="residual",  sls=0.05, aic=0, pr=FALSE)
> val.boot
            index.orig training    test optimism index.corrected    n
Dxy             0.2654   0.2848  0.2366   0.0482          0.2172 1000
R2              0.0622   0.0772  0.0517   0.0255          0.0367 1000
Intercept       0.0000   0.0000 -0.1819   0.1819         -0.1819 1000
Slope           1.0000   1.0000  0.8607   0.1393          0.8607 1000
Emax            0.0000   0.0000  0.0679   0.0679          0.0679 1000
D               0.0377   0.0478  0.0308   0.0171          0.0206 1000
U              -0.0053  -0.0053  0.0013  -0.0066          0.0013 1000
Q               0.0429   0.0531  0.0295   0.0236          0.0193 1000
B               0.1545   0.1525  0.1568  -0.0043          0.1588 1000
g               0.4857   0.5409  0.4428   0.0980          0.3876 1000
gp              0.0821   0.0894  0.0743   0.0151          0.0670 1000
```

If we want to have a graphical representation of the normal tissue complication probability (NTCP) as a function of dose (here the absolute bladder dose-surface, DSH, value at 8.5 Gy/week), which is obviously the most interesting predictor when considering dose-response model, and of an important clinical predictor, we can have a new 2-variable fit and then plot the resulting curves. The resulting plot is presented in Figure 8.11: NTCP as a function of bladder dose and of use of drugs for cardiovascular diseases.

```
> lrm.fit4<-lrm(delta10.f~DSH+cardiodrug*DSH,data=dataex,method="lrm.fit",
model=T,x=T,y=T,se.fit=T)
> summary(lrm.fit4)
```

```
Logistic Regression Model

lrm(formula = delta10.f ~ DSH + cardiodrug * DSH, data = dataex,
    method = "lrm.fit", model = T, x = T, y = T, se.fit = T)
```

		Model Likelihood Ratio Test		Discrimination Indexes		Rank Discrim. Indexes	
Obs	380	LR chi2	11.98	R^2	0.049	C	0.607
0	303	d.f.	3	g	0.390	Dxy	0.214
1	77	Pr(> chi2)	0.0074	gr	1.477	gamma	0.218
max |deriv|	2e-09			gp	0.068	tau-a	0.069
				Brier	0.156		

	Coef	S.E.	Wald Z	Pr(>|Z|)
Intercept	-1.8960	0.3081	-6.15	<0.0001
DSH	0.0060	0.0035	1.72	0.0850

```
cardiodrug         -0.4720 0.5786 -0.82  0.4146
DSH * cardiodrug   0.0117 0.0073  1.62  0.1061
```

```
> sigmoid1 = function(x) { 1 / (1 + exp(-(-1.8960+x*0.0060)))}
> sigmoid2 = function(x) { 1 / (1 + exp(-(-1.8960+x*0.0060+x*0.0117-0.4720)))}
> xxx = seq(40, 300, 1)
> plot(xxx,sigmoid1(xxx),main="Normal Tissue Complication Probability",
xlab="DSH values",  ylab="toxicity (delta IPSS>=10", pch=1,xlim=c(0,300),ylim=c(0,1.0),col=4)
> par(new=T)
> plot(xxx,sigmoid2(xxx),main="Normal Tissue Complication Probability",
xlab="DSH values", ylab="toxicity (delta IPSS>=10",  pch=1,xlim=c(0,300),ylim=c(0,1.0),col=4)
>
```

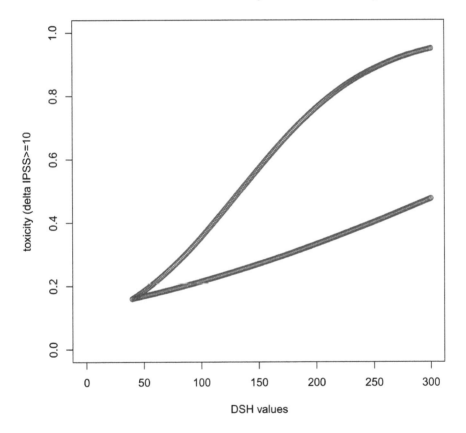

Figure 8.11 Normal tissue complication probability (NTCP) as a function of bladder dose (here the absolute bladder dose-surface, DSH, value at 8.5 Gy/week) and of use of drugs for cardiovascular diseases (superior NTCP curve for patients using selected drugs and inferior curve for patients not using drugs) .

The library PredictABEL allows calculation of Hosmer-Lemeshow test and display of calibration plot using the same groups of the Hosmer-Lemeshow test. A frame of data is created starting from predictions (pred3) and observed toxicity events (delta10). The plotCalibration function calculates HL test (number of groups determined by argument groups, 10 groups is the following example) and calibration plot, by considering the observed events (cOutcome to be read in the data frame) and the predicted risk (predRisk to be read in the data frame).

```
> library(PredictABEL)
> mydata <- data.frame(pred3,delta10)
> cOutcome <- 2
> rangeaxis <- c(0,1)
> risk=mydata[ ,1]
> groups <- 10
> plotCalibration(mydata, cOutcome=cOutcome, predRisk=risk,
groups=groups, rangeaxis=rangeaxis)
$Table_HLtest
                  total meanpred meanobs predicted observed
[0.0855,0.132)     38     0.122   0.105     4.62       4
[0.1318,0.144)     38     0.137   0.132     5.21       5
[0.1436,0.154)     38     0.148   0.079     5.62       3
[0.1540,0.163)     38     0.159   0.237     6.03       9
[0.1630,0.177)     38     0.170   0.263     6.46      10
[0.1770,0.193)     38     0.185   0.132     7.02       5
[0.1930,0.217)     38     0.205   0.184     7.78       7
[0.2173,0.249)     38     0.231   0.263     8.78      10
[0.2489,0.293)     38     0.272   0.237    10.33       9
[0.2930,0.762]     38     0.399   0.395    15.15      15

$Chi_square
[1] 6.878

$df
[1] 8

$p_value
[1] 0.5498
```

In this example, p-value for HL test is 0.55, thus indicating acceptable goodness-of-fit. The resulting calibration plot is depicted in Figure 8.12.

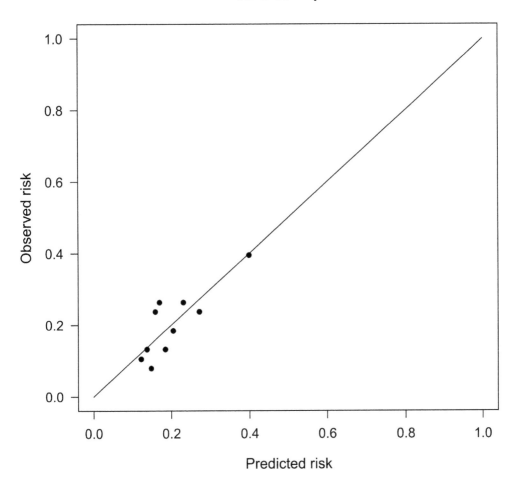

Figure 8.12 Calibration plot for the 3-variable logistic model described in the text and using the ten equal-sized grouping considered by the Hosmer-Lemeshow test.

8.7   CONCLUSIONS

We have described classical multivariable methods for statistical modeling of outcomes. These methods based on linear and logistic regression are among the commonly used in oncology. We have presented their theoretical background and provided examples for their application including a model parameter selection process and statistical tests to avoid pitfalls of over-fitting.

CHAPTER 9

Data driven approaches II: Machine Learning

Sarah Gulliford, PhD

CONTENTS

ABSTRACT

Machine learning techniques provide a powerful tool for predicting outcomes. Their role has risen to more prominence in the era of big data. In this chapter, we present different flavors of machine learning techniques, many of which have been successfully applied for outcome modelling in oncology. We discuss issues related to feature reduction, appropriate selection of machine learning algorithms, fitting pitfalls, and available resources.

Keywords: outcome modeling, machine learning, feature selection, NTCP modeling.

9.1 INTRODUCTION

The subject of Machine Learning can seem slightly mystical. Often used in complex situations and a buzz word related to Big Data [94, 24, 9, 370], the reality is a group of algorithms that have features that extend the flexibility of fitting data without a priori information about the functional form of the relationship between the variables and outcomes. Machine learning is a field in computer science devoted to algorithms where computers learn from examples with the intention of eventually making decisions from unseen cases, mimicking to an extent the way the human brain learns. A classic example would be car number plate recognition. Many different example images would be given to the algorithm as training data, each one matched with correct interpretation of letters and or numbers (depending on international variation). The examples are presented iteratively in the same way a small child learns by repetition. Once trained on a broad enough selection of examples the computer could be used to process data from a speed camera or toll booth. Machine learning approaches are non-linear allowing maximum flexibility for learning. This flexibility is what makes machine learning appealing as an extension to classical statistical approaches.

The field of machine learning dates back to the 1940's when McCuloch and Pitts [371] published their paper "A Logical Calculus of the Ideas Immanent in Nervous Activity which describes a computational algorithm known as a perceptron based on the function of the human brain." Constructed as a network of interconnected nodes, weights between the nodes were optimized to relate input data to output data. This is analogous to the neurons and synapses of the human brain. When making an outcome prediction, the intention is to either make a classification (select the correct group for a case) or regression (a continuous predictive value such as Normal Tissue Complication Probability (NTCP)). If the data can be described as linearly separable, *i.e.*, a straight line can be used to separate the classes when the data is plotted, then machine learning is not necessary as a simple statistical model is sufficient.

To define linearly separable, consider the classic machine learning dataset known as the Wisconsin Breast data (see Section 9.6). The database kept by Dr William H Wolberg, University of Wisconsin hospital [372] over a period of years included patients referred with a suspect breast lesion, and gave details of 9 cytological features from fine-needle aspirations and the corresponding diagnosis of benign or malignant. The features were uniformity of cell shape and size, marginal adhesion, single epithelial cell size, bare nuclei, bland chromatin, normal nucleoli and mitosis. Input data were presented in a format suitable for a multilayer perceptron (MLP) with all information grouped numerically with values in the range 0 to 1. The output data were classified as benign or malignant. There were 700 cases in the datasets of which 241 were malignant. Figure 9.1 is a representative example of how 2 of the variables are related to outcome. It is obvious that no straight line can separate the outcome classes (x (benign) vs o (malignant)). For some cases the same input values correspond to different outcomes.

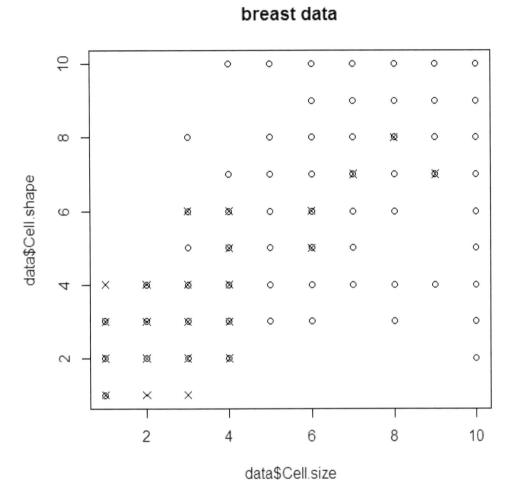

Figure 9.1 Demonstrating the concept of linear separability of classes using two variables from the classic "breast data" machine learning data set, Cell size and Cell shape. The known outcome of the fine needle aspiration biopsy is coded x for benign, o for malignant.

In situations where the relationship between input data and output data cannot be described as linearly separable, machine learning comes into its own. Since machine learning approaches are developed without a priori information regarding the type of relationship between the data, the quality of the training data is fundamental to the ability of the model to produce acceptable results. Optimizing the model may be computationally intensive however since trained models do not require iterative calculations, results are produced relatively quickly. Many machine learning approaches are also suitable for datasets where the input data are known to be correlated. This is true of the bins of a dose-volume histogram commonly used in modelling radiation outcomes. Sometimes it is useful to look at only the input data in order to establish clustering or patterns in the data, this is known as unsupervised learning. The corollary, where the corresponding outcomes of example cases are integrated in model design is known as supervised learning.

Machine learning is a broad subject and a comprehensive discussion is beyond the scope of a single chapter, more detailed description in given in the text by [16]. However, this chapter will explore the potential advantages of machine learning for specific outcome modelling by introducing commonly used methods, considering when it is appropriate to use them and highlighting examples in the context of radiotherapy outcome modelling [373]. Some of the practicalities which are common to different approaches to machine learning (and generally to statistical modelling) will be included.

9.2 FEATURE SELECTION

There are two components to building a predictive model. The first is to decide which variables are relevant, known as feature selection, and the second is to model the strength of the relationship between the variables and the outcome, often referred to as parameter optimization.

Feature selection is implicit in any model as there is a finite selection of input variables introduced within the dataset. However, within these variables it is assumed that the contribution of individual variables will vary in the final predictive model. Feature selection can be used as a pre-processing step to reduce the number of variables in a model to those with the most relevance. Feature selection can be integrated within optimization of model parameters rather than as a separate process. Testing the contribution of each variable to a final model is often referred to as the 'leave one out method'. If no change in results is observed it can be said that the input is redundant. Redundant input variables can then be removed as part of the development process. A more common approach to assessing the contribution of individual variables to the outcome of the model during development is to add or remove variables sequentially and assess the resultant outcome metrics of the model. This is the stepwise approach classically used in logistic regression. However, neither this approach nor leave one out considers interactions between variables. *i.e.*, where for two separate variables adding them individually to a model may make only minimal improvement, but when both are incorporated properly then the model improves substantially.

The overall objective of feature selection is to find the best set of input parameters however using a larger number of variables may be computationally expensive and in clinical practice models with fewer variables are easier to implement. Optimization criteria such as Akaike information criteria (AIC) and Bayesian information criteria (BIC) [374] penalize overly complex models (see Chapter 4).

9.2.1 Principal Component Analysis (PCA)

An alternative approach to feature selection by removing variables is to pre-process the input data to retain useful information whilst reducing the number of variables. Principal component analysis (PCA) is an unsupervised methodology often used for feature selection.

The intention is to find a smaller set of uncorrelated orthogonal vectors which describe the variance in potentially correlated input variables of a dataset. The principal components are linear combinations of the original data with the first principle component describing the largest amount of the variance in the dataset. The coefficients of the principal component are the coefficients of the eigenvector of the covariance matrix. Each subsequent principle component augments the description of the variance by describing a smaller amount of the variance. Often only the first 2 or 3 principal components are required to provide a good description of the data. The results of the PCA analysis can either be used to cluster the input data to show a relationship with the corresponding outcomes or used as inputs to a supervised model where a few principal components replace a larger original set of inputs.

9.2.1.1 When should you use them?

For outcome modelling in radiation oncology, the classic example would be dose-volume histogram reduction, where many, correlated, histogram bins are described using just a few principal components. However PCA can be used for feature selection/ reduction on any data type. PCA is often used to assess if data is linearly separable.

9.2.1.2 Who has already used them?

There are a reasonable number of papers that detail the use of PCA in radiotherapy outcome modelling. The study by Bauer *et al.* [375] gives a comprehensive description of the method used to differentiate the rectal DVHs of patients with/without rectal bleeding. Dawson *et al.* [154] demonstrate how PCA can be used to separate the DVH of patients with toxicity from those without toxicity using the principal components to cluster data. Using data for two different organs at risk (parotid gland and liver), it was shown that describing the variance in the input data was reasonable as a method to relate input data to outcomes.

The concept of describing the dose-volume histogram as a function rather than discretized data has been explored by Benadjaoud *et al.* [376]. The discrete bins of differential rectal DVH of patients treated with prostate radiotherapy were described as continuous functions by fitting the DVH to probability density functions. The resultant probability density functions were then processed in an analogous approach to PCA using functional principal component analysis [377] by decomposing the function into eigenfunctions. In addition to the functional PCA data clinical and patient-related factors were incorporated in to a customized functional logistic regression model and compared to conventional PCA, logistic regression and the LKB model [378] (or see Chapter 7 in this book). Generally, the model performance (assessed using AIC and AUC) was similar with much larger variation depending on the variables chosen rather than the model used. The 1st and 2nd functional principal component were interpreted to have a radiotherapy based rationale, discrimination between 2 and 2.5 Gy for the former and intersection of 4-field and 2-field regions of the 4 field box radiotherapy technique for the latter.

Although PCA and functional PCA are extremely helpful for reducing the number of input features to a model, neither relates the input data to the outcomes only looking at the variation between input data (unsupervised learning). The concept can be extended using functional partial least squares (PLS) [379]. Here the functional description of the input data, are uncorrelated covariates which account for the relationship between predictor variables and corresponding output data. Again the resultant FPLS components were modelled using logistic regression. A comparison between functional PCA, functional PLS and more conventional penalized logistic regression was made using radiotherapy data to predict acute toxicity endpoints (mucositis and dysphagia) following head and neck radiotherapy. There were only slight differences in the results between methods when considering AUC, however, the difference was more marked when performance was assessed using calibration curves. For dysphagia particularly, FPCA and FPLS were much closer to the ideal slope and intercept values (1 and 0) respectively than PLR (0.79 and -0.04(0 FPLS) vs 2.5 and -0.96). Higher fractional doses were associated with both toxicity endpoints whilst cisplatin was significantly associated with dysphagia by both functional models.

9.3 FLAVORS OF MACHINE LEARNING

9.3.1 Artificial Neural Networks

9.3.1.1 The basics

An Artificial Neural Network (ANN) is a high-level mathematical abstraction of the process of learning in biological neural systems. The brain learns by developing connections between neurons. Similarly an ANN is composed of processing units known as nodes. The most commonly known ANN is the multilayer perceptron (MLP). The original concept of the perceptron was developed with the introduction of a hidden layer and a non-linear activation function to increase flexibility.

The input layer is provided with the variables which are to be included in the model and the output layer is the corresponding known outcome. This can be one node which makes a continuous prediction between 0 and 1 or a number of nodes to represent classification groups. There are often one or two hidden layers in the model with no predefined values. These hidden layers add flexibility in forming the relationship between input and output. The number of nodes in the hidden layer can be optimized to ensure the neural network learns efficiently and generalizes successfully. Information is propagated from the input layer through hidden layers to the output by weighted connections between the all the nodes in adjacent layers (analogous to synapses). This is known as a fully connected neural network (Figure 9.2).

During training, the inputs for a particular case (n) are fed forward through the network (left to right). The sum on each hidden node $v_j(n)$ is calculated as the connection weight w_{ji} between node i in the input layer and node j in the hidden layer multiplied by the value on the connected input node y_i. This is summed over all the nodes (m) which are connected to the node in question.

$$v_j(n) = \sum_{m=0}^{m} w_{ji}(n) y_i(n) \tag{9.1}$$

In addition to the connected nodes, the input from a bias node is included. The bias input can be regarded as an extra node in each of the input and hidden layers with a fixed value of 1 and a connection weight developed in the same way as all the others. A nonlinear activation function (*e.g.*, sigmoid) is applied to the sum on each processing node to introduce non-linearity into the network. There are a variety of functions that can be

Input layer (i) Hidden layer (j) Output layer (k)

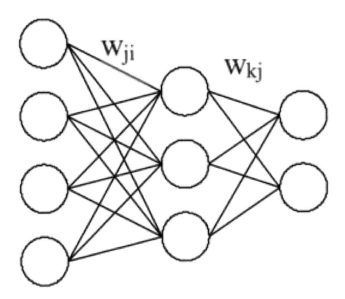

Figure 9.2 Schematic diagram of the architecture of an MLP with 4 input nodes, 1 hidden layer with 3 nodes and 2 output nodes.

used as the activation function. A sigmoidal activation function is common.

$$\varphi_j\left(v_j\left(n\right)\right) = \frac{1}{1 + e^{-(v_j(n))}} \tag{9.2}$$

The output of each node is known as the activation. Cases are presented to the network iteratively and the weights updated to minimise the error between predicted and actual outcome. The weights are updated using the concept of backpropagation of errors [380], where the error is used to guide the size of the change of the values of the weights. This is achieved by calculating the error gradient for both the hidden $(\delta_j(n))$ and output layers $(\delta_k(n))$ and adjusting the weights by propagating a fraction of the error values back through the network (Figure 9.3). The magnitude of the change is determined by the learning rate η, which has a value less than unity. The algorithm can be improved by the addition of a momentum term α which augments consecutive changes in the same direction by including a fraction of the change from the previous iteration. This is known to speed up computation and smooth the optimization process. This optimization scheme is known as the generalized delta rule 9.3 and is calculated for a node in the hidden layer j at iteration n as:

$$\Delta w_{ji}\left(n\right) = \alpha \Delta w_{ji}\left(n - 1\right) + \eta \delta_j\left(n\right) y_i\left(n\right) \tag{9.3}$$

MLPs have a reputation of being a black box since the complex relationships formed between the inputs and outputs within the network can be hard to interpret. However, the simplest way of analyzing how the input information is used to produce the outputs of the MLP is to use the 'leave one out' method. If no change in results is observed, it can be said

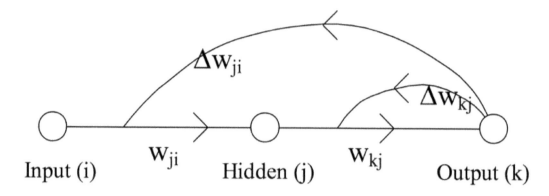

Figure 9.3 Summary of the backpropagation of weights through a Multilayer Perceptron.

that the input in question is redundant and is removed as part of the development process.

More recently, feedforward neural networks have been developed to contain many layers. In these neural networks the layers are not necessarily fully connected. To extending the concept of how neurons are connected in the brain, Convolutional Neural Network (CNN) was inspired by understanding of how the visual cortex processes image data using cells which only process information from a small sub-region of the visual field. The analogous CNN uses many locally connected layers. Often the local weights are identical in each separate region, acting as a filter for the incoming data. The CNN is a form of deep learning, a term which is synonymous with Big Data.

9.3.1.2 When should you use them?

ANNs are very useful for datasets with large numbers of input variables which may or not be correlated. Useful for both classification or regression, the flexibility to increase the number of layers and customize the connections between layers make neural networks an appealing choice for machine learning. CNN and other deep learning approaches are ideal for image-based inputs.

9.3.1.3 Who has already used them?

As a classic form of machine learning there are a reasonable number of publications that utilize ANN for outcome prediction in oncology [381, 382, 383, 384, 385]. Recently, Nie *et al.* [386] used a classic 3 layer MLP to optimize the choice of image-based metrics derived from MRI to determine the response to chemo-radiation of rectal tumours. The authors were able to show that voxel-based parameters were more predictive than volume-averaged parameters. Sun *et al.* [387] published a study using CNN to classify breast lesions detected using mammography as benign or malignant. Since CNN require large amounts of data to optimize the many layers a semi-supervised learning approach was used where images of lesions with unknown labels (classification) were used to augment the cases by matching with labelled cases of images of lesions which had been classified by a radiologist. The CNN approach was evaluated with/without labelled data and against Support Vector Machines (SVM) and conventional ANN. The CNN with unlabelled data resulted in the highest accuracy and area under the ROC curve (AUC).

9.3.2 Support Vector Machine

9.3.2.1 The basics

A popular alternative to artificial neural networks is support vector machines (SVM). Also defined as a supervised learning algorithm where both the inputs and corresponding outputs of a dataset are used for training, an SVM uses a kernel-based mapping to translate a non-linear dataset in to a higher dimensional space, where the two outcome classes become linearly separable, by defining a hyper-plane. Commonly the kernels used to map the data are either polynomials or a radial basis function of the form:

$$K\left(x, x'\right) = e^{\left[-\frac{\|x - x'\|^2}{2\sigma^2}\right]} \tag{9.4}$$

where σ is the width of the radial basis function. The function is then optimized to maximize the separation between the outcome classes and choose vectors describing the hyperplane (boundary), which discriminates between the outcomes classes with the largest possible margin between them. Only a subset of the cases is used to define the hyperplane. The corresponding vectors are known as the support vectors and represent cases that have crossed in to the margin of the boundary. One of the flexible attributes of SVM is that it is accepted that support vectors may be on the incorrect side of the hyperplane (see figure 9.4).

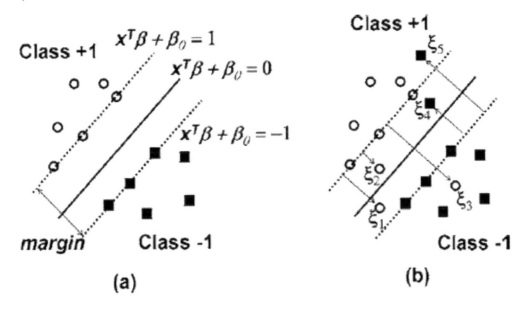

Figure 9.4 describing the support vectors which define a hyperplane which can separate 2 outcome classes. Taken from Investigation of the support vector machine algorithm to predict lung radiation-induced pneumonitis [388] .

9.3.2.2 When should you use them?

SVM are ideal for non-linear classification problems with a binary outcome.

9.3.2.3   *Who has already used them?*

SVM are overtaking ANN in publications related to radiation outcome modelling. Seminal publications include the work by Chen et al [388] who used SVM to predict radiotherapy-induced pneumonitis and El Naqa *et al.* [160], which describes using PCA to assess non-linearity of example datasets and then compare SVM with ANN and logistic regression. More recently Klement *et al.* [389] published a study using SVM to predict local recurrence following Stereotactic Body Radiotherapy (SBRT) to stage 1 lung tumours. Available input data included dose to the tumour described using biologically effective dose (BED) and patient characteristics including Karnofsky and age. Lung function information was included using forced expiratory volume in 1 second (FEV1). Since 88% of patients had not recurred after 6 months the outcome classes were unbalanced. Two strategies for under-sampling were considered. The first was to under sample the majority group (no recurrence) so that the incidence was balanced with the recurrence group. As the authors point out this is rather wasteful as many cases are discarded. A less wasteful strategy is synthetic minority over-sampling technique (SMOTE) [390] whereby synthetic copies of the minority data are made by choosing a nearest neighbour case and permuting the input variables randomly by multiplying the original values by a random number between 0 and 1. A combination of these 2 techniques was found to result in the highest AUC for the SVM model. Following on from this Zhou *et al.* [391] modelled incidence of distant failure in a similar patient population (early stage non small cell lung cancer (NSCLC) treated with SBRT. Three methods for feature selection were considered, clonal selection algorithm [392], sequential forward selection and a statistical analysis based method (backward stepwise logistic regression). The clonal selection algorithm approach was found to produce the best results.

9.3.3   Decision Trees and Random Forests

9.3.3.1   *The basics*

Interpreting the relationship between input variables and outcome predictions can be tricky for some machine learning approaches. This cannot be said of decision trees where a diagrammatic representation of the results is common. Analogous to a real tree, a decision tree is constructed from an initial input data set, bifurcating at nodes according to splitting criteria. This technique is known as recursive partitioning analysis and is repeated iteratively until terminal nodes (leaves) are reached. There are two types of decision tree, classification, where each terminal node represents an outcome class or regression, where the node value is a continuous probability. The general principle is to find criteria which maximally differentiate the data. Figure 9.5 demonstrates a decision tree constructed from a simple dataset with two variables and three output classes. The decision tree can be visualized as the partitioning of a rectangular space (the concept can be developed for further dimensions). The variables are split at a threshold which minimizes the misclassification of the data.

The architecture of the tree can be determined using a number of algorithms with different cost functions to choose the best split (known as the impurity function). For classification trees a common function is the Gini index G [394], which for a specific split criteria with k possible classes is given by:

$$G = \sum_{k=1}^{k} \widehat{p}_k (1 - \widehat{p}_k) \qquad (9.5)$$

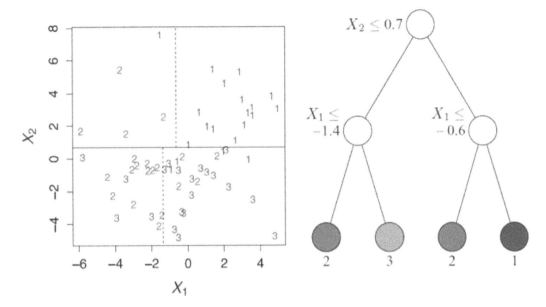

Figure 9.5   A classification decision tree taken from [393] .

where $\widehat{p}_k$ is the proportion of patients classified in class k by the split criteria. The tree is constructed by testing all possible splits for each variable at each node with the lowest value of the *Gini* index being chosen. This is a computationally greedy approach. If a regression tree is being constructed an alternative impurity function is sum squared error within a split group (each rectangle defined in Figure 9.5. Trees are grown until the number of cases to be split is less than a predefined threshold. This is to mitigate against overfitting. Pruning can be performed after growing by assessing the contribution of individual nodes to the overall prediction. So far the description has been for a single decision tree however the use of ensemble of trees known as a Random Forest has become more popular. Ensembles of decision trees are created and initialized using a randomly generated (bootstrapped) subset of the available data cases. Each tree is then constructed independently. The final result is aggregated from the contributions of each tree. For classification this will be the most votes, *i.e.*, class chosen by the most trees or for regression the outcome will be averaged across all the trees.

9.3.3.2   *When should you use them?*

Decision trees and random forests are more commonly used for classification. They are very helpful for clinical decision making as the splits on the nodes can be easily tested as a set of rules. Looking at figure 9.5 that would be if $X_2 \leq 0.7$ and $X_1 \leq -1.4$ then outcome class is class 2 and so on. It is important as with all statistical models to ensure that the model has been tested for predictive accuracy and generalisability.

9.3.3.3   *Who has already used them?*

In recent times random forests have become popular for building predictive models for radiation outcome modelling. Since the data is often large and complex with subtle relationships between input and output data, a forest constructed of a 1000 trees may well be better suited to the task than an individual decision tree. Ospina *et al.* [395] developed a random forest (RF) to predict the risk of rectal toxicity five years after conformal prostate radiotherapy. A comprehensive number of dosimetric variables were included in an initial random forest. Each tree was built with a dataset that was created from a bootstrapped sample of the original database. 75% of patients were used for training with the rest used for validation. Once the important dosimetric variables had been established, then, a further random forest was trained with additional, patient, tumour, and treatment characteristics. (n.b., this may not be optimal if relationships between dosimetric variables and treatment variables such as technique exist). The random forest results were compared to logistic regression and Lyman-Kutcher-Burman (LKB) models and were shown to be consistent in terms of both prediction and feature selection. A more recent paper by Dean *et al.* [379] compared logistic regression, SVM and RF to predict acute mucositis following head and neck radiotherapy using both conventional and spatial descriptors of the dose distribution. Results were assessed using both discrimination (AUC) and calibration with RF slightly superior in terms of calibration.

9.3.4   Bayesian approaches

9.3.4.1   The basics

An alternative approach in machine learning is to develop a probabilistic model based on Bayes Theorem. Bayesian networks have conditional independence statements that can be used to calculate joint conditional probabilities which describe the contribution of individual variables to an outcome. Bayesian Networks [396] are presented graphically using Directed Acyclic Graphs (DAG) (see figure 9.6). The representation of the model in a graphical format enables easy interpretation. Variables are represented as nodes with connections between nodes described as edges. The direction of the relationship is shown using arrows with parent nodes feeding in to child nodes. The probability of a node is conditional on the nodes feeding in to it and the graphs are not permitted to include loops. Each node has an associated conditional probability table.

As with other forms of machine learning, Bayesian Networks can be constructed from the available examples/cases. It is also possible to structure the network with known relationships. Learning the structure of the network becomes increasingly complex with an increasing number of variables; to overcome this computational challenge methods such as Markov Chain Monte Carlo [397] are commonly implemented. This method involves sampling from the multidimensional solution space to build up an accurate representation of the full solution. Once the structure has been optimized expectation maximization is often used to optimise the probabilities/weights of individual nodes.

9.3.4.2   When should you use them?

Bayesian Networks are a versatile approach to outcome modelling with easy interpretability. They are particularly useful for modelling interactions between variables and are also suitable for datasets with missing data. Using an ensemble of networks minimises the risk of choosing a biased model.

9.3.4.3 *Who has already used them?*

Jayasura *et al.* [398] capitalized on the fact that Bayesian Networks are inherently capable of dealing with missing data to develop a model to predict 2 year survival for lung cancer patients who received radiotherapy. It had been previously shown that adding information regarding positive lymph node status obtained from pre-treatment PET imaging improved predictive accuracy. However, this information is not always available. The Bayesian network that was developed is shown in Figure 9.6. Three external datasets from other institutions were used to test the predictive accuracy of the model. For comparison a support vector machine was also developed. Models were assessed using AUC and the Bayesian Network was shown to be superior to the SVM and robust to missing data.

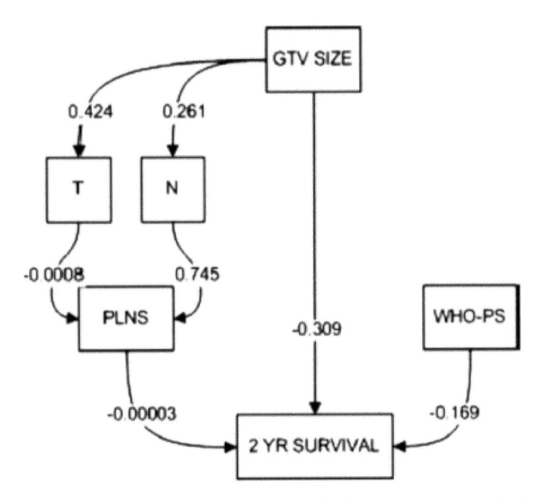

Figure 9.6 Example of a directed acyclic graph of a Bayesian network to predict 2 year survival in lung cancer patients treated with radiotherapy [398]

More recently, Lee *et al.* [255] used a Bayesian network ensemble to predict radiation pneumonitis by combining dosimetric, clinical and biomarker variables. The concept of a Markov blanket [153] was used for feature selection whereby it is assumed that only nodes with immediate connections (parents, children and children's other parents) influence the node. The rest of the nodes in the network are shielded. Despite the small dataset included in the study, the Bayesian network resulted in an AUC of 0.83, selecting a small subset from the biomarkers, clinical and dosimetric variables originally considered.

9.4 PRACTICAL IMPLEMENTATION

9.4.1 The data

Any model is only as good as the data on which it is constructed. It is important to ensure that the dataset is of good quality i.e. ideally prospectively collected and carefully curated for accuracy. It is also important to have an understanding of the data available and consider, for example, correlation between variables. However, this is much less of an issue for machine learning than in conventional multivariable logistic approaches. Where missing data is unavoidable, a reasonable strategy may be implemented or if only affecting a few cases consider removing incomplete data from the data set (checking to ensure this will not introduce bias to the data.) It is important to ensure that there are sufficient cases in the dataset, the rule of thumb for machine learning is that there should be at least 10 cases for each input variable. Although this is not a strict rule it is helpful to consider the concept of the curse of dimensionality whereby for high-dimensional data the available cases may be rather sparse. In addition to considering which variables should be included in a model, it is important to consider format/type of data. If there are orders of magnitude of difference between two variables for example Age and Prostate Specific Antigen (PSA) then the models may be biased by a numerically dominant variable. It is therefore good practice to normalize each variable either by scaling the input to the range 0 to 1 or standardizing to zero mean and unit variance. The concept of "Garbage in, Garbage out" is simple but powerful when considering data preparation. The corollary is also true whereby a good quality dataset with associations between input variables and outcomes should result in good models using different (but appropriate) types of machine learning. Many publications compare results using different flavors of conventional and machine learning models [399]. Agreement between these methods can be considered helpful although not necessarily an endorsement of accuracy.

9.4.2 Model fitting and assessment

Strategies for ensuring a good fit have been discussed in previous Chapters 4, 6, 7, 8; using all available data to train a model will result in the model fitting the training data 'too well'. Each type of machine learning algorithm has a cost function which is used to measure the error between the outcomes of the model and the known outcomes of the training data. With successive iterations the total error on the training cases should decrease. However, if the model trains too efficiently on the training cases it will lose the ability to generalize to new (unseen) cases. To avoid this problem, techniques such as cross-validation [400] should be implemented in the training procedure. In this method the available cases are split in to equal sized groups. Commonly 10 fold cross-validation is implemented by repeating the model fit using each sub group of the data as the validation cohort for the model in turn. Overall results are then averaged over the 10 model fits and variability between folds used as a metric for generalization. An alternative approach of bootstrapping where each dataset is constructed by sampling with replacement may also be appropriate for some machine learning techniques. Internal validation is important to avoid overfitting and therefore ensure generalizability. Classically, machine leaning studies have often removed a subset of data before model fitting to act as a test set. However, the recent tripod guidelines [26] would recommend using all available data to perform model fitting with internal validation. Saving external validation for a truly independent test set (for example from another institution or, if compatible, from another cohort). The classic metric to assess predictive model performance is area under the curve (AUC). While this is a helpful indication of the discriminative abilities of the model, more detailed analysis using calibration curves can be more informative. Model assessment is discussed in more detail in Chapter 4.

9.5 CONCLUSIONS

Machine learning methods have become front and center in the era of big data. This chapter was intended to demystify machine learning. Many of the examples studies from radiation outcome modelling cited within the chapter provide more detailed information regarding implementation. The chapter closes with a list of resources including on-line tutorials which will also provide practical information. Machine learning is a powerful extension to classic statistical with a rigorous mathematical basis. Their role ion outcome is likely to grow particularly in the area of outcome modelling given the rapid increase in patient-specific data from clinical, physical and biological sources as discussed in Part I of this book. As such they are incredibly useful for teasing out the complex treatment responses in radiation oncology or oncology in general.

9.6 RESOURCES

Most of the major software environments provide packages/libraries for machine learning. Where possible, using a familiar environment will make developing initial machine learning approaches easier. Commonly used platforms include R (various open source packages), Matlab (machine learning tool box) and Python (`scikit-learn package`).

Test data sets can be found in the UCI Machine learning dataset repository, some are preloaded in machine learning packages.
`https://archive.ics.uci.edu/ml/about.html`

Many online tutorials exist to lead beginners through machine learning implementations, these include:
`http://machinelearningmastery.com/start-here/`
`https://medium.com/@ageitgey/machine-learning-is-fun-80ea3ec3c471`.

III

Bottom-up Modeling Approaches

Stochastic multi-scale modeling of biological effects induced by ionizing radiation

Werner Friedland

Pavel Kundrat

CONTENTS

ABSTRACT

An overview is given on bottom-up stochastic modeling of radiation-induced effects on sub-cellular and cellular scales with the PARTRAC suite of Monte Carlo codes. PARTRAC simulations start from modeling particle tracks by following individual interactions of the primary as well as all secondary particles with the traversed medium, typically liquid water as a surrogate for biological material. Dedicated modules then convert these energy deposits to reactive species and follow their diffusion and mutual reactions. Multi-scale models of DNA and chromatin structures in human cell nuclei are implemented, representing DNA double-helix, nucleosomes, chromatin fiber, domains and chromosome territories. By overlapping with the simulated tracks, these target structures are used in assessing initial radiation-induced DNA and chromatin damage both via direct energy depositions and indirectly through attacks of reactive species. DNA fragmentation patterns can be analyzed.

In a subsequent module, cellular repair of DNA double-strand breaks by non-homologous end-joining is simulated. Followed is the enzymatic processing as well as spatial mobility of individual broken chromatin ends; correct rejoining, misrejoining, and formation of chromosome aberrations are scored. The chapter is focused on the underlying principles of radiation action in biological systems on subcellular scales and their stochastic modeling with PARTRAC. The pieces of experimental information represented in this tool are highlighted, together with numerous independent benchmarking tests against observed data and trends. Future developments are discussed, directed towards extending PARTRAC scope and predictive power, but also towards applications in medical physics and radiation protection.

Keywords: Ionizing radiation, particle tracks, radiation quality, cross sections, reactive species, water radiolysis, radiation-induced DNA damage, DNA damage response, non-homologous end-joining, chromosome aberrations, Monte Carlomodeling, bottom-up model, computational biology.

10.1 INTRODUCTION

Ionizing radiation is widely used in medicine, science and technology. Its industrial applications include, *e.g.*, quality control of different goods, food conservation, or X-ray imaging in security checks. Its medical applications serve both diagnostic and therapeutic purposes: Diagnostic methods based on ionizing radiation include X-ray imaging, computed tomography, positron or single photon emission tomography and other techniques. Therapeutic purposes include external radiotherapy using photon, electron, proton or ion beams or brachytherapy with short-range radiation sources placed into or close to the cured region, for the curative treatment of cancer or palliative treatment of painful diseases.

In optimizing these and future applications, a balance must be found between benefits and potential long term health risks of radiation exposure. The health risks follow from the fact that ionizing radiation elicits a number of adverse biological effects [401], including its pro-carcinogenic action or recently documented enhancement of cardiovascular diseases [402].

Detailed understanding of the mechanisms and processes that underpin the biological effects of ionizing radiation is needed in order to allow the benefits of radiation applications to be fully exploited and at the same time its negative effects being reduced as far as possible. Mechanistic modeling plays an important role towards achieving these goals [331, 263, 403]. Modeling serves as a useful complement to experimental studies: it integrates the knowledge gained from the experiments. It allows alternative hypotheses on the underlying processes and mechanisms to be tested. Modeling results inform experimental research on conditions and endpoints where additional experiments should be performed to foster the existing knowledge or rule out alternative hypotheses. Modeling also allows interpolations between measurements and extrapolations to conditions not measured yet or experimentally not accessible at all.

Ionizing radiation affects biological systems over multiple spatial as well as temporal scales; see Figure 10.1. These scales are linked with objects that undergo modifications by certain processes, and usually specific modeling approaches are adopted to describe the underlying mechanisms. The fastest events with temporal scales down to 10^{-20}s or even

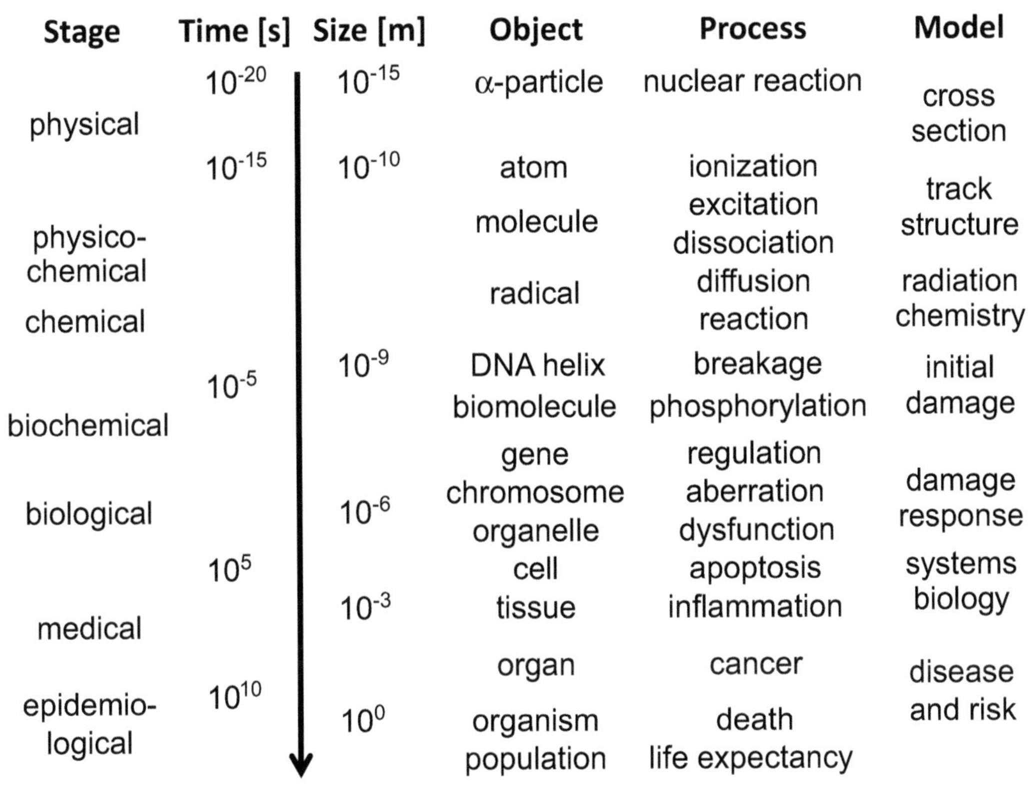

Stage	Time [s]	Size [m]	Object	Process	Model
physical	10^{-20}	10^{-15}	α-particle	nuclear reaction	cross section
physico-chemical	10^{-15}	10^{-10}	atom	ionization	track structure
			molecule	excitation dissociation	
chemical			radical	diffusion reaction	radiation chemistry
biochemical	10^{-5}	10^{-9}	DNA helix biomolecule	breakage phosphorylation	initial damage
biological		10^{-6}	gene chromosome organelle	regulation aberration dysfunction	damage response
	10^{5}		cell	apoptosis	systems biology
medical		10^{-3}	tissue	inflammation	
			organ	cancer	disease and risk
epidemio-logical	10^{10}	10^{0}	organism population	death life expectancy	

Figure 10.1 Scales of radiation effects in biological context. The development of radiation effects proceeds in multiple stages that cover distinct time scales and are related to objects of generally increasing size. The timescales, objects and processes are largely exemplary items. The underlying processes are theoretically assessed by various model approaches.

shorter are processes which involve the atomic nucleus, such as fragmentation of an incident carbon ion upon hitting another nucleus. This spatial scale is also related to radioactive decay events that lead to the emission of energetic helium nuclei (alpha particles), electrons or positrons (beta particles), photons (gamma rays), or other particle types in further decay modes. Neither natural nor induced radioactivity is addressed further in this chapter. However, time scales of these events cover essentially the whole range shown in Figure 10.1, indicating that the listed relation of objects and processes to time scales is not mandatory. Key events setting the scene for the biological effects of ionizing radiation are ionizations and excitations of individual atoms and molecules within the traversed medium; these processes occur on spatial scales given by the size of atoms and small molecules, 10^{-10}m, and temporal scales of 10^{-15}s. Within the subsequent physico-chemical phase up to 10^{-12}s, reactive species are produced from excited and ionized water molecules, which are major constituents of most biological materials. During the following chemical phase, up to 10^{-6}s, biologically important macromolecules are damaged via attacks of radicals (primarily hydroxyl radicals, •OH) in addition to the damage resulting from direct energy deposits. The target molecules include, first of all, DNA within the cell nucleus, the carrier of the cell's genetic information. Although the total length of a human DNA molecule is a few meters, it is highly compacted within a cell nucleus into chromosomes with micrometer dimensions. The initial radiation damage on sub-cellular scales is induced within microseconds to seconds, but the cell possesses a number of dedicated repair pathways that aim at removing the damage and restoring the original functionality; these repair processes typically operate on timescales of seconds to hours. Unremoved damage to biologically important macromolecules and/or the shift in redox balance upon radiation exposure may then manifest over the timescales of hours to days in malfunctioning of organelles, induction of mutations, or even cell death. At the tissue level, inflammation or diverse types of tissue damage can be observed, corresponding to spatial scales of millimeters to decimeters and a broad range of temporal scales, hours to years. On even larger and longer scales, radiation-induced effects include organ dysfunction and enhanced or accelerated induction of cancer or cardiovascular diseases. Spatial and temporal scales may be even further extended when risks for populations are considered, *e.g.*, in the framework of nuclear waste disposal strategies. All the mentioned effects depend on the quality of radiation (particle type and energy), applied dose, dose rate, and in the context of medical applications on the selected fractionation scheme.

Being a multi-scale issue, biological effects of ionizing radiation require multi-scalemodeling approaches. Model approaches addressing particular processes and topics are listed in the last column of Figure 10.1. Cross section models provide a dataset of interaction probabilities for the traversed medium in dependence on the incident particle type and its energy. These are then used as an input for track structure models that aim at reproducing the detailed pattern of energy deposition and its stochastic nature. The initial radiation damage is assessed then by combining the resulting track structure with the structure of the relevant target; some models work with a short piece of DNA, while more advanced approaches model the total genomic DNA inside a single cell nucleus. In nanodosimetric studies, the target DNA and chromatin structures are approximated simply by small volumes, and the numbers and sizes of ionization clusters in these volumes are studied in relation to biological effects such as DNA damage [404]. To consider also the indirect damage due to radical attacks, the physical track structures have to be complemented by a chemistry model, and used with a model of the target and corresponding assumptions on the susceptibility of the target to damage induction. During the past three decades a variety of track structure models have been developed and used for calculations of radiation-induced DNA damage [405]; individual models differ in particular in how the indirect damage is considered, what incident particles and energies are covered, and how sophisticated the underlying DNA model is. In some models, endpoints such as cell killing have been linked with initial damage in a semi-phenomenological manner, using suitable parameterizations [291, 406]. A mechanism-oriented, dynamic simulation of DNA damage response has been addressed by a few research groups only [407, 296, 408]. The semi-phenomenological approaches offer simplicity and robustness needed for applications in treatment planning; the mechanistic approaches are computationally more expensive, but allow one, for example, to estimate mutations in specific genes and hence can be more easily integrated into complementary top-down approaches of biology-based models of carcinogenesis [409].

Tools for stochastic bottom-up simulation of track structures and resulting radiation-induced biological effects on subcellular and cellular scales have been developed in various frameworks, such as KURBUC [13], RETRACKS & RITRACKS [410], the Geant4-DNA project [411] and PARTRAC [265]. This chapter is focused on PARTRAC as a representative example of such simulation methods. PARTRAC possesses a modular structure corresponding to the underlying processes sketched above. It starts from modeling particle tracks, i.e. spatial patterns of excitations and ionizations induced by photons, electrons, protons, alpha particles or heavier ions. This physical structure is then followed by pre-chemical and chemical modules, in which excitations and ionizations are converted into chemical species, whose diffusion and mutual reactions are traced. Multi-scale models of DNA and chromatin structures in human cell nuclei are overlaid with the track structures to score DNA and chromatin damage from direct effects through energy deposits to DNA, to assess the impact of chromatin structures on radical chemistry in the surrounding water, and to determine indirect effects mediated by the attacks of reactive species. DNA fragmentation from single and multiple tracks can be analyzed in a separate module. DNA damage response is modeled step-by-step in time, including the misrejoining of broken DNA ends and the induction of chromosomal aberrations.

The stochastic nature of the processes is reflected in PARTRAC by using the Monte Carlo simulation technique. Typically, a number of alternative or competing processes are considered. The processes to take place (e.g., whether a water molecule is ionized or excited) and their detailed characteristics (e.g., the energy and direction of a secondary electron liberated in the ionization process) are sampled, using computer-generated (pseudo-)random numbers, from corresponding databases derived from experimental data and theoretical research. Output from a preceding module is used as an input for the subsequent module, e.g., DNA repair is simulated starting from the modeled initial DNA damage patterns.

The principles of individual PARTRAC modules, the pieces of knowledge integrated, and the simulation results are discussed to greater detail below. For further information the reader is referred to [265, 412] and references therein.

10.2   PARTICLE TRACKS: PHYSICAL STAGE

The setting scene for the biological effects of ionizing radiation is given by the fast physical stage; within femtosecond time frames, ionizing particles interact with the traversed medium and induce excitations and ionizations to its constituting atoms and molecules. PARTRAC follows a semi-classical, non-quantum picture of particle trajectory: A particle is characterized by its position and velocity, and is considered as a point object without any internal structure. The actual charge state of the particle is traced for protons and alpha particles, which at low energies are often present as hydrogen or singly charged or neutral helium atoms, respectively. For heavier ions (at low energies including neutral atoms), neither excited states nor charge states are considered explicitly, but the model works with energy-dependent mean charge states of these incident particles.

In general, biological material is approximated by liquid water in track structure-based studies, without considering the particular material composition and/or heterogeneity inside the cell and cell nucleus. One exception was an investigation of photon-induced effects taking into account the heterogeneous target structure of the DNA helix, DNA bases and histones inside the cell nucleus [413]. In that study, pronounced effects were found to be limited to ultra-soft X-rays between carbon (278 eV) and oxygen K-shell energies (539 eV); this justifies the use of liquid water as proxy for biological material. Hence, in all studies reported in this chapter, energetic particles and all their secondary and higher-order particles are followed event-by-event inside a certain simulation region ("world region") that is filled with liquid water and surrounded by vacuum.

Figure 10.2 illustrates an exemplary track structure calculation for an energetic particle such as a proton, for instance. Within the physical stage (panel a), the proton is started with defined energy in a random or pre-defined direction from a randomly selected location inside a source volume or area. The 'world region' encompasses the source. The particle moves a certain distance corresponding to the arrow, until an interaction within the medium occurs (circle). Energy deposition events inside the region of interest ('target region'), a sub-volume of the world region, are stored for further analysis. They also serve as initial events for the physico-chemical stage (panel b). In this stage, inelastic events give rise to the production of reactive species, and slow electrons come to rest and become finally hydrated. These new species provide the starting condition for the subsequent chemical stage for which the original locations of inelastic events are no longer of interest.

In the physical stage, primary and secondary particles are traced as long as they do not leave the world region. The margin between the world and target regions has to be selected large enough to include essentially all particles that may leave and re-enter the target region as depicted in the bottom-right part of panel a, but small enough in order not to waste the computing time for unnecessary simulations. The characteristics of the track, such as path lengths, energy deposits, secondary electron energies and directions, etc., are determined by total and differential cross sections discussed below. Ionization and stripping events of the primary particle liberate secondary electrons. After oxygen K-shell ionizations, an additional Auger electron is released (bottom-right part of panel a). All these electrons are traced on an event-by-event basis within the world region until their energy has dropped below the threshold for excitations. Finally, the electrons are transported as sub-excitation electrons in random direction to the position where they come to rest (dotted arrows).

Some particular features of the PARTRAC approach are indicated in Figure 10.2a: Angular scattering of ions and atoms by elastic or upon inelastic interactions is presently not considered in PARTRAC. These heavy particles are thus assumed to travel along straight lines, contrary to electrons that suffer from large scatter, especially at low energies. Nevertheless, the proton track shows a slight shift (delocalization) of individual energy deposition events from a straight line, which accounts for the density limit on ionizations and excitations corresponding to the density of water molecules; multiple events in the same water molecule are not considered in PARTRAC. The electron transport direction is conserved upon excitations if its energy stays above the excitation threshold. To reduce computing time, for electrons below 30 eV the transport to the next inelastic event is approximated (dashed arrow in panel a), lumping together the numerous elastic events which are up to 500 times more frequent than inelastic ones.

In cases where the material or its density changes during the path between two interactions, the particle transport is restarted with the same direction and energy at the intersection point with the mean free path of the then traversed material.

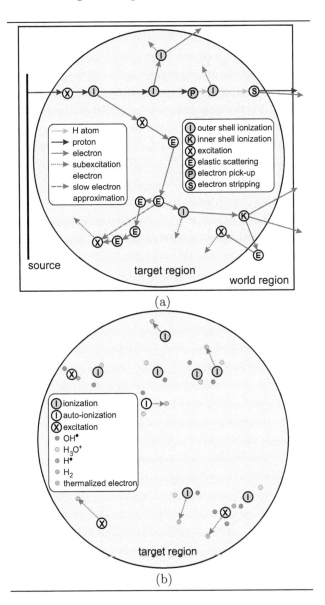

Figure 10.2 Sketch of an exemplary track structure calculation. Panel a: physical stage, panel b: physico-chemical stage. The species in panel b describe the initial condition of chemical stage; types and locations of the processes in the physical stage are then no longer of relevance. Further information see text..

The prerequisite for simulating the track structure within the physical stage is the knowledge of overall probabilities and detailed characteristics of alternative interaction processes in dependence on the type and energy of the impinging particles and on the traversed medium. Such interaction probabilities are quantified by so-called cross sections : In a classical analogue, the cross section corresponds to the target area that the impinging particle has to hit for the process to take place. Total cross sections cover all considered types of events. Process-related cross sections refer to a certain subset of processes such as ionizations or specifically oxygen K-shell ionizations of water molecule. Further, differential cross sections describe the details of the selected process; *e.g.*, cross sections differential in energy and angle of a secondary electron liberated from an ionized water molecule describe the probabilities of an electron being emitted with particular energy and direction. The cross sections reflect the wealth of knowledge from particle and nuclear physics experiments and theories; they do account for the quantum nature of the processes.

Total macroscopic cross sections, *i.e.*, the sum σ of cross sections σ_i for all possible processes i multiplied by the density of water molecules N, provide information on typical transport distances between individual interactions: the mean free path μ is the inverse of the total macroscopic cross section, $\mu = 1/(N\sigma)$. The actual transport distance t between two interactions is a stochastic quantity; it is sampled by $t = -\mu\ln(r)$, where r is a uniformly distributed (pseudo-)random number, $0 \leq r \leq 1$, determined by the Mersenne twister algorithm [414]. Mean free path values of electrons and several light ions are presented in Figure 10.3.

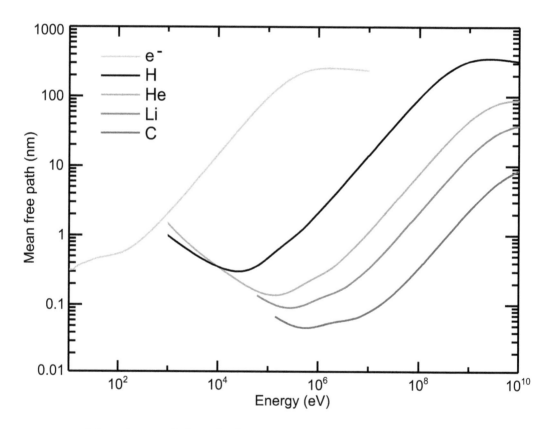

Figure 10.3 Mean free path lengths for electrons and ions liquid water. Data for H, He, Li and C ions are averaged over charge states during slowing down; they include the contributions from neutral atoms that are particularly relevant at low energies. Based on cross sections from [415, 416, 412].

The process to take place is selected as illustrated in Figure 10.10 for proton transport in liquid water. The event-specific cross sections [416] are represented as tables of cumulative probabilities, for a sufficiently dense grid of particle energies (log-equidistant distribution with 100 values per decade in PARTRAC). For a given case, a (pseudo-)random number is drawn and the intersection point with particle energy determines the interaction type; *e.g.*, drawing 0.6 means for 0.1 MeV protons that an ionization of type $3a_1$ is selected. In a similar manner, further interaction details are sampled, such as the energy and direction of flight of the liberated electron. Then the transport distance to the next event is sampled, the particle moved, the next event type and details selected, and so on. Followed is not only the primary particle but also all secondary and higher-order ones with all their interactions.

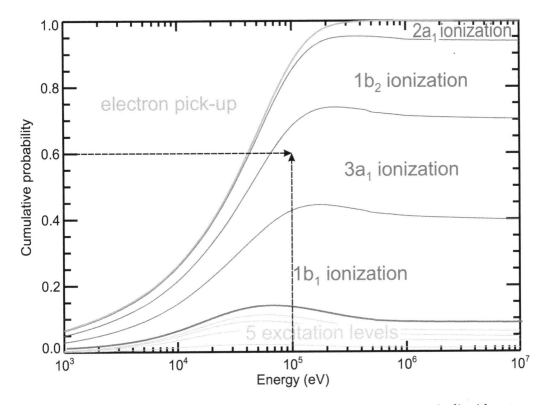

Figure 10.4 Cumulative probabilities of proton interaction processes in liquid water. The contribution of oxygen K-shell ionizations is not visible (0.2% at 1 MeV, 0.5% at 10 MeV). For a selected random number between 0 and 1 taken as cumulative probability (here 0.6), the intersection point with the proton energy (here 0.1 MeV) determines the event type (here 3a1 ionization).

From the above-outlined scheme it is clear that cross sections play a crucial role in Monte Carlo simulation of particle tracks. In PARTRAC, dedicated cross section databases are available for photons [417], electrons [415], hydrogen (protons) [416] and helium (alpha particles) [418]. Hydrogen cross sections are scaled to provide cross section data of heavier ions [412]. Essentials on these data bases are provided below; additional details can be found in the references.

Photon cross sections in PARTRAC are based on the EPDL97 database [417]; they cover photon energies from 1 eV to 100 GeV. The interaction processes taken into account are coherent scattering, photoelectric effect, Compton scattering, and pair production. Relaxation processes through Auger electron or fluorescence photon emissions are considered, too [419]. These libraries provide atomic cross sections, which are then converted to those for water or other biologically relevant material by the additivity rule and density scaling. This approach accounts for abundances of different atoms in the traversed medium under the implicit assumption of a homogeneous structure.

Electron interaction cross sections in the energy range between 10 eV and 10 MeV are based on calculations within the plane-wave Born approximation, with a semi-empirical model for the dielectric response function of liquid water with relativistic and semi-empirical corrections [415]. Five distinct excited states and five ionization shells of water molecules are considered. Electrons below 10 eV are assumed to deposit their energy at the point of their last inelastic interaction and then transported a certain thermalization distance [420] to the location where they become hydrated.

Cross sections for protons and neutral hydrogen atoms in liquid water [416] are available in PARTRAC from 100 eV but due to neglecting nuclear interactions not adopted below 1 keV; the upper energy limit is 10 GeV for protons and 10 MeV for hydrogen (above 10 MeV, virtually all hydrogen is ionized). The proton/hydrogen cross sections account for ionizations, excitations and charge changing processes (i.e. electron capture by proton and electron loss by neutral hydrogen atom). The data for energies above about 500 keV are based on the first Born approximation and Bethe theory. At relativistic energies, corresponding corrections and the Fermi density effect are taken into account. Below 500 keV, semi-empirical approximation and corrections are adopted, considering also the resulting stopping cross sections. Elastic scattering of these heavy particles is neglected and modifications of the transport direction upon ionization events due to momentum conservation are not considered.

Cross sections for helium cover interactions of doubly charged ions (He^{2+}), singly charged ions (He^{+}), and neutral helium atoms (He^{0}) in liquid water over energies of 1 keV – 10 GeV [418]. Similarly to the case of protons, helium cross sections are based on the plane-wave Born approximation and Bethe theory, with semi-empirical models and corrections at low energies. Elastic scattering and deviations from the initial transport direction due to inelastic processes are not considered.

Cross sections for ions with atomic number Z > 2 are scaled from the ones for protons at the same velocity or, at energies below 1 MeV/u, the corresponding equilibrium mixture of protons and neutral hydrogen atoms [421, 412]. The scaling algorithm is applicable for energies from 10 GeV/u down to about 10 keV/u; at lower energies nuclear interactions not considered in PARTRAC actually play an important or even dominating role. The scaling is based on the Barkas term for the effective charge Z_{eff} as a function of β, the particle velocity in units of the speed of light: $Z_{eff} = Z(1 - \exp{-125\beta Z^{-2/3}})$.

It is assumed that total cross sections of particles with different atomic numbers Z, including hydrogen/protons, scale according to their relation of Z_{eff}. The scaling law is assumed to govern all electronic cross sections; this means that effectively only the mean free paths are scaled, while the type of interaction, the amount of energy transferred, and angular distributions are determined as for proton/hydrogen at the same speed. This results in a corresponding scaling for stopping powers, which have been validated against established databases recently [412]. For high energies where the particle is fully ionized, the given formula reduces to scaling with the ion charge squared, Z^2. At intermediate energies, the full charge is replaced by the effective charge, which accounts for the expected state of the particle at the given energy (or velocity); this is the standard Barkas scaling approach [422].

In addition to having started from cross sections that represent available experimental data and reasonably agree with other models and theories, it is important to benchmark the results on physical tracks against independently obtained data on more macroscopic endpoints. To illustrate such benchmarking, PARTRAC has been successfully validated against literature data on stopping powers of light ions (Figure 10.5), and on radial dose profiles along oxygen and helium ion tracks in liquid water (Figure 10.6).

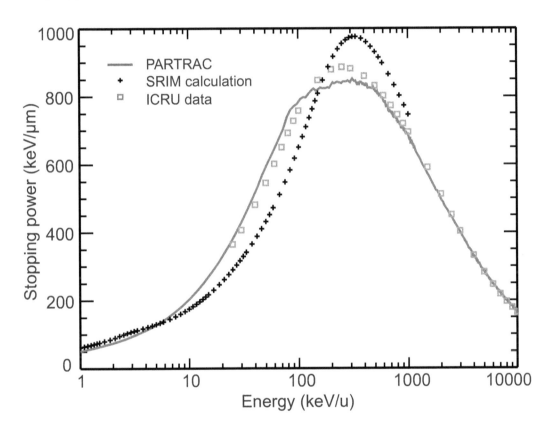

Figure 10.5 Stopping power of carbon ions. The values of stopping power from event-by-event track structure simulations in PARTRAC based on scaled cross sections [412] are compared to the reference data from ICRU [423] and to the results of calculations with the SRIM code [424].

The track structure simulations allow taking views that are much more detailed than what can be reached experimentally. To illustrate this, exemplary simulated tracks of a proton, an alpha particle and a carbon ion are shown in Figure 10.7. The viewpoint is close to the track ends in the foreground on the left hand side. The decreasing size of the spheres and the repeated rulers in the background on the right hand side corresponds to the increased distance from the start of the tracks there. The three ions have the same initial energy

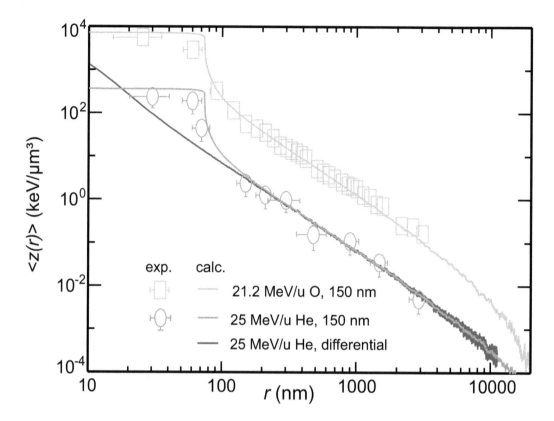

Figure 10.6 Radial distribution of energy depositions around ion tracks. Data on mean specific energy per ion for 21.2 MeV/u oxygen and 25 MeV/u helium ions are shown. Symbols: experimental data for 150 nm simulated site size (Schmollack *et al.* 2000); lines: PARTRAC calculations for 21.2 MeV/u oxygen (cyan) and 25 MeV/u helium (orange), both for 150 nm site size, and the underlying radial distribution for 25 MeV/u helium (violet). The specific energy inside the track core (1 nm radius) is about 106 keV/μm^3 (not shown in the figure).

per nucleon of 1.25 MeV/u and travel with about 5.8% of the speed of light. Captured are all inelastic interactions of the primary particle as well as of secondary and higher-order electrons are also the moving particles themselves. Most secondary electrons possess low energies only, and together with the interactions of the primary particle constitute the track core with a high ionization density. Some secondary electrons are, however, emitted with relatively large energies (> 2 keV), and travel quite far from the primary track, forming the penumbra region with lower mean/expected energy deposition. Note, however, the high inhomogeneity of energy deposition in the penumbra: large parts of the penumbra region do not see any energy deposit at all, but the energy deposition density in the tracks of energetic secondary electrons is quite high. Due to the stochastic nature of radiation interaction with matter, each track looks differently; however, for massive charged particles such as protons, alphas or especially heavier ions, there are always the high-density core and low-density penumbra regions. Also note that ions travel along straight lines since lateral scattering of these particles is neglected in PARTRAC; nevertheless, the delocalization from a straight line is visible for alpha particles and carbon ions. On the contrary, electron tracks are curly, especially at low energies, *i.e.*, towards track ends. The number of electrons moving during slowing-down with more than 10 eV is relatively low for all ion tracks.

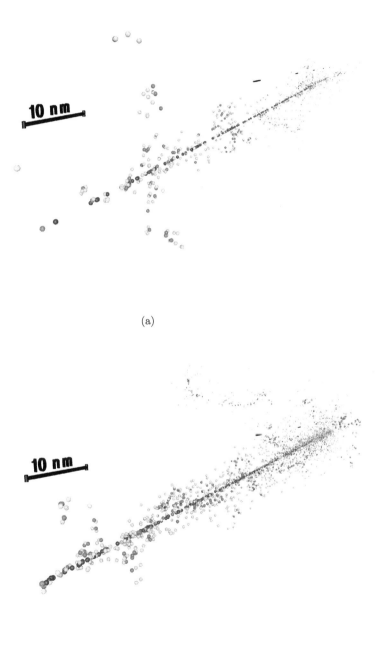

(a)

(b)

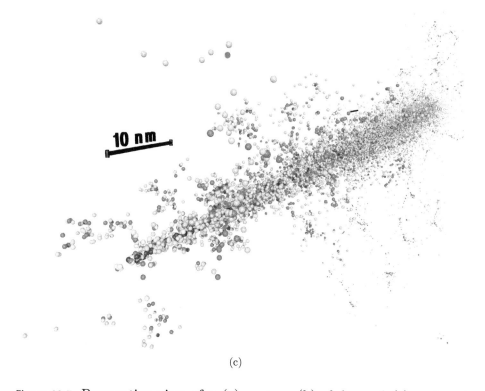

(c)

Figure 10.7 Perspective view of a (a) proton, (b) alpha and (c) carbon ion tracks calculated by PARTRAC. The snapshots have been taken 32 fs after the particles have started with 1.25 MeV/u initial energy at the right-hand end of the tracks. During the passage of nearly 0.5 μm to the left-hand end, they have lost 17 keV (p), 47 keV (α) and 306 keV (C) due to inelastic interactions; thus, the unrestricted LET values in keV/μm are twice these numbers. Indicated by colored spheres (0.6 nm diameter) are: the moving primary particle (pink), moving secondary electrons above 10 eV (yellow) and below 10 eV (white), thermalized electrons (green), locations of ionized (red) and excited (cyan) water molecules from secondary electrons and due to primary particle interactions (blue). Apparent sizes of the spheres are inversely proportional to the distance from the viewpoint close to the track end on the left-hand side in 55 nm distance from the track core; to show the wide-angle perspective, the 10-nm-ruler is repeated as a black line every 100 nm. On the left-hand side, the tracks are still developing due to further interactions of moving particles; on the right-hand side, the subexcitation electrons have largely stopped and shine up as a green cloud. The secondary electrons of the three tracks differ in their density corresponding to the LET values but have otherwise the same characteristics (*e.g.*, energy distribution) since the primary particles possess the same energy per nucleon. Visualization by POV-Ray^{TM} ray tracer software (Persistence of Vision Raytracer Pty. Ltd., Williamstown, Victoria, Australia).

10.3 PARTICLE TRACKS: PHYSICO-CHEMICAL AND CHEMICAL STAGE

Within the physico-chemical stage of track structure development during the time interval from 10^{-15} s to 10^{-12} s after the passage of the ionizing particle, excited and ionized water molecules relax or dissociate to reactive species. Likewise, secondary electrons that have been followed by the physical stage module down to sub-excitation energies and transported over a thermalization distance are converted to hydrated (solvated) electrons, eaq-. The simulated pattern of particle interactions with the water medium (Figure 10.2a) is thus replaced at the beginning of a chemistry module by a pattern of reactive species (Fig. 10.2b): Ionized water molecules are replaced by a pair of hydroxyl radical, $^{\bullet}$OH, and hydronium (also called oxonium) cation, H_3O^+, which are products of a reaction of the ionized molecule with its unaffected neighbor, $H_2O^+ + H_2O \rightarrow\, ^{\bullet}OH + H_3O^+$. Excited water molecules are assumed to dissociate also with formation of atomic hydrogen radicals ($H^{\bullet}$) and atomic oxygen bi-radicals $^{\bullet}O^{\bullet}$; the latter is assumed to interact with a nearby water molecule and produce two $^{\bullet}$OH radicals and a hydrogen molecule (H_2). Important parameters for the simulation of the physico-chemical stage are the branching ratios of the excited states into the different decay channels including also auto-ionization and relaxation without formation of species, and the initial positioning of these species. The knowledge on these settings is very limited. Since they significantly influence the reaction dynamics during the chemical stage, an agreement between simulation and experiment in the yields of species in the course of time and as a function of the radiation quality has to be sought; in turn, the agreement with the data can be taken also as a validation of the settings for the physico-chemical stage [425].

In the chemical stage module, the initially formed reactive species are allowed to diffuse and undergo mutual reactions. Using diffusion coefficients of the species that are known from chemistry [426, 425], a step-by-step simulation is performed. Starting from their initial positions, the species are randomly displaced by $\Delta x = (6D\Delta t)^{1/2}$ at each time step Δt, where D denotes the diffusion coefficient. Reactions are assumed to happen whenever two reaction partners get close to each other, within the reaction radius, R, related to the corresponding reaction rate constant, k, according to the theory of diffusion-limited reactions [427]: $k = 4\pi D'R^2/(R + (\pi D'\Delta t)^{1/2}$, where D' is the relative diffusion coefficient of the reaction partners, $i.e.$, the sum of their diffusion coefficients. Nine reactions between reactive species are explicitly modeled in PARTRAC [426, 425]: Hydrated electrons produce hydrogen radicals with hydronium or form hydroxide anions (OH^-) as new species with other reaction partners; $^{\bullet}$OH recombine with hydrogen radicals or produce hydrogen peroxide (H_2O_2) as new species upon reaction with each other; two hydrogen radicals combine to molecular hydrogen; and hydronium cations recombine with hydroxide anions. A so-called jump-through correction is employed to correct for reactions that might have taken place but were jumped over due to finite displacements Δx [428]. To save computational expenses, the time steps are gradually increased during the chemical stage simulation; this is enabled by the diminishing concentrations of reactive species.

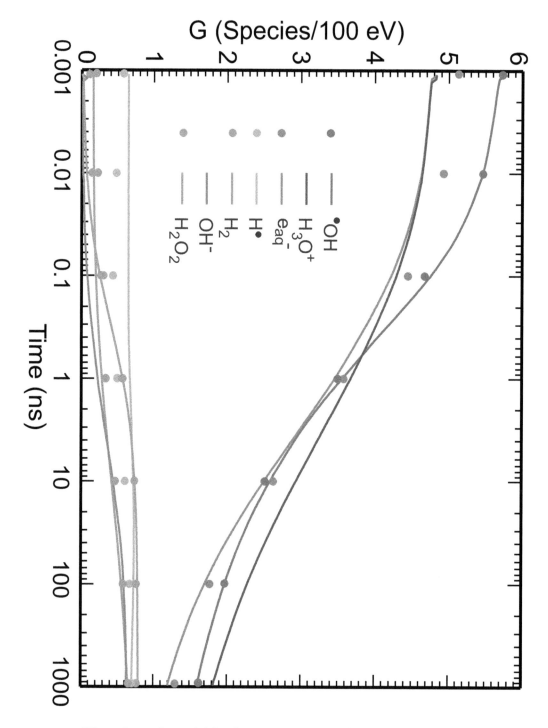

Figure 10.8 Time-dependent yields of reactive species after 5 MeV proton irradiation. Lines: results from PARTRAC calculations [425], symbols: results from another Monte Carlo Study [429].

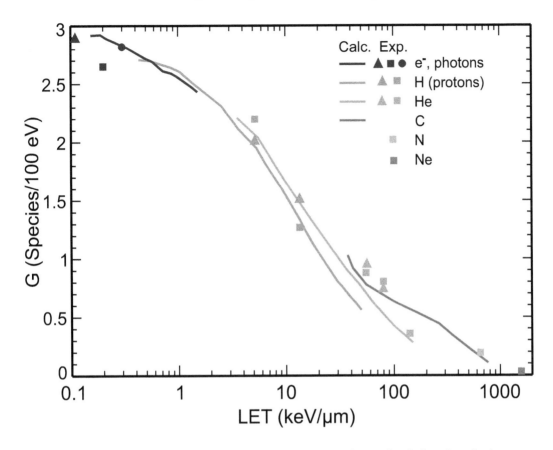

Figure 10.9 LET-dependent yield of OH radicals at the end of the chemical stage. PARTRAC calculations (lines) after 1 μs for electron and ion irradiation [425] compared to experimental results; triangles: [430], squares: [431], circle: [432].

In Figure 10.8, PARTRAC results on the yields of the considered species in the course of time from 1 ps up to 1 μs upon irradiation with protons of 5 MeV are shown, together with the results of another Monte Carlo simulation study [429]. The yields are presented as the numbers of species per 100 eV energy deposit. Both calculations are in good mutual agreement; they show the consumption of the initially highly abundant species (eaq$^-$, $^{\bullet}$OH and H_3O^+) and the production of other species (OH$^-$, H_2O_2 and H_2). As a validation of the simulations against experimental results, the yields of $^{\bullet}$OH at the end of the chemical stage are presented in Figure 10.9 from literature data and for PARTRAC calculations in dependence on radiation quality, captured here by the linear energy transfer (LET) that reflects the ionization density: With increasing LET the $^{\bullet}$OH radicals and their reaction partners are formed closer together and may frequently recombine; the reduction of the yields during the chemical stage (see Figure 10.8) become more pronounced. The simulated data agree very well with the measured data. For additional information on radiation chemistry simulation in PARTRAC and its benchmarking, we refer to specific studies [426, 425].

Exemplary visualizations of chemical track structures in their spatial and temporal development are presented in Figure 10.10. Shown are the results of a simulation based on the physical track structure of the 5 MeV alpha particle from Figure 10.7b, at an early (10 ps), an intermediate (1 ns) and a later (10 ns) stage after the particle traversal. Note that with increasing time, not only the yields of the initially dominant species ($^{\bullet}$OH, eaq^{-} and H_3O^{+}, see Figure 10.8) decrease and the yields of other species increase, but also the track structure gradually disappears and a largely homogenized picture starts to arise after about 10 ns.

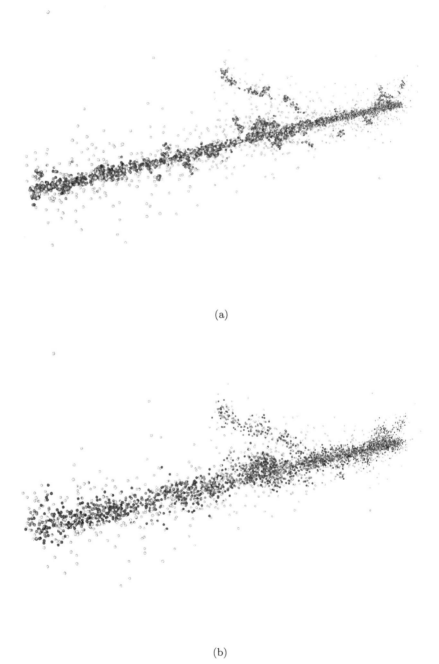

(a)

(b)

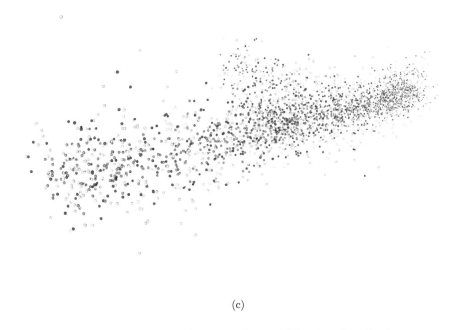

(c)

Figure 10.10 Perspective view of track structures within the chemical stage resulting from the passage of an alpha particle. The ion has started with 5 MeV initial energy on the right hand side; its track structure within the physical stage is shown in Figure 10.7b in a different perspective (viewpoint in 0.4 μm distance from track core). Reactive species are shown by colored spheres of 2 nm diameter; their apparent sizes are inversely proportional to the distance from the viewpoint. Green: eaq$^-$, red: $^\bullet$OH, blue: H$_3$O$^+$, yellow: H$^\bullet$, cyan: H$_2$, white: OH$^-$, grey: H$_2$O$_2$. The snapshots have been taken after 10 ps (panel a), 1 ns (panel b) and 10 ns (panel c); after 100 ns the branch due to the high-energy secondary electron is no longer visible. Initially, pairs of $^\bullet$OH and H$_3$O$^+$ in close vicinity are found along the tracks surrounded by a cloud of eaq$^-$, finally, the track structure is almost lost and similar amounts of various species are present overall (not shown). Visualization by POV-Ray^{TM} ray tracer software (Persistence of Vision Raytracer Pty. Ltd., Williamstown, Victoria, Australia).

10.4   MULTI-SCALE DNA AND CHROMATIN MODELS

DNA damage, in particular DNA double-strand breaks (DSB) that may lead to chromatin breaks, represent critical damage to cells. In PARTRAC, DNA damage is assessed by superimposing the simulated tracks with multi-scale models of DNA and chromatin structures. Physical track structures, *i.e.*, patterns of individual energy deposition events, are taken to model direct radiation effects due to energy deposits to DNA by such superposition. Chemical track structures, *i.e.*, time-dependent patterns of reactive species, are used to determine indirect radiation effects mediated by radical attacks onto the sugar-phosphate backbone or bases of the DNA macromolecule. Locations of DNA bases and sugar moieties are considered as non-diffusing reaction partners of $^{\bullet}$OH and eaq$^-$ with certain reaction rate constants, in competition with all other species. Reactions of $^{\bullet}$OH with sugar lead with high probability to DNA strand breaks from indirect effects, reactions with bases to indirect base damage. Otherwise, the merged atoms of the chromatin model define a volume not filled with liquid water and thus not accessible to reactive species; those entering it are assumed to be scavenged and removed from further water radiolysis simulation.

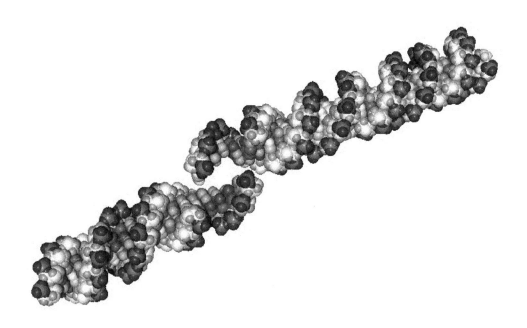

Figure 10.11 DNA double-helix model with double-strand break. Spheres represent individual atoms of phosphate group (blue), deoxyribose (white, gray), adenine (yellow), cytosine (pink), guanine (red) and thymine (green). Visualisation by POV-Ray^{TM} ray tracer software (Persistence of Vision Raytracer Pty. Ltd., Williamstown, Victoria, Australia).

The target structures of DNA and chromatin in PARTRAC start with an atomic model representation of the DNA double-helix (Figure 10.11) and of a histone molecule based on X-ray crystallography data [433]. A DNA segment wrapped around a histone octamer serves as a model of the nucleosome, the most elementary repeating motif in chromatin. The next level of chromatin organization, the 30 nm fiber, is obtained by positioning one nucleosome after the other into a corresponding virtual tube and by connecting them with a linker DNA helix. The arrangement of subsequent nucleosomes determines the overall chromatin fiber structure. This is still a debated issue: Chromatin structures seem not to be uniform and regular, but have to be viewed in the context of specific biological functions [434]. The two-start zigzag topology and the type of linker DNA bending that defines solenoid models may be simultaneously present in a chromatin fiber with a uniform 30 nm diameter [435]. The DNA model in PARTRAC has the capability to describe regular or stochastic arrangements of nucleosomes in solenoidal as well as zigzag or crossed-linker topology [436]. In order to cope with the amount of DNA of about 6 Gbp (6×109 base pairs) in a human diploid cell, short pieces of a 30-nm chromatin fiber are used as building blocks for the entire genome. Five types of basic fiber elements are defined within cubic voxels of 40 or 50 nm edge length in which the chromatin fibre connects the bottom plane with one of the other five walls of the cube. For a seamless connection of these building blocks, the interface structure has to be identical. The interface may be reduced to a single nucleosome positioned between two boxes (see Figure 10.12); this allows consideration of two basic sets of building blocks with different compaction so as to describe hetero- and euchromatic structures [271]. When several nucleosomes are cut at the interface [290], the chromatin fiber may become very compact, with the underlying building block structure hardly discernible. The building blocks are generated by a special trial-and-error algorithm using the Monte Carlo technique: under given boundary conditions of fiber structure parameters and the already positioned nucleosomes, an additional nucleosome is placed, the linker DNA determined, and all the newly positioned atoms are tested for spatial overlap with the formerly arranged ones. In order to setup the building blocks for a highly compact chromatin fiber (with about 6 nucleosomes per 10 nm fiber length), the above trial-and-error procedure has to be executed several million times.

With a given set of building blocks for the fiber structure, the next higher-order chromatin structure is provided by a path of linked voxel elements on a regular three-dimensional grid with side length corresponding to the building blocks, combined with an algorithm for tracing their orientation. This path is established based on data provided by a 1-Mbp spherical chromatin domain model [290]. The chromatin domain centers are taken as anchor sites of chromatin loops and in the DSB repair model as fixed nuclear attachment sites: the chromatin path periodically departs from the center and approaches it again until about 1 Mbp is allocated by the formed loops; the path is then directed to the next chromatin domain. The length and territorial organization of the chromosomes within the G0/G1 phase is provided by the number and locations of the chromatin centers. Figure 10.13 shows the chromatin fiber path model of the chromosomes in an elliptical fibroblast nucleus.

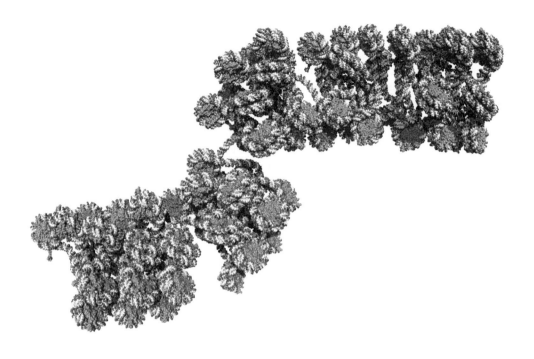

Figure 10.12 Chromatin fibre model. Five basic fiber elements are seamlessly connected. The elements are cubic boxes with 40 nm side length. A linear element on the left hand side is in sequence with the four types of bent elements. The single nucleosome in the interface layer allows linkage with other sets of elementary boxes, either compact ones representing heterochromatin (shown) or loose ones with lower numbers of nucleosomes that correspond to euchromatin (not shown). Spheres represent individual atoms of phosphate group (blue), deoxyribose (white), adenine (yellow), cytosine (pink), guanine (red), thymine (green) and histone (turquoise). Visualization by POV-Ray^{TM} ray tracer software (Persistence of Vision Raytracer Pty. Ltd., Williamstown, Victoria, Australia).

Figure 10.13 Model of human fibroblast nucleus. Basic chromatin fibre elements are represented by small straight or bent cylinders, forming tangled-up rope-like structures; individual chromosomes are distinguished by their colours. Visualisation by POV-Ray™ ray tracer software (Persistence of Vision Raytracer Pty. Ltd., Williamstown, Victoria, Australia).

10.5 INDUCTION OF DNA AND CHROMATIN DAMAGE

Direct effects of ionizing radiation on DNA are in PARTRAC scored from energy deposits to this macromolecule. Following experimental information on the sensitivity of DNA to irradiation with low-LET radiation, a strand break is scored with a probability that depends on the energy deposited to the sugar-phosphate backbone of DNA: energy deposits below 5 eV are ignored, all deposits above 37.5 eV are converted to strand breaks, and between these two points a linear increase of breakage probability with the deposited energy is assumed. Indirect effects are scored from attacks of •OH radicals onto DNA. 65% of such interactions onto a sugar are recorded as strand breaks. This probability results in a fraction of about 13% of •OH interactions with DNA that leads to strand breaks, since attacks to bases are about 4 times more frequent than those to the sugars. DNA double-strand breaks (DSB) are scored whenever breaks on opposite strands of the DNA molecule occur within 10 bp genomic distance. Obviously, a DSB can be induced by two direct breaks, two indirect ones, or a combination of both. In addition, 1% probability is assumed for a conversion of single-strand breaks (SSB) into DSB, to account for experimentally observed DSB induction by very low-energy (10 eV) photons and electrons and other findings supporting DSB induction by single events [413]. In addition to strand breaks, also damage to DNA bases can be investigated with PARTRAC; in particular base lesions in a close vicinity to a DSB have been taken into account in the DSB repair simulation.

The results of PARTRAC simulations on DNA damage induction have been thoroughly benchmarked against experimental data. Regarding the chemical stage, PARTRAC simulations for SSB induced in linear DNA under varied lifetimes of hydroxyl radicals have been found in accord with measured data for different concentrations of dimethyl sulfoxide (DMSO), an efficient scavenger of these radicals; similarly good agreement with the data has been obtained for SSB and DSB induced in SV40 minichromosomes [437]. As illustrated in Figure 10.14, experimentally observed yields for DSB induction after irradiation with light ions of diverse energies (and thus diverse values of linear energy transfer, LET, which is a measure of ionization density and, hence, of radiation quality) are well reproduced by PARTRAC simulations. The agreement of simulations with the data is even better when plotted in terms of relative biological effectiveness (RBE) , since the simulations regularly somewhat overestimate the measurements. The reported simulations take into account the experimental limitations in detecting short DNA fragments. Actually, the simulations predict that DNA fragment distributions after high-LET radiation are heavily skewed towards short fragments [438] that could not have been detected by the experimental protocol. Figure 10.15 shows that the simulations are also in good agreement with measured DSB distributions [439] on genomic scales of about 20 kbp – 5 Mbp. On short genomic scales, PARTRAC simulations have reproduced the observed 78 bp and 190 – 450 bp peaks in fragment size distributions [440], related to the nucleosome and short-scale arrangement in chromatin fiber [413]. Taken together, these benchmarks indicate that both track and chromatin structures are represented properly in PARTRAC, since DNA fragmentation patterns naturally reflect both of these structures.

In the following, simulation results are presented on initial DNA damage induced by a variety of light ions at energies from 250 down to 0.25 MeV/u [412]. Detailed information on the simulation setup, methods and further results can be found in that reference. The yields are presented in dependence on the LET of the particles; the LET increases during particle slowing down up to a maximum value occurring around or below 0.5 MeV/u, after which the ion comes to rest within less than 10 μm. In Figure 10.16, the calculated induction of DNA strand breaks due to direct and indirect effects is shown in dependence on the LET of the ion irradiation. The yield from direct effects is almost constant over a wide range of LET values. The yields from indirect effects, about 60% higher than those from direct effects at low LET values, decrease with increasing LET to values below the contributions from direct effects. At higher energies, *i.e.*, lower LET values the yields of strand breaks do not depend on ion type, but at low energies, *i.e.*, approaching the maximum LET values

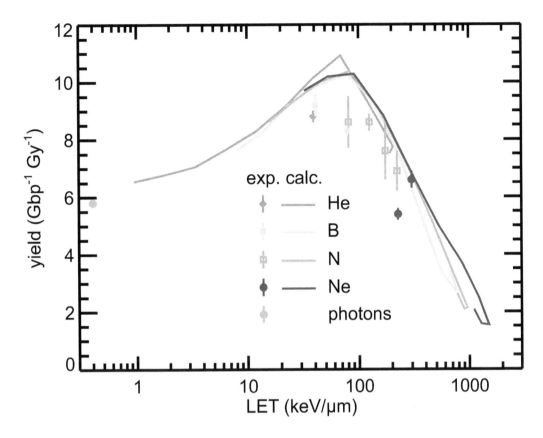

Figure 10.14 LET-dependent yield of DSB after photon and ion irradiation. Symbols: experimental results from pulsed-field gel experiments [441, 442], lines: PARTRAC simulations that disregard DSB linked to fragments shorter than 5 kbp. Adapted from [412].

for a given ion, both direct and indirect effects tend to be lower than for heavier ions at the same LET. High-LET radiation tends to induce multiple DSB in the close vicinity. Such DSB cluster, equivalent to DSB^{++} in the notation by Nikjoo and coworkers [295, 13], are likely processed by the cell as a single complex lesion. Thus such clustered DSB have been merged into a broken site (DSB site) counted only once in the analysis of PARTRAC damage simulation; also a separated DSB is counted as one DSB site. The LET-dependent yield of DSB sites in Figure 10.17 is largely independent of particle type, and shows a pronounced maximum around about 200 keV/μm LET. Similar patterns have been reported in studies on cell survival upon exposure to radiations with diverse qualities [443]. The downward trend in simulation results for LET > 200 keV/μm follows from scoring multiple clustered DSB as a single DSB site only; this is similar to disregarding fragments < 5 kbp in Figure 10.14. The mentioned qualitative agreement with cell survival data indicates that the cells likely process DSB clusters together, rather than repairing individual DSB in a cluster independently. The simulations allow one to look at the DSB clusters in more detail: The LET-dependent formation of DSB clusters shown in Figure 18 culminates in a maximum around 500 keV/μm LET for heavier ions or at the maximum LET value for lighter ions; however, the maximum yields decrease with increasing mass (or charge) of the ion. The

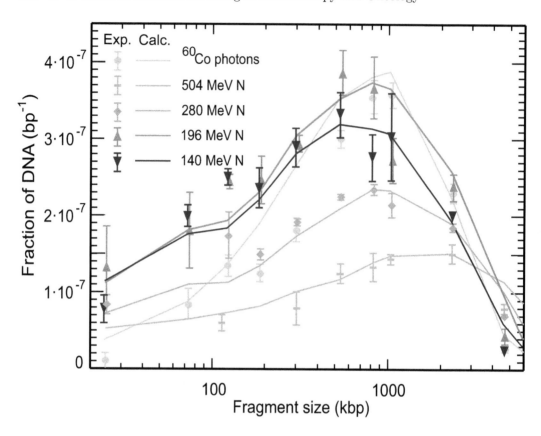

Figure 10.15 Size distributions of DNA fragments after photon and nitrogen ion irradiation. Symbols: Experimental results for 100 Gy (photons), 5 ions/μm^2 fluence (504 MeV, 280 MeV, 196 MeV) and 5.5 ions/μm^2 (140 MeV) [439]; lines: PAR-TRAC calculations for the same irradiation conditions.

decrease in the yield of DSB clusters at the highest LET values is linked to an increase in the mean number of DSB per cluster [412], as eventually smaller clusters merge to a larger one. Taken together, these results provide a detailed picture on the interplay between the frequency of initial DNA damage and its complexity on nm-scales during ion slowing down.

10.6 DNA DAMAGE RESPONSE

Cells possess a number of pathways that aim at recognizing and repairing damage to DNA as the key macromolecule in which genetic information is stored. Radiation-induced DNA damage is no exception, and is processed by the same mechanisms as similar damage induced endogenously or by other stressors. Following their recognition, base damage and single-strand breaks can be repaired by base excision repair, nucleotide excision repair, and mismatch repair systems. These are relatively error-free and quick processes; most of radiation-induced base damage and single-strand breaks are commonly considered as easily and quickly reparable.

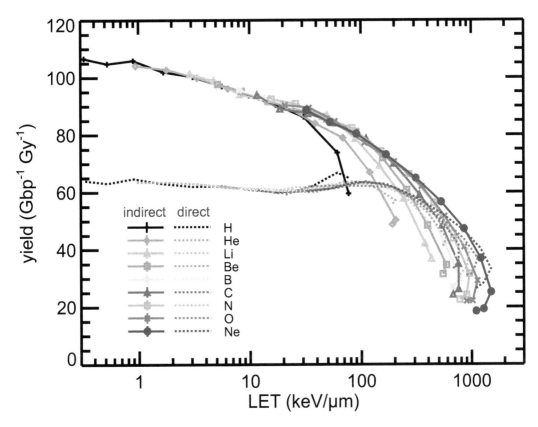

Figure 10.16 Calculated LET-dependent yield of DNA strand breaks after ion irradiation. Symbols connected by solid lines: contribution from indirect effects, dotted lines: contribution from direct effects.

Double-strand breaks, on the other hand, represent a severe class of DNA damage since not only one but both DNA strands are affected; thus, some genetic information as well as the connection of the broken ends may be lost. However, even DSB are repaired quite efficiently; human cells typically repair more than 90% of radiation-induced DSB within a day post irradiation. Non-homologous end-joining (NHEJ) is the dominant DSB repair pathway in eukaryotic cells in the G1/G0 cell cycle phase; it aims at restoring the DNA integrity in the absence of information on the original sequence, and as such is prone to errors. Following DNA replication in the S phase, in the G2 phase the sister chromatid can be used as a template for restoring the original sequence in an error-free way by the homologous recombination pathway. Conceptually in between these two pathways is a third one, microhomology-mediated end-joining, which is based on using 5 − 25 bp microhomologous sequences downstream or upstream of the break to align the broken strands and try to restore the original sequence; this pathway operates in the S-phase, and is not free of errors.

The PARTRAC tools include a dedicated module to simulate DNA damage response. The framework of the model has been used so far to model DSB repair via NHEJ, since DSB are the most relevant lesions and the underlying initial DNA damage refers to the G0/G1 phase of the cell cycle where NHEJ is the dominant rejoining process. The processing of both induced broken DNA ends as well as their motion is traced individually, so that not

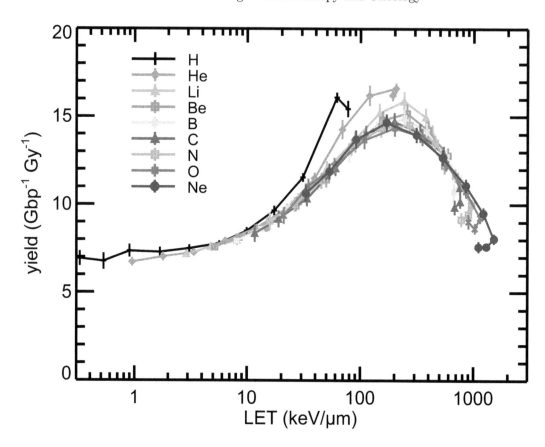

Figure 10.17 Simulated LET-dependent yield of DSB sites after ion irradiation. Isolated DSBs as well as multiple DSBs in clusters are counted as one DSB site. Adapted from [412].

only correct rejoining but also misrejoining and even the formation of chromosome aberrations can be assessed.

The following key steps of the NHEJ pathway [444] for the repair of DSB are modeled in PARTRAC (Figure 10.19): After a quick damage recognition and chromatin remodeling step, DNA-dependent protein kinase (DNA-PK) has to attach to the induced DNA end. DNA-PK consists of two subunits, Ku70/80 and the catalytic subunit DNA-PKcs, whose attachments are actually modeled as two subsequent steps. Two DNA ends in close proximity, both with attached complete DNA-PK, form then a synaptic complex. Following the cleaning steps in which nearby base lesions and/or additional single-strand breaks are removed, the final ligation takes place. The enzymatic processing is represented in the simulation as transferring the DNA end from one to another state, $e.g.$, from a state with attached Ku70/80 to a state with complete DNA-PK. These transitions are described as processes in first order kinetics; for each process, the transition rate is a single parameter that enters the model. For the early phase the transition rates between individual states have been estimated from corresponding experimental data for Ku80 and DNA-PK attachment and turnover [445, 446]. The actual states and transitions of individual ends are simulated by Monte Carlo modeling analogous to the chemistry module; diffusive transport of DNA ends

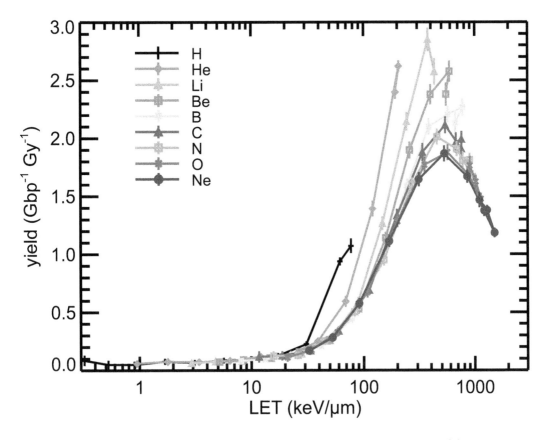

Figure 10.18 Simulated LET-dependent yield of DSB clusters (DSB^{++}) after ion irradiation. DSB clusters includes two ore more DSB within not more than 25 bp distance. Adapted from [412].

alternates with potential changes of end states. The adopted time steps in typical calculations are of the order of 0.1 s. The delayed repair of dirty DNA ends due to cleaning steps in the model structure in Figure 10.19, combined with increasing fractions of dirty ends with increasing LET of the irradiation, leads to repair kinetics depending on the radiation quality that corresponds to measured data after irradiation with photons or nitrogen ions of different LET, as shown in Figure 10.20.

In parallel to the enzymatic processing, the mobility of the DNA ends during repair is accounted for by a stochastic model of DNA end mobility, conceptually similar to the one used for reactive species in the chemical track structure module. However, the diffusive motion of the DNA ends is confined by chromatin fibres being assumed to be attached at nuclear attachment sites and, for short fragments, by a maximum distance to the other end of the fragment corresponding to the genomic length of the fragment. The fibre compactification is gradually released, i.e. the radius of the sphere in which the end may diffuse around the nuclear attachment site is increased with time. With these mechanisms the observed sub-diffusive mobility of telomeres and induced DNA ends [447] is reproduced by the calculations [448, 271].

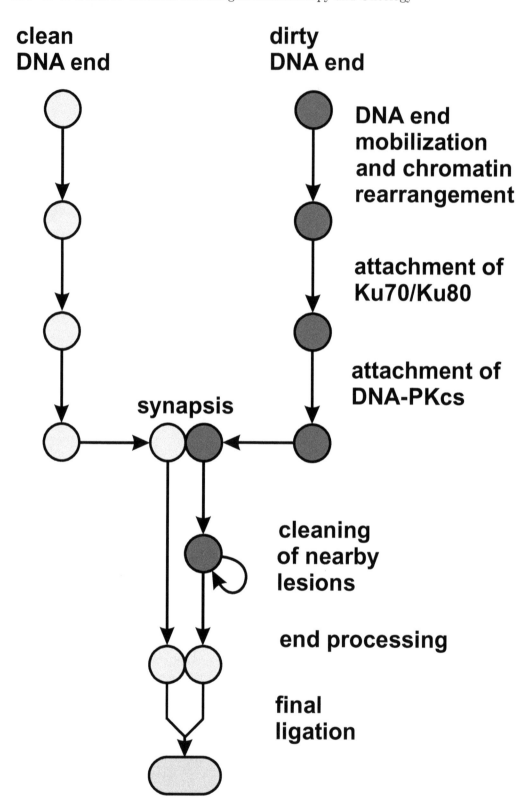

Figure 10.19 Simplified scheme of DSB repair model in PARTRAC.

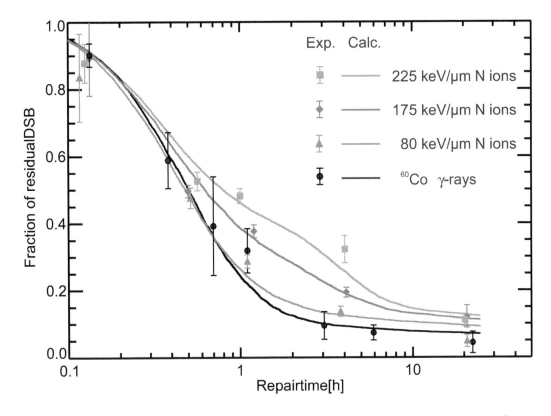

Figure 10.20 Kinetics of DSB repair after photon and nitrogen ion irradiation. Symbols: experimental data [449], lines: PARTRAC calculations.

Throughout this procedure, individual DNA ends induced by the DSBs are followed independently and in a stochastic manner. Hence it may happen that one end has already attached a molecule of DNA-PK but its original partner not yet (or has diffused away). If another DNA end with attached DNA-PK is found in close proximity, the two ends form a synaptic complex and, if the complex does not break, finally get ligated. Through this interplay of enzymatic processing and spatial mobility, not only correct rejoining but also misrejoining is simulated. If the two ligated segments used to belong to different chromosomes, a chromosomal aberration is scored. Upon having complemented the simulations with information on the location of centromeres, PARTRAC has been used formodeling the induction of dicentric chromosomes, i.e., misrejoined chromosomes that contain fragments with centromeres from two different chromosomes. Without adjustment of model parameters, the simulations have correctly reproduced the dose dependence, although the absolute yields of dicentrics for gamma and alpha-particle irradiation were overestimated [271].

The repair module in PARTRAC provides a powerful framework that can be extended in many respects. One example is simulating the action of individual molecules of repair enzymes that mediate certain steps in the repair process instead of adopting first order kinetics with a given rate constant [450]. The drawback of such a detailed approach is the increase in the number of parameters to describe enzyme production, mobility, reaction radius, time constant for repopulation, nonlinear features such as feedback or upregulation etc., which are largely not known and can hardly be derived from their impact on the calculated kinetic and outcome of the repair process. Another example are changes in DSB end states that are allowed only under certain boundary conditions, such as its overhang length, which may also restrict the formation of a synaptic complex between DSB ends. Consideration of a maximum spatial distance of the ends of short DNA fragments forces the DNA ends to stay in proximity during diffusion; however, without further restriction this leads to high numbers of small rings: about 75% of the rings are smaller than about 10 kbp [271]. To account for the stiffness of the DNA helix that is seen in the linear short fragment structures due to DSB in plasmid DNA, the formation of rings can be suppressed below a certain genomic size, usually a few kbp.

The main aim of the model is to test hypotheses on key mechanisms during DNA damage response, such as the origin of misrejoining of broken DNA ends, the reason for slowly repairing DSB or the cause for unrejoined DSB ends after long repair times. For a solid test it is not sufficient to yield an agreement under prevailing hypotheses but to show also the failure of alternative hypotheses. However, the calculated modeling endpoints may not depend in a pronounced way on the tested mechanisms. Such result has been obtained for different hypotheses on the origin of misrejoining where only DNA end mobility turned out to be trivially indispensable [296]. It is generally accepted that the complexity of initial DNA damage has a high impact on biological effects of radiation. However, the relative biological effectiveness for reproductive cell death culminates around LET values of 100 – 200 keV/μm [443], whereas the formation of DSB clusters and their complexity is still rising with LET. This highlights the need for a more detailed evaluation of DNA damage distribution on different scales.

In this situation innovative microbeam experiments have provided unprecedented data for DNA damage response model testing and development. These experiments offer a separation of the nanometre-scale clustering of radiation-induced damage from its clustering on micrometre scales, which are otherwise closely interrelated in conventional experiments with different radiation qualities [451]. Specifically, high-LET irradiation by carbon ions of 55 MeV (LET: 310 keV/μm) delivered on a quadratic matrix with 5.4 μm side length has been mimicked by bunches of 117 protons of 20 MeV (LET: 2.6 keV/μm) or 5 lithium ions of 45 MeV (LET: 60 keV/μm) focused to spots of sub-μm size and delivered on the same matrix; secondly, the matrix size has been varied with related change of ions per spot to maintain a constant dose to the cells. The induction of dicentrics has been determined as the biological endpoint: an increase with the LET of the ions but also a pronounced effect of focusing for protons and lithium ions has been found; both were initially not adequately reproduced by PARTRAC model calculations [452]. After extensive parameter studies, relatively good agreement between experiment and model calculations has been obtained, see Figure 10.21, regarding the focusing effect (except the largest matrix size) and the LET effect of the incident ions. In this simulation, misrejoining events result primarily from a small fraction of DSB ends which lose their proximity to the corresponding DSB end due to high mobility in the early phase of DSB repair. Including other possible mechanisms leading to misrejoining, such as more frequent break-up of synaptic complexes or exchange of DNA ends upon interaction of two such complexes, has reduced the conformity with the data; yet alternative hypotheses continue to be tested aimed at finding a scenario with an improved agreement with the measured data and expectations on endpoints such as unrejoined breaks after one day repair time . In spite of these limitations, the induction of chromosomal exchange aberrations has turned out to correlate closely with the probability of clonogenic cell death.

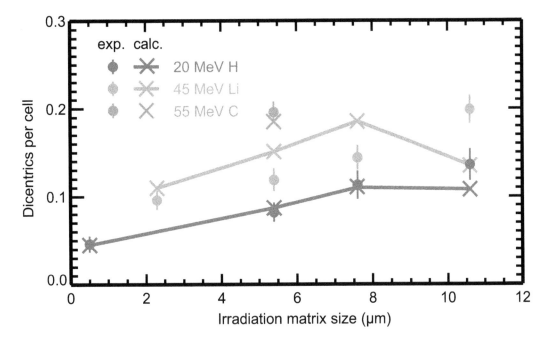

Figure 10.21 Induction of dicentrics after spot matrix irradiation by ion microbeam. Alternative numbers of ions have been focused to sub-μm spots (1, 117, 234 and 468 for H, 1, 5, 10 and 20 for Li, 1 for C) and spots applied on a regular matrix pattern, all corresponding to a unique dose of about 1.7 Gy. Circles: Dicentric chromosomes scored in hybrid human-hamster AL cells [452]. Crosses and lines: PARTRAC simulations.

10.7 MODELING BEYOND SINGLE-CELL LEVEL

As is generally the case in modeling complex multi-scale systems, when moving towards a larger scale, only simplified characteristics of the processes underpinning a smaller scale can be represented. Models on tissue and organ level thus hardly can account for full details of track structures, DNA damage and its repair included in PARTRAC as described above. Yet, PARTRAC has been successfully used in pilot studies on intercellular communication. Radiobiological experiments on so-called bystander effects have demonstrated that unirradiated cells exposed to medium transferred from irradiated cells show similar responses as if they were irradiated directly, including DNA damage induction or cell killing [453]. PARTRAC simulations have been used to study alternative signal emission scenarios consistent with an observed threshold-like behaviour of bystander effects, which show an onset at doses around 2 mGy of gamma irradiation [454]. Signal emission scenarios assuming above-threshold energy deposits to mitochondria and signal amplification by neighbor cells have been found consistent with the data [455]. Independent of PARTRAC simulations, non-linear response of bystander cells to the signals has been modelled [455, 456].

Another experimental system of intercellular communication whose radiation-induced modifications have been studied in detail is the selective removal of precancerous cells by apoptosis upon signaling by their normal neighbors. This intercellular induction of apoptosis likely represents a natural anti-carcinogenic mechanism [457]; mechanistic modeling suggests that this process is capable of stopping the growth of a population of precancerous cells, leading to a kind of dormancy [458]. Cell-culture experiments have shown that ionizing radiation may enhance the underlying signalling and the resulting removal of precancerous cells by apoptosis [459]. However, for in vivo conditions of early carcinogenesis, where a small population of precancerous cells is challenged by a large, dense population of normal cells and the signaling species possess shorter lifetimes, the modeling predicts that this anti-carcinogenic process be reduced upon radiation exposure [460]. This modeling prediction would thus remove the discrepancy between the well-known epidemiological evidence that ionizing radiation induces cancer and the mentioned radiobiological experiments. To facilitate the urgently needed experimental benchmarking of this prediction, conditions have been formulated under which such inversed behavior is predicted already and which, contrary to *in vivo* conditions, are achievable in cell-culture experiments; the modeling predictions thus await experimental verification.

10.8 CONCLUSIONS

PARTRAC represents a state-of-the-art tool for stochastic multi-scale modeling of the biological effects of ionizing radiation on subcellular and cellular scales. It integrates knowledge on radiation physics, chemistry and biology. It provides means to test alternative hypotheses on the processes and mechanisms that underpin the observed biological effects. It also helps interpret the data and interpolate between or even extrapolate outside measured regimens, even to conditions and/or to the level of detail hardly accessible experimentally.

Future activities in PARTRAC development shall include linking it with a macroscopic radiation transport code to facilitate its applications in medical physics and to enable addressing biological effects over spatial scales up to decimetres relevant in therapeutic applications of radiation. It is also planned to propose a module to simulate cell killing based on track structure calculations for initial DNA damage and subsequent modeling of DNA damage response. On the one hand, such a detailed mechanistic model would serve as a benchmark to existing, rather descriptive models; on the other hand, it would extend PARTRAC applicability to an endpoint of utmost relevance to medical physics and radiation therapy. It is also desirable to address the biological variability between different cell types, in particular regarding their capability to repair radiation-induced DNA damage. Likewise, one shall continue extending the scope of mechanistic modeling beyond the single-cell level. Last but not least, efforts shall be made to reduce the computational expensiveness, enhance the predictive power and enable a widespread use of PARTRAC.

CHAPTER 11

Multi-scale modeling approaches: application in chemo– and immuno–therapies

Issam El Naqa

CONTENTS

ABSTRACT

Multi-scale modeling of cancer growth and treatment response would allow better understanding of the different biophysical and biochemical interactions at the molecular, cellular, tissue, and organ levels. Different mathematical models ranging from continuum, discrete, to hybrid have been proposed and evaluated in different theoretical and experimental settings. These process models not only can help explain the complex tumor-treatment system but can also be used to personalize treatment single or combined modality regimens and predict their efficacy. In this chapter, we will present examples of such modeling approaches and discuss its ability in the cases of chemotherapy drugs and immunotherapy agents.

Keywords: outcome models, multi-scale models, chemotherapy, immunotherapy.

11.1 INTRODUCTION

M ULTI-SCALE modeling is a form of mathematical models that is routed in physical sciences as a tool that allows solving complex problems at multiple temporal and spatial scales. The complexity of cancer and its dynamical "carcinogenesis" processes and their interaction with their host immune system and therapeutic agents at varying scales invite the application of such advanced modeling schemes. These models can guide and optimize response to therapy and would allow making quantitative predictions concerning their functional behavior and the physiochemical parameters that influence cancer response [461, 462]. Multi-scale models can complement mathematical tools already used in clinical and cancer biology studies and enrich the library of treatment outcome prediction models by challenging current paradigms and shaping future research in cancer outcomes [463, 464]. The multi-scale framework fro modeling cancer dynamics is presented in Figure 11.1 [464].

Cancer is a diverse family of diseases (about 100 known cancers) that is characterized by an abnormal growth of cells that acquired the ability of uncontrolled progression and invasion of surrounding tissues and disrupting their vital functions and leading to patient death. Biologically, cancer cells comprise of six major *tumor hallmarks* that underlie their genetic characteristics: sustained proliferative signaling, evasion of growth suppressors, resistance to cell death, replicated immortality, induction of angiogenesis, and activation of invasion and metastasis [465], which are the subject of active multi-scale modeling. Among these we will focus on the first four in our modeling description below. The dynamics of mutation accumulation in cancer are typically modeled in computational biology by using a fixed-size stochastic *Moran process* while metastasis dynamics are modeled by stochastic *branching process* [462].

Patients diagnosed with cancer have multiple treatment options to choose from depending on their disease type and its status. Typically, a patient with localized and early stage disease would receive surgery, benefit from watchful waiting or active surveillance as in prostate cancer [466], or the use of medication for palliative purposes, rather than curative purposes, as in certain types of leukemia [467]. However, patients at more advanced stages of disease would receive chemoradiotherapy, molecular targeted therapy, immunotherapy, or a combination of these treatment modalities. In a previous chapter (Chapter 10), the focus was primarily on on multi-scale modeling in radiation oncology using Monte Carlo (MC) techniques; therefore, here will present examples from chemo- and immuno-therapies. However, the ideas could be directly generalized to other treatment modalities. Furthermore, it is currently an open question on how best to combine a specific immunotherapy agent with chemotherapies, targeted therapies, surgery or radiation, in which both the timing, the concentration profile and the synergistic combinations need to be identified [463].

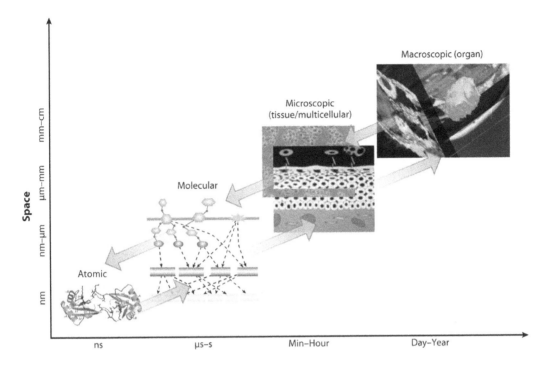

Figure 11.1 Schematic illustration of the biological scales of significant relevance for multi-scale cancer modeling including atomic, molecular, microscopic (tissue/multicellular), and macroscopic (organ) scales. Different scales represent different spatial and temporal ranges; the methods for modeling these distinct scales differ as well. Multi-scale cancer modeling requires the establishment of a linkage between these scales.

11.2 MEDICAL ONCOLOGY TREATMENTS

11.2.1 From chemotherapy to molecular targeted agents

There are different categories of chemical agents that aim to eradicate tumor cancers [468]. Among the most common ones are alkylating agents, which substitute an alkyl groups (hydrocarbon) for a hydrogen atom of an organic compound including DNA (*e.g.*, Temozolomide). There are also antibiotics (*e.g.*, Doxorubicin, Blemoycin). Other common agents are antimetabolites (*e.g.*, Methotrexate, 5-Fluorouracil,Taxanes,vinca alkaloids). Another agents that do not fall into any of these classes include platinum compounds (Cisplatin) and topoisomerases DNA winding enzymes inhibitors. Recently, more signaling pathway molecular targeted agents have been developed such as anti-epidermal growth factor receptor (EGFR) (*e.g.*,Cetuximab or Erbitux) [469]. In addition to the agent type and dosage, the timing of the administration of the agent can influence treatment response. Chemotherapy could be administrated after the completion of the local treatment such as radiation and is called *adjuvant* chemotherapy; if administered before the local treatment and is called *induction* chemotherapy, or, if given during local treatment, it is called *concurrent* chemotherapy. In particular, concurrent chemoradiation has been demonstrated to be effective in the treatment of several cancers, in which the chemotherapy agent can act as a radiosensitizer by aiding the destruction of radiation resistant clones or act systematically and potentially eradicate distant metastases [470]. However, it is noted that despite therapeutic advances in molecular targeted therapy (*e.g.*, EGFR/ALK inhibitors in lung cancer), the vast majority of cancer patients do not benefit from such treatment and would still receive generic cytotoxic agents [471]. A timeline of chemotherapy agents is shown in Figure 11.2 [468].

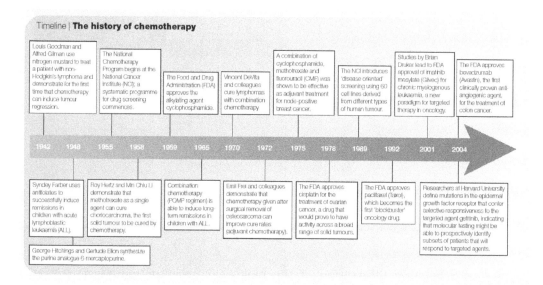

Figure 11.2   Timeline — The history of chemotherapy [468].

11.2.2   Immunotherapy

Immunotherapy by checkpoint blockade (CPB) has emerged as a promising therapeutic alternative for these tumors with a high degree of heterogeneity often caused by high somatic mutation burden (as in tobacco associated lung cancers). Immune tolerance is mediated in part, by cytotoxic T-lymphocyte-associated antigen 4 (CTLA-4) or expression of programmed death-ligand 1 (PD-L1) by the tumor and infiltrating immune cells. These checkpoints act as negative regulators of T-cell immune functions. The roles of CTLA-4 and PD-1 in inhibiting immune responses, including antitumor responses, are largely distinct. CTLA-4 is thought to regulate T-cell proliferation early in an immune response, primarily in lymph nodes, whereas PD-1 suppresses T cells later in an immune response, primarily in peripheral tissues as shown in Figure 11.3 [472]. In case of PD-1, for instance, PD-L1 binds to its cognate receptor, programmed death-1 (PD-1), on T cells and macrophages, profoundly altering their proliferation and function [473]. Blocking PD-1 or PD-L1 promotes anti-tumor immunity by promoting a tumor-reactive adaptive and innate immune response [474]. Recently, published clinical trials of immune checkpoint inhibitors in metastatic lung cancer patients induced improved overall response rate of 15-20% with a median survival of 9-12 months, for instance [475]. However, prognostic and predictive models are not clearly established, and an inability to precisely identify the patients who will benefit, limits CPB promise [476] and places patients at higher risk for immune related toxicities [477]. Therefore, more sensitive and specific models that can explain and predict immunotherapy responsiveness would be particularly useful.

Lymph node　　　　　　　　　　　　　　　　　　　　　　　　　**Tumor site**

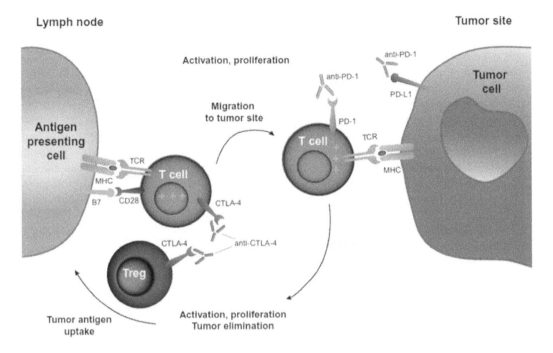

Figure 11.3　CTLA-4 and PD-1 pathway blockade. CTLA-4 blockade allows for activation and proliferation of more T-cell clones, and reduces Treg-mediated immunosuppression. PD-1 pathway blockade restores the activity of antitumor T cells that have become quiescent. A dual pathway blockade could have a synergistic effect, resulting in a larger and longer lasting antitumor immune response. CTLA-4 indicates cytotoxic T-lymphocyte-associated antigen 4; MHC, major histocompatibility complex; PD-1, programmed death 1; PD-L1, programmed death ligand 1; TCR, T-cell receptor; Treg, regulatory T cell [472].

11.3　MODELING TYPES

Models for cancer could be categories based on how the tumor tissue is represented into continuum models, discrete cell-based models, or a combination of both (hybrid) using multi-scale settings. There are several excellent textbooks and review articles dedicated to this subject [478, 463, 462] that can provide the interested reader with more details beyond the scope of this chapter.

11.3.1　Continuum tumor modeling

Continuum tumor models are based on the reaction-diffusion differential equations, which attempt to provide a description of the tumor cell density, the extracellular matrix (ECM), the matrix degrading enzymes (MDEs), and the concentrations of cell related substrates (*e.g.*, glucose, oxygen, and growth factors and inhibitors) [478]. These methods employ an ordinary differential equation (ODE) or partial differential equation (PDE) to represent the tumor and its microenvironment. The models could be single-species (*e.g.*, single-phase tumors) or multiphase (mixture) models , which have been developed to describe detailed interactions between multiple solid cell species and extra- or intracellular liquids [478]. These models further try to model the two main types of tumor growth, namely; *avascular* (with no blood vessels) and *vascular* (with blood vessels [angiogenesis]). These models are typically confined to a spherical geometry (spheroids) to simplify numerical calculations. The basic tumor growth (avascular) model would represent tumor cells proliferating in such a way that they form a small sphere-like structure without direct access to the vasculature. During this avascular growth, the tumor cells receive oxygen, nutrients, and growth factors via the diffusion process from the surrounding host tissue. Mixture models incorporate more detailed biophysical processes than can be accounted for in single-phase models, and they provide a more general framework for modeling solid tumor growth with avascular or vascular capabilities. Phenotypic and genetic heterogeneity (*e.g.*, mutations) in the cell population can be represented mathematically, and this is a critical improvement towards more realistically simulating mutation-driven heterogeneity [478, 462].

As a representative example of this category, we will briefly describe the model developed by Wise *et al.* [479]. The model follows from the laws of conservation of mass and momentum, and consists of a fourth-order nonlinear advection-reaction-diffusion equations (of the *Cahn-Hilliard*-type) that describe the growth of various tumor-cell species and diffusion reaction equations representing the time evolution of the environmental substrate concentrations that affect tumor growth. The evolution equation of the total tumor volume fraction model (ϕ_T) is given by:

$$\frac{\partial \phi_T}{\partial t} = M\nabla(\phi_T \nabla \mu) + S_T - \nabla \cdot (\mathbf{u_S}\phi_T), \mu = \acute{f}(\phi_T) - \epsilon^2 \nabla^2 \phi_T, \qquad (11.1)$$

where $M > 0$ is the mobility constant related to phase separation between tumoral and healthy tissues; S_T is the net source of tumor cells; $\mathbf{u_S}$ is the tissue velocity $f(\phi) = \phi^2(1 - \phi)/4$, which is a double-well potential; $\epsilon > 0$ is the parameter specifying the thickness of the interface between healthy and tumoral tissues. A nonlinear multigrid scheme using an adaptive-mesh and finite difference methods is used to solve equation 11.1 efficiently for complex tumor morphologies [479]. A 3D simulation example is shown in Figure 11.4.

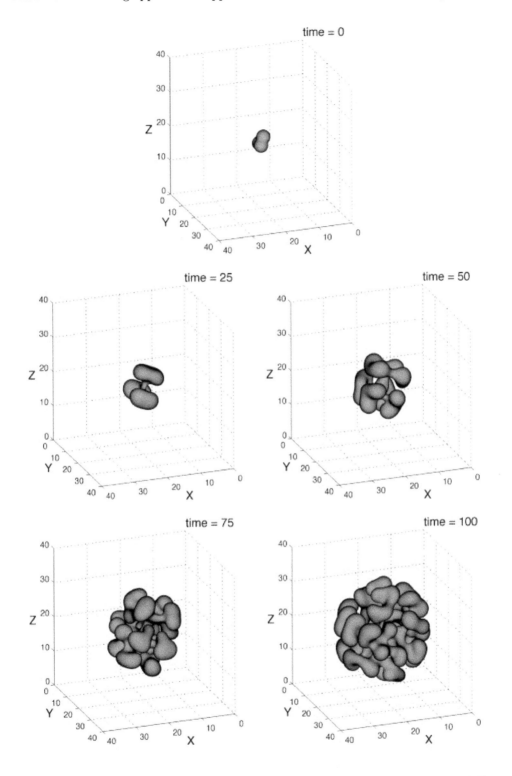

Figure 11.4   A 3D simulation during the growth of an asymmetrical tumor at different time points [479].

11.3.2   Discrete tumor modeling

Discrete modeling has enjoyed a long history in applied mathematics and biology since the seminal work of von Neumann in applied lattice crystals. These models are divided into two categories: lattice-based (*e.g.*, cellular automata) and lattice-free (*e.g.*, agent-based). Both approaches track and update individual cells according to a set of biophysical rules. Typically these models involve a composite discrete-continuum approach in the sense that the microenvironmental variables (glucose, oxygen, extracellular matrix, growth factors, etc.) are described using continuum fields while the cells are discrete. In *lattice-based* modeling , the cells are confined to a 2D or 3D lattice and each mesh point in the lattice is updated in time according to deterministic or stochastic rules derived from physical conservation laws and biological constraints. Cellular automata (CA) is considered a special case, where a mesh point represents an individual cell. In case of *lattice-free* (*agent-based*) models , they do not restrict the cells' positions and orientations in space. This allows more flexibilitty and accurate coupling between the cells and their microenvironment providing the individual cells in the model with more freedom in terms of motility, death, or survival. This is especially useful in modeling cases like angiogenesis, carcinogenesis, immune system attacks on tumor cells, and metastasis [480]. However, this freedom may come at a hefty computational cost when compared with continuum approaches, especially when dealing with large systems with millions of cells.

As an example of these models, we will consider a lattice-based approach using an extended Q-Potts model, which originated in statistical mechanics to study materials. The most common approach utilizes the Q-Potts model to simulate cell sorting through differential cell-cell adhesion and is known as the Graner-Glazier-Hogeweg (GGH) or cellular Potts model (CPM) model [481] . In this approach, each cell is treated individually and occupies a finite set of grid points within a Cartesian lattice; space is divided into distinct cellular and extracellular regions. Each cell is deformable with a finite volume. Cell-cell adhesion is modeled with an energy functional (Hamiltonian). A Monte Carlo algorithm is used to update each lattice point and hence change the effective shape and position of a cell. Although the description of the cell shape is less detailed than in the continuum (ODE/PDE) approaches described above, finite-size-cell effects are still incorporated.

Like any statistical mechanics system, the Hamiltonian is the main component to define in the GGH/CPM model and is determined by the configuration of the cell lattice. The original Hamiltonian proposed by Graner and Glazier included adhesion energies and area (volume) constraints for 2D (3D) tumors [481]:

$$H = \sum_{(i,j)neighborhood} J(\tau(\sigma_i), \tau(\sigma_j))(1 - \delta(\tau(\sigma_i), \tau(\sigma_j)) + \lambda \sum_i (v(\sigma_i) - V(\sigma_i))^2, \quad (11.2)$$

where $\tau(\sigma)$ is the cell type associated with cell σ; $J()$ is the surface energy determining the adhesion between two cells; δ is the Kronecker (tensor) product; $v(\sigma)$ is the volume of cell σ, $V(\sigma)$ is the target volume; and λ is a Lagrange multiplier determining the strength of the volume constraint. The Hamiltonian has been modified to control cell behaviors such as chemotaxis, elongation and haptotaxis by using other sub-lattices containing information such as the concentrations of chemicals as shown in Figure 11.5 [482].

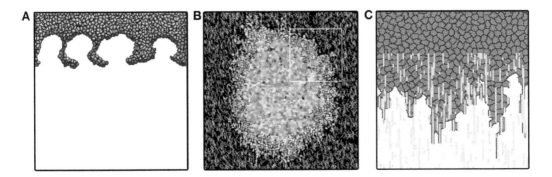

Figure 11.5 Tumor invasion with homogeneous and heterogeneous ECM. (A) Invasion front of tumor penetrating the stroma via viscous fingers. (B) Avascular tumor model of Rubenstein and Kaufman, exploring invasion along ECM fibers. (C) Invasion front of persistently moving cancer cells penetrate in fingers along ECM fibers describe penetration dynamics even without cell division [482].

11.3.3 Hybrid tumor modeling

These models combine the advantages of continuum and discrete approaches in order to better simulate multi-scale multi-body complex tumors problems. They can provide more realistic descriptions of microscopic mechanisms while efficiently evolving the entire system to obtain macroscopic observations. These models typically contain one or more continuum and discrete components; each component is described by a set of equations that satisfy the conservation of *mass* and *momentum*. Using statistical mechanics, the connection between the continuum and discrete mass and momentum could be established and the coupling mechanism could be derived, this could be further generalized across the different scales and incorporate further available molecular and/or imaging information as depicted in Figure 11.6 [478].

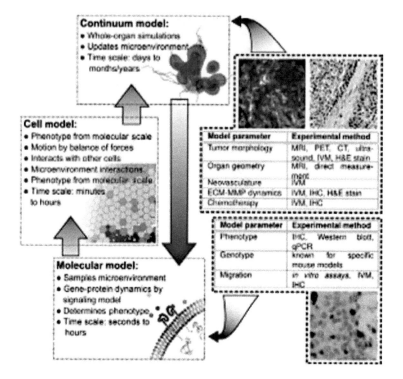

Figure 11.6 Hybrid multi-scale cancer modeling incorporating data flow across the physical scales of a tumor [478].

11.4 MODELING EXAMPLES

11.4.1 Modeling of chemotherapy

The response to chemotherapy like in the case of radiotherapy, tends to globally follow a sigmoidal curve (S-shape). Different approaches have been used to model response to chemotherapy (systematic or targeted), which are typically coupled with compartmental pharmacokinetic/pharmacodynamic (PK/PD) models to identify optimal drug administration schedules. For instance, a model based on the stochastic birth and death process was developed to design optimized dosing strategies for EGFR-mutant in non-small cell lung cancer (NSCLC). The model was trained on tumor cell populations *in vitro* and incorporated pharmacokinetic data from human clinical trials with the drug *erlotinib* to ensure that the drug doses proposed were clinically achievable in humans [483]. Sample prediction results are shown in Figure 11.7 [483].

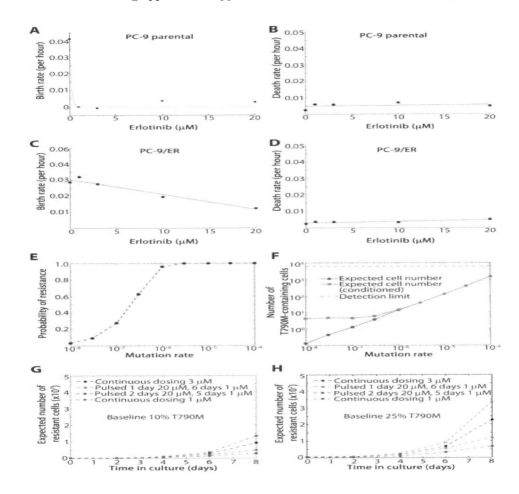

Figure 11.7 Targeted chemotherapy cancer modeling predictions to delay the development of resistance. (A) PC-9 cell birth rate at erlotinib concentrations of 0, 1, 3, 10, and 20 μM. (B) PC-9 cell death rate as a function of increasing erlotinib concentration. (C) PC-9/ER cell birth rate in the presence of erlotinib (0, 1, 3, 10, and 20 μM). (D) PC-9/ER cell death rates as a function of increasing erlotinib concentration. (E) Probability of preexisting T790M-harboring cells in a population of 3 million cells initiating from one cell harboring just a drug-sensitive EGFR mutation that grew in the absence of drug for a range of mutation rates (10^4 to 10^8 per cell division). (F) Expected number of resistant cells present in the population, both averaged over all cases and averaged only over the subset of cases, where at least one resistant cell is present. (G) An initial population of 750,000 cells, 10% of which harbor T790M, treated with continuous low-dose erlotinib (1 and 3 μM) selects for the emergence of T790M-harboring cells (green and black lines). The addition of one or two high-dose erlotinib "pulses" (20 μM) followed by 1 μM for the remaining days of a 7-day cycle decreases the expected number of resistant cells (red and blue lines). (H) Analogous results as in panel (G) starting with an initial population with 25% T790M-harboring cells.

In another study, the efficacy of a cytoxic chemotherapy drug (*Doxorubicin*) was simulated using a continuum approach with a vascular tumor model (cf. 11.3.1)[484]. The drug distribution was represented by a diffusion-convection PDE. The toxicity of the drug targeted to the tumor cells is determined by the drug concentration and tumor cell activity, with higher concentrations of drug that are more likely to induce cell cycle transitions from active to quiescent or necrotic. The intravascular drug concentration can be predicted using body level PK/PD models as shown in Figure 11.8 [484].

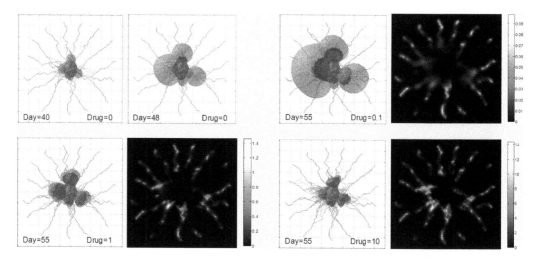

Figure 11.8 Tumor morphology and chemotherapy drug (doxorubicin) distribution following three different drug administration concentrations (0.1, 1, and 10 mol/m^3), [484].

11.4.2 Modeling of immunotherapy

Modulating the interplay between the immune system and cancer cells via immunotherapy alone or on in combination with other therapeutics is becoming one of the most promising cancer treatment approaches. Due to the complexity of this interplay, computational modeling methods are playing a pivotal role [463]. A summary of such efforts using different modeling approaches is shown in Figure 11.9 [485]. For instance, an agent-based (cf. Section 11.3.2) model was used to study the effects of a specific immunotherapy strategy against B16-melanoma cell lines in mice and to predict the role of CD137 (tumor necrosis factor (TNF) receptor family) expression on tumor vessel endothelium for successful therapy [486]. Another study applied a discrete-time pharmacodynamic model to illustrate the potential synergism between immune checkpoint inhibitors (PD1-PDL1 axis and/or the CTLA-4) and radiotherapy. The effects of irradiation were described by a modified version of the linear-quadratic (LQ) model [487]. The model was able to explain the biphasic relationship between the size of a tumor and its immunogenicity, as measured by the *abscopal effect* , *i.e.*, an induced non-localized immune response. Moreover, it explained why discontinuing immunotherapy may result in either tumor recurrence or a durably sustained response indicating potential for better synergistic scheduling [487].

A continuum type model was developed to combine therapeutic effects from chemotherapy and immunotherapy was proposed by Pillis *et al.* by including the influence of several immune cell effector subpopulations (tumor antigen-activated $CD8^+$ T cells, natural killer (NK) cells and total circulating lymphocytes), in addition to the concentrations of inter-leukin (IL)-2 and chemotherapy drug in the bloodstream [488]. However, such combined modeling approaches remains an active area of research today.

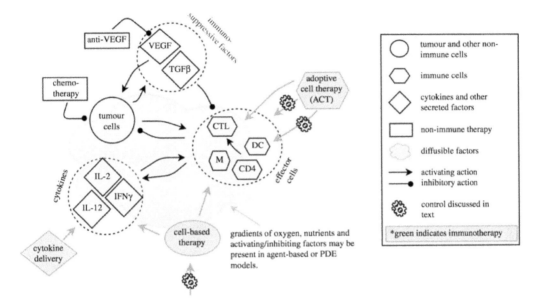

Figure 11.9 Summary of modeling efforts in immunotherapy [485].

11.5 SOFTWARE TOOLS FOR MULTI-SCALE MODELING

There are several software tools for multi-scale modeling in the literature. Among the commonly used ones is *CompuCell3D*. CompuCell3D is an open-source simulation environment for multi-cell, single-cell-based modeling of tissues, organs and organisms. It uses the Glazier-Graner-Hogeweg (GGH) model as described above [489]. The software is written in C++ programming language and is constructed with a friendly object-oriented (modular) design. Models are described at a basic level using XML, but far greater versatility is afforded by use of custom extensions written in the Python scripting language, through which much of the model architecture can be accessed (a Simplified Wrapper and Interface Generator (SWIG) interface exposes the C++ modules to Python). An example demonstration is shown in Figure 11.10. Other software tools that exist include *EPISIM* [490], *Chaste* [491], and *PhysiCell* [492].

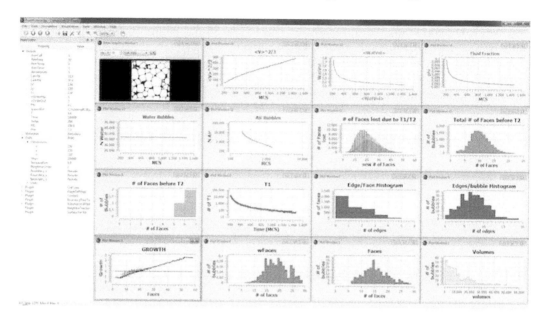

Figure 11.10   COMPU3DCELL simulation software demonstration.

11.6   CONCLUSIONS

Multi-scale modeling techniques (continuum, discrete, or hybrid) provide a useful and complementary role to other treatment outcome modeling schemes based on statistical or machine learning techniques. Moreover, it allows for better mechanistic understanding of the interplay between the tumor and various therapeutic agents. However, with increased complexity (parameterization in this case), further experimental and clinical data is required to meet proper fitting requirements of these methods. This would demand closer collaborations among modelers, basic scientists and clinicians to achieve the potential of such a powerful mathematical modeling approach to improve our understanding of cancer treatment outcomes, and address current open questions to how to prescribe a certain treatment regimen or optimize synergy of timing and dosing of multiple treatment modalities for localized or metastatic cancers.

IV

Example Applications in Oncology

CHAPTER 12

Outcome modeling in treatment planning

X. Sharon Qi

Mariana Guerrero

X. Allen Li

CONTENTS

ABSTRACT

The goal of radiotherapy (RT) is to deliver the therapeutic radiation dose to the targets
while minimizing potential complications to the adjacent normal structures. Physical quantities, such as dose and dose-volume metrics, have been used to optimize a treatment plan
and evaluate the quality of the plan. Such physical quantities are unlikely to predict patient
biological response, which is usually characterized by a nonlinear dose response relationship.
Several dose-response models have been proposed and are expected to more closely reflect
clinical goals and quantitatively predict treatment outcome. In recent years dose-response
models have been included in commercial treatment planning systems and play an increasingly important role in RT planning. In this chapter, we summarized the most commonly
used dose-response models and biological quantities, followed by the application of the models in the treatment planning process, both for plan evaluation and/or plan optimization for
major commercial treatment planning platforms. The platforms are Monaco (Elekta, Maryland Heights, MO, USA), Pinnacle (V8.0h, Philips Medical Systems, Andover, MA, USA),
Eclipse (V10.0, Varian Medical Systems, Palo Alto, CA, USA), and RaySearch (RaySearch
Laboratories AB, Stockholm, Sweden).

Keywords: radiotherapy treatment planning, dose-volume metrics, biological models, treatment outcome, treatment planning platforms.

12.1 INTRODUCTION

The ultimate goal of radiation treatment planning is to generate plans to deliver the therapeutic radiation dose to the targets while minimizing the complications to the adjacent normal tissues. Historically, due to the lack of understanding of the underlying biological mechanisms and the sparsity of available clinical data, physical surrogates such as dose-volume endpoint have often been adopted to judge the quality of a radiation treatment plan. However, such dose and dose-volume quantities are unlikely to predict patient specific biological response, usually characterized by a nonlinear dose-response relationship. Biological models, on the other hand, have been proposed for many decades and are expected to more closely reflect clinical goals and quantitatively predict treatment outcome [493, 152].

Treatment planning tools that use biological models for plan optimization and/or evaluation are available for clinical use. A variety of dose response models and quantities along with a series of organ-specific model parameters are included in these tools [152]. In fact, the major treatment planning platforms, such as Monaco (Elekta, Maryland Heights, MO, USA), Pinnacle (V8.0h, Philips Medical Systems, Andover, MA, USA), Eclipse (V10.0, Varian Medical Systems, Palo Alto, CA, USA), and RaySearch (RaySearch Laboratories AB, Stockholm, Sweden), have implemented a biological-based treatment planning module into a routine treatment planning process for plan evaluation and/or plan optimization.

12.1.1 Review of the history and dose-volume based treatment planning and its limitations

Current dose-volume based treatment planning is a result of the history of radiotherapy. In the 1960-70s, treatment planning was in its primitive stage prior to the application of computers [493]. In the 1980s, 3-dimensional conformal radiotherapy (3D-CRT) technology became available allowing investigators to correlate the dose-volume information with clinical outcome data. This so-called 'forward planning' is no longer feasible for the emerging intensity modulated radition therapy (IMRT) as there are far too many degrees of freedom in IMRT planning.

Modern techniques such as IMRT often require some kind of optimization using powerful computer software (so-called "inverse planning"). The user specifies the single or multiple treatment goals and lets the computer system find the best possible solutions. The treatment goals are often defined in terms of physical dose limits to different volumes, which are based on clinical studies that demonstrate correlation between tumor control/complication incidence and particular dose-volume metrics. For example, V20 (percentage of lung volume receiving at least 20 Gy) is used to gauge the probability of a plan causing grade ≥ 2 or grade ≥ 3 radiation pneumonitis [494]. However, there are a number of limitations associated with the dose-volume based treatment planning approach as summarized below:

1. Dose-volume based treatment planning typically utilizes one or more organ-specific DVH constraints (*e.g.*, V5, V10, V20, mean lung dose) that correlates with the complication based on previous experience. This correlation is typically specific to the treatment delivery technique, *i.e.*, IMRT or 3DCRT, beam arrangements, dose fractionation scheme, etc. Caution should be taken when the limits are clinically applied but were derived from other institutions, treatment techniques and/or planning systems.

2. Optimization with dose-volume constraints requires substantial practice and experience in selecting objectives and relative weights to meet planning goals. The trial-and-error method is generally adopted, which may be time consuming for some challenging cases and is highly dependent on the planner experience level.

3. Specifying multiple dose-volume constraints increases computational complexity of the inverse treatment-planning problem. Moreover, cost functions based on dose-volume constraints can lead to multiple local minima [495, 496], potentially leading to less favorable dose distributions [152].

4. Most current dose-volume based plan optimizations lack tools for routinely evaluating biologically based metrics alongside DVH metrics. Since dose distributions for plans driven by biological methods constraints may differ substantially from those driven by dose-volume point constraints, evaluation tools are important as a basis for progression to preferentially adopting biological methods in optimization.

12.1.2   Emerging dose-response modeling in treatment planning and advantages

Dose-response based treatment planning, or biological-based treatment planning, however, is an emerging development that can further improve radiation plan quality when used in conjunction with the dose-volume based treatment planning for plan evaluation and optimization. The advantages of biological based treatment planning can be summarized as follows:

1. Biological model based optimization is designed to reflect more realistic dose-response relationships for the targets and OARs and quantitatively estimate TCP/NTCP probabilities in the selection of treatment plans. Instead of uniform dose distribution to the target, a more inhomogeneous treatment plan may be expected compared to the current dose-volume based optimization [497, 498].

2. Biological model based optimization may be more effective in OAR sparing. While multiple dose-volume points are normally selected for dose-volume based inverse treatment planning optimization via trial-and-error method, a biological based objective function, *i.e.*, single EUD-based cost function with appropriate choice of a volume effect parameter, may lead to a better treatment plan in a more efficient way since the biological cost function controls greater space in the dose-volume domain.

3. Biological based plan evaluation provides additional biological metrics such as EUD, TCP, NTCP, UTCP, that are otherwise unavailable in dose-volume based plan evaluation tools. Estimates of biological metrics are useful in the selection of rival plans that may spare one OAR over the other or have crossing DVHs for a critical structure. These situations are illustrated in Figure 12.1(a) and (b). Figure 12.1(a) shows a rectum DVH for two competing prostate plans. Both plans fulfill the clinical goals but the dose distributions are quite different so the knowledge of NTCP values could be useful in deciding between the competing plans. Figure 12.1(b) shows the DVHs for esophagus and ipsilateral lung for two competing lung plans. One plan benefits the esophagus at the cost of more lung dose and vice versa with both plans having similar target coverage and other OAR DVHs. Again, NTCP modeling could help make better clinical decisions in such cases.

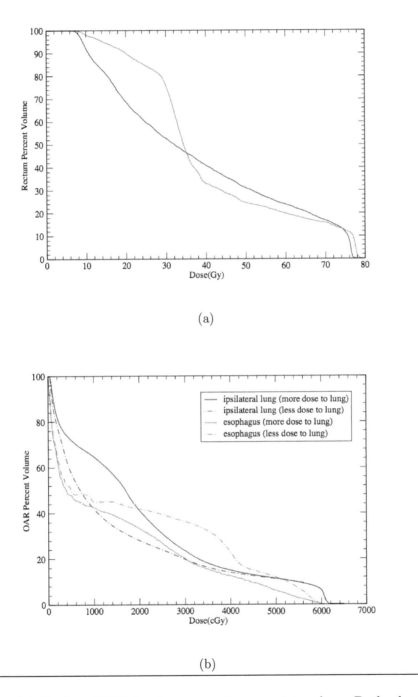

(a)

(b)

Figure 12.1 (a) Rectum DVH for two competing prostate plans. Both plans fulfill the clinical goals but the dose distributions are quite different so the knowledge of NTCP values could be useful in deciding among plans. (b) DVHs for esophagus and ipsilateral lung for two competing lung plans.

12.2 DOSE-RESPONSE MODELS

12.2.1 Generalized equivalent uniform dose (gEUD)

The concept of generalized equivalent uniform dose (gEUD) assumes that any two dose distributions are equivalent if they cause the same biological effect [499, 151]. More specifically, the gEUD is defined as the uniform dose that gives the same biological effect as a given inhomogeneous dose distribution and can be calculated based on the power law dependence of dose response for both tumor and organ-at-risk (OAR):

$$gEUD = \left(\sum_i \nu_i D_i^a \right)^{1/a} \tag{12.1}$$

where ν_i is the fractional organ volume within the target or OAR irradiation to dose D_i. The parameter a is an organ-specific biological parameter. Figure 12.2 shows a set of DVHs with various degrees of inhomogeneity of the corresponding dose distributions that has the same EUD.

The gEUD model is parameterized by the single biological parameter a, which is specific for a pre-defined and quantifiable biological endpoint for given tumor and/or OAR. The parameter a in gEUD and the Lyman model parameter n are inversely related as $a = 1/n$. A gEUD can be calculated directly from the dose calculation points or from the corresponding dose-volume histograms (DVHs). For $a \to \infty$, gEUD approaches the minimum dose; thus negative values of parameter a are generally used for tumors to penalize target cold spots.

Due to its simplicity, specifically the same gEUD formalism being applied to both targets and OARs with a single parameter capturing different dosimetric characteristics of the biological responses, the concept of gEUD is often utilized in plan evaluation and optimization.

12.2.1.1 Serial and parallel organ models

For normal organs, the concept of functional sub-units (FSU) is used to model the radiation response. FSUs may be arranged in serial, parallel, or a combination of both. Serial organs involve FSUs in a line, meaning the organ fails if a single FSU is susceptible to high point doses and therefore the volume effect is small and the response depends mostly on the maximum dose. Since gEUD approaches the maximum dose as $a \to \infty$ a large value is often used to characterize serial organs. The spinal cord and the gastrointestinal tract are typical serial organs.

On the other hand, each FSU in a parallel organ is able to function independently of the others. Loss of a single FSU leads to a slight decrease in function of the organ. Parallel organs are more sensitive to volume effects and complications are assumed to occur after a substantial fraction of FSUs are damaged. A small value of the parameter a (close to 1) in gEUD is typically used for parallel organs, and therefore gEUD approaches the mean dose. Typical parallel organs include the liver and the kidneys.

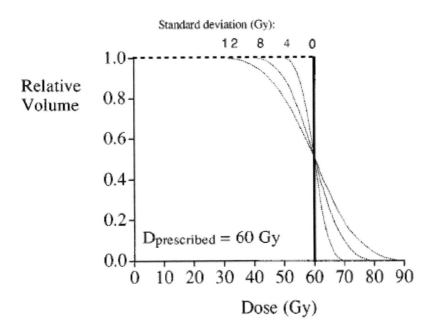

Figure 12.2 A set of DVHs (thin lines) with various degrees of inhomogeneity of the corresponding dose distributions that has the same EUD (thick line) for $a = 1$.

12.2.2 Linear-Quadratic (LQ) Model

The linear-quadratic (LQ) formalism [285, 500, 322] is the most commonly used model to describe cell survival in radiotherapy. The LQ model provides a biologically plausible and experimentally established method to quantitatively describe the dose-response to irradiation in terms of clonogenic survival. In the basic LQ formula, the clonogenic-surviving fraction (SF) following a radiation dose D (Gy) is described by an inverse exponential approximation:

$$SF = e^{-E} = e^{-\left(\alpha D + \beta G(D)D^2\right)} \tag{12.2}$$

where,

$$E = \alpha D + \beta G(D)D^2 \tag{12.3}$$

where α and β are tissue specific model parameters that govern the linear and quadratic elements of the dose response and $G(D)$ is the generalized Lea-Catcheside protraction factor [285]. The parameters α and β are experimentally derived for the linear and quadratic terms, respectively. The general expression for the protraction factor $G(D)$ for any generic dose-rate $\dot{D}(t)$ is (Sachs et $al.$ [285] and reference therein):

$$G(D) = \frac{2}{D^2} \int_{-\infty}^{\infty} \dot{D}(t)\,dt \int_{-\infty}^{t} e^{-\mu(t-t')}\dot{D}(t')\,dt' \tag{12.4}$$

where $\mu = ln(2)/T_{rep}$ is the repair rate of the tissue and T_{rep} is the characteristic repair time. The protraction fraction G is 1 for an acute dose (negligible treatment time) but in general is less than 1 for a protracted irradiation, showing an increase in cell survival due to sub-lethal repair. For a continuous irradiation with constant dose-rate $\dot{D}_0$ and treatment time T the protraction factor is:

$$G\left(D, \mu T\right) = \frac{2}{D^2}\left(\mu T - 1 + e^{-\mu T}\right) \tag{12.5}$$

where T is the treatment time $T = \frac{D}{\dot{D}_0}$. For a series of n acute dose fractions of size d, $G = 1/n$ so that

$$E = \alpha D + \frac{1}{n}\beta D^2 = \alpha D \cdot \left(1 + \frac{d}{\alpha/\beta}\right) \tag{12.6}$$

This expression assumes that there is complete repair in between fractions, which is true when the time interval between fractions is much larger than T_{rep}.

12.2.3 Biological effective dose (BED)

The biological effective dose (BED) is used to understand tumor and normal tissue response across different treatment modalities and fractionation schemes. The basic form of BED for fractionated radiotherapy with n fraction of nearly instantaneous doses d is defined as:

$$BED = \frac{E}{\alpha} = nd\left(1 + \frac{d}{\frac{\alpha}{\beta}}\right) \tag{12.7}$$

This is the simplified model for which cell repopulation and sublethal-damage repair can be ignored during the irradiation.

A more general BED model that is often used includes cell repopulation as follows:

$$BED = \frac{E}{\alpha} = nd\left(1 + \frac{d}{\frac{\alpha}{\beta}}\right) - \frac{\ln(2)}{\alpha T_{pot}}(T - T_k) \tag{12.8}$$

where T_{pot} is the potential doubling time and T_k is the delay of the onset of rapid repopulation.

To account for different fractionation schemes (other than 2 Gy per fraction), normalized total dose (NTD) is a commonly used radiobiological dose-fractionation model that converts the biological effect referred to the standard fractionation of 2 Gy also known as the equieffective dose at 2 Gy (EQD2). The NTD equation is based on the LQ model and can be expressed in the following:

$$NTD_{2Gy} = D_x\left(\frac{\frac{\alpha}{\beta} + x}{\frac{\alpha}{\beta} + 2}\right) \tag{12.9}$$

where NTD_{2Gy} is the normalized total dose to 2 Gy fractions and D_x is the actual dose for the fractionation scheme of x Gy per fraction.

12.2.4 Tumor control probability (TCP) models

The TCP models describe of the probability of killing all tumor clonogenic cells in the defined tumor volume following irradiation to a certain total dose. Common TCP models are often built following the Poisson distribution, assuming that all the clonogenic cells need to be eradicated to achieve tumor control. Tumors are consisting of non-clonogenic cells with a limited lifetime and clonogenic cells proliferating exponentially, basically uncontrolled. The Poisson TCP model is generally expressed as:

$$TCP = e^{-K \cdot S} \tag{12.10}$$

where K is the initial number of clonogenic cells and S is the surviving fraction. Theoretically, all clonogens need to be eradicated to achieve tumor control since one remaining clonogenic cell is able to repopulate the entire tumor.

$$P(D_i) = e^{-Ke^{-\alpha D_i - \beta D_i^2}} = e^{-e^{e\gamma - \alpha D_i - \beta D_i^2/n}} \tag{12.11}$$

where $P(D_i)$ is the probability of controlling the tumor or including a certain injury to an organ that is irradiated uniformly with a dose D_i. Here, the number of clonogens K is approximated by $e^{e\gamma}$, and γ is the maximum normalized value of the dose-response gradient. The expression of $K = e$ is exact when $\beta = 0$ or while the dose per fraction remains constant [152]. If the clinical dose-response parameters γ and the dose at 50% response rate D_50 have been derived from treatments where a fixed dose per fraction was applied, the dose-response relation can be described by the linear-Poisson model :

$$P(D_i) = e^{-e^{e\gamma - (\frac{D_i}{D_{50}})(e\gamma - \ln(\ln(2)))}} \tag{12.12}$$

Thus, the values of α and β for each organ can be derived from the corresponding values of D_50 and γ by equating Eqs. 12.11 and 12.12:

$$\alpha = \frac{e\gamma - \ln(\ln(2))}{D_{50}(1 + \frac{2}{\alpha/\beta})} \tag{12.13}$$

$$\beta = \frac{e\gamma - \ln(\ln(2))}{D_{50}(\alpha/\beta + 2)} \tag{12.14}$$

The logistic function is often adopted to describe the sigmoid shape dose-response TCP model with the following form [501]

$$P(D_i) = \frac{e^{(D_i - D_{50})/k}}{1 + e^{(D_i - D_{50})/k}} \tag{12.15}$$

where k is related to the normalized dose-response gradient according to $k = D_{50}/(4\gamma)$. The empirical log-logistic function often expressed as:

$$P(D_i) = \frac{1}{1 + (D_{50}/D_i)^k} \tag{12.16}$$

where k determines the slope of the curve. This formulation is recommended in AAPM 137 [502] for use in brachytherapy treatments.

12.2.5 Normal Tissue Complication Model (NTCP) models

12.2.5.1 Lyman-Kutcher-Burman (LKB) model

The Lyman model [503] was proposed to estimate the complication probability of a normal tissue structure from a dose-volume histogram for uniformly irradiated whole or partial organ volumes. In the Lyman model, the volume dependence is represented by a power-law relationship:

$$D(V) = TD(1)/V^n \qquad (12.17)$$

where $TD(V)$ is the tolerance dose for a given partial volume (V). $TD(1)$ is the tolerance dose for the full volume and n is a fitted parameter that describes the magnitude of the volume effect using a power-law relationship between the tolerance dose and irradiated volume [503]. The dose dependence is represented by the integral of a normal distribution (a sigmoid shape curve) via

$$NTCP = \frac{1}{\sqrt{2\pi}} \int_{-x}^{t} e^{-t^2/2} dt \qquad (12.18)$$

where

$$t = (D - TD_{50}(V))/\sigma(V)) \qquad (12.19)$$

The TD_{50} is the dose that would result in 50% complication probability after 5 years. The complication probability of a normal organ can be defined by the three parameters: $TD_{50}(1)$, n determines whether the structure is "serial" (n close to 0) or parallel (n close to 1), and m is related to the curve steepness.

Small values of n correspond to small volume effects for serial organs such as cord and rectum. Large values (close to 1) correspond to parallel organs, such as lung and liver.

Among other models, the effective volume method [504, 505, 149], the combined formalism Lyman-Kutcher-Burman (LKB) model [149] is the most commonly used complication model to convert a heterogeneous dose distribution into a uniform partial or whole organ irradiation resulting in the same NTCP. The LKB model is calculated using the following equations:

$$NTCP = \frac{1}{\sqrt{2\pi}} \int_{-infinity}^{t} e^{-\frac{x^2}{2}} dx \qquad (12.20)$$

$$t = \frac{D_{eff} - TD_{50}}{m \cdot TD_{50}} \qquad (12.21)$$

$$D_{eff} = \left(\sum_i \nu_i D_i^{1/n} \right)^n \qquad (12.22)$$

Where D_{eff} is the dose that, if given uniformly to the entire volume, will lead to the same NTCP as the actual non-uniform dose distribution. TD_{50} is the uniform dose given to the entire organ that results in 50% complication risk, m is a measure of the slope of the sigmoid curve, n is the volume effect parameter and v_i is the fractional organ volume receiving a dose D_i. Note that D_{eff} is conceptually equivalent to the gEUD with parameter $a = 1/n$. Small values of n correspond to small volume effects (for serial organs) and large values correspond to large volume effects (for parallel organs).

12.2.5.2 Relative seriality (RS) model

The relative seriality model or the s-model [149] describes response of an organ with a mixture of serial- and parallel-arranged functional subunits (FSUs). The relative contribution of each type of architecture is described by the parameter s, which is equal to unity for a fully serial organ and zero for a fully parallel organ. NTCP is given by the following equations [149]:

$$NTCP = \left\{ 1 - \prod_{i=1} [1 - (P(D_i))^s]^{\nu_i} \right\}^{1/s} \tag{12.23}$$

$$P(D_i) = 2^{-e^{\left(e\gamma(1 - \frac{D_i}{D_{50}}) \right)}} \tag{12.24}$$

where $P(D_i)$ is the probability of complication due to the irradiation of the relative volume v_i at the dose D_i described by an approximation of Poisson statistics. The RS model contains three parameters: D_{50}, γ, s. D_{50} has the same meaning as for the LKB model, γ is a slope that describes the steepness of the sigmoid shape dose-response curve, and s is a generalized parameter that describes the relative seriality of organ/tissue under consideration (the ratio of serial subunits to all subunits of the organ). Large values (≈ 1) of s indicate a serial structure and small values ($\ll 1$) indicate a parallel structure. The Poisson model for $P(D_i)$ often used in treatment planning systems (TPSs), *i.e.*, Eclipse, for calculation of NTCP.

12.2.5.3 Model parameters and Quantitative Analysis of Normal Tissue Effects in the Clinic (QUANTEC)

One of the most challenging tasks in radiation treatment is to minimize the risk of normal tissue toxicity while delivery therapeutic dose distribution to the target. Knowledge of how dose distributions affect normal tissue outcomes is needed to guide planning/delivery goals. In 1991, Emami *et al.* [27] published a comprehensive review of the available dose-volume and outcome data, along with expert opinion where data were lacking. Over the past 20 years, there have been numerous new studies available providing dose/volume/outcome data. AAPM and ASTRO, therefore, funded the QUANTEC effort, wherein over 60 volunteer physicists and radiation oncologists collaborated in an expert review of normal tissue tolerance data and modeling in the literature from the "3DCRT" era. Consensus dosimetric guidelines for 16 normal organs were published in early 2010 (International Journal of Radiation Oncology, Biology and Physics V76, 3 supplement). For some complications, the QUANTEC recommendations are in general concordance with the Emami guidelines, but for others there are notable differences [31]. The QUANTEC [7] review summarizes the currently available three-dimensional dose-volume and outcome data and updates and refines the normal tissue dose/volume tolerance guidelines in Emami *et al.* [27]. The QUANTEC provide a practical clinical guidance and excellent resources to assist physicians and treatment planners in determining acceptable dose/volume constraints. Since most of the data in the QUANTEC is derived based on conventionally fractionated radiation schemes, caution needs to be taken when applying QUANTEC criteria to stereotactic body radiotherapy (SBRT) regimen.

Mathematical/biological models have been used to relate 3D dose-volume data to clinical outcome along with associated model parameters, limitations and uncertainties. The readers are referred to the following literatures for detail [30, 31, 135, 134].

12.2.6 Combined TCP/NTCP models –Uncomplicated tumor control model (UTCP or P+)

Uncomplicated tumor control probability model (UTCP), also called P+, is defined as the probability of complication-free tumor control [506]. It combines TCP and NTCP models together to express the treatment plan quality in a single value. By maximizing UTCP, the dose distribution that gives optimal probability of complication-free tumor control according to the used biological models is obtained. The trade-off between tumor control and normal tissue complication is controlled by the biological models instead of using penalty functions in the dose domain.

In the simplest form, if TCP and NTCP are assumed to be statistically independent processes, the UTCP model can be given by:

$$UTCP = TCP \times (1 - NTCP) \tag{12.25}$$

If TCP and NTCP are assumed to be totally correlated processes, UTCP is given by

$$UTCP = TCP - NTCP \tag{12.26}$$

The UTCP is ranged between 0 to unity. Higher UTCP value suggests the superior radiation treatment plan. Statistically independent responses are assumed during optimization.

12.3 DOSE-RESPONSE MODELS FOR STEREOTACTIC BODY RADIOTHERAPY (SBRT)

The LQ model has been successfully used for at least two decades to quantify radiation effects at standard dose fractionation schemes, *i.e.*, 1.2–3.0 Gy. Validity of the LQ model for large doses per fraction encountered in radiosurgery (SRS) and SBRT has become a matter of ongoing debate [507, 508, 317]. The issue in contention is that the LQ formalism predicts a continuously bending survival curve while experimental data clearly demonstrate that at large doses the surviving fraction becomes an exponential function of dose; *i.e.*, follows a straight line on a semi-log plot. Hybrid solutions accounting for this effect have been suggested [330]. Despite these controversies, the LQ model remains a tool of choice for isoeffect calculations in conventionally fractionated photon beam therapy.

12.3.1 Linear-Quadratic (LQ) model applied to SBRT

The LQ model is the most commonly used radiation induced cell-killing model [509, 272]; it has been useful in predicating and understanding biological effects in conventional fractionated RT schemes in which small dose per fraction (*i.e.*, 1.8–4 Gy) is delivered over the course of 4-7 weeks. With the emerging stereotactic radiosurgery (SRS) and SBRT techniques being increasing utilized in which larger doses per fraction (*i.e.*, 10–20 Gy) are delivered over the course of 5 fractions or less, the need for SBRT dose-response biological models are becoming prominent.

The LQ model predicts a continuously bending curve in the high-dose range, while experimentally measured data have shown a linear relationship between the dose and surviving clonogens [510]. In other words, the LQ model overestimates the effect of radiation on clonogenic cells survival in high dose ranges such as those normally used for SBRT. Several extensions and/or alternatives to the LQ model have been proposed to address this shortcoming. In this section, we describe three possible alternative models.

12.3.2 Universal survival curve (USC) model

The universal survival curve (USC) model is a hybrid model proposed by Park *et al.* in 2008 [330]. The USC model combines two classical biological models, the LQ model in the low-dose range (1.8 – 4.0 Gy) and the multi-target model in the high-dose range. For the conventional fractionated radiotherapy (CFRT), the LQ model has been shown to have a good description and understanding of the biological effects, but in the high-dose range, the LQ model overestimates the effect of radiation on clonogenic cell survival [510].

The multi-target model provides an alternative description of the clonogenic survival as a function of radiation dose. The multi-target model assumes there are $\bar{n}$ targets that need to be hit to destroy the cell's clonogeneity. The surviving fraction is given by:

$$S = e^{-d/d_1} \cdot \left\{ 1 - (1 - e^{d/D_0})^{\bar{n}} \right\} \tag{12.27}$$

where d_1 and D_0 are the parameters that determine the initial and final slopes of the survival curve. In the high-dose range, where $d \gg D_0$, the multi-target model survival curve approaches an asymptote:

$$\ln S \approx \frac{-1}{D_0} d + \ln(\bar{n}) = \frac{-1}{D_0} d + \frac{D_q}{D_0} \tag{12.28}$$

where $\ln(\bar{n})$ is the y-intercept of the asymptote, and can be rewritten in terms of its slope $1/D_0$, and x-intercept D_q. The fundamental basis of the multi-target model is not favored as the mechanism of underlying radiobiological processes; however, the multi-target model is still valuable because it fits the empirical data well, especially in the high-dose range [511].

The USC takes advantage of the simplicity of the LQ model in the low-dose range and enhanced data fit of the multi-target model in the high-dose range. The USC model can be expressed as:

$$\ln S = \begin{cases} -(\alpha \cdot d + \beta \cdot d^2), & d \leq D_T \\ \frac{-1}{D_0} d + \frac{D_q}{D_0}, & d \geq D_T \end{cases} \tag{12.29}$$

where D_T is a transition dose point that the LQ model smoothly transitions to the asymptote of the multi-target model. The D_T can be mathematically calculated as the following function:

$$D_T = \frac{2 \cdot D_q}{1 - \alpha \cdot D_0} \tag{12.30}$$

Since the asymptotic straight line of the multi-target model is tangential to the LQ model parabola at D_T, the β can be solved and written as

$$\beta = \frac{(1 - \alpha \cdot D_0)^2}{4 D_0 \cdot D_q} \tag{12.31}$$

The USC model was shown to describe the measured cell survival curve for the H460 survival curve [31] in high-dose range beyond the shoulder. However, the USC model does not offer any mechanistic explanation for the complex biology that underlies cell ablation and loss of clonogenicity for SBRT treatment [512, 330]. The applicability of the USC model to in vivo tumor control probabilities needs to be further validated through prospectively designed trials.

12.3.3 Linear-Quadratic-Linear (LQL) model

The LQL (originally denoted the MLQ, "modified" LQ) model was first proposed to address the problem of biological effective dose calculations (BED) for tumor and normal tissues when the dose per fraction d is large (greater than 5 Gy or so) [328]. While the LQ model is well established for standard fractionations, i.e., fraction dose of 1–3 Gy, the behavior at larger doses may sometimes deviate from the LQ prediction as shown in in-vitro cell survival experiments [328] where the survival versus dose curve becomes linear at high doses instead of the quadratic behavior derived from the LQ model. In his milestone article [513], Barendsen reviewed a large number of animal in-vivo tolerance doses and pointed out deviations from the LQ predictions at large doses per fraction and derived the ratios based on an average of the high dose and low dose behavior. However, for clinical practice where the ratios are derived from data for low doses only, it is important to at least investigate the impact of possible deviations from the standard LQ model.

The LQL model [328] considered the cell survival for the LQ model of a single dose exposure D where the dose-rate effects are included in the protraction factor (a.k.a the Lea-Catcheside factor)

$$G(x) = 2(x + exp(-x) - 1)/x^2 \qquad (12.32)$$

$$S_{LQ}(D) = exp(-\alpha D - \beta G(\mu T)D^2) \qquad (12.33)$$

where the α term represents the direct creation of non-repairable (lethal) lesions, and the β term represents lethal lesions arising from the interaction of a pair of sub-lethal lesions. The constant μ is the repair rate of the tissue and T the delivery time. The LQL introduced an additional parameter (here denoted δ) to shift the function G and match the lethal-potentially lethal (LPL) [283] model at high doses. The LPL is a kinetic model that has also been used to interpret *in vitro* experiments. At low doses, the LPL can be shown to exhibit the LQ model behavior, but at high doses, it predicts a significantly lower level of cell kill. By shifting $G(\mu T) \to G(\mu T + \delta D)$ and choosing δ appropriately, the LQL matches the LPL essentially at all clinically relevant doses and all LPL parameters can be mapped into the LQL parameters. With this modification, the survival fraction expression becomes for continuous irradiation:

$$S_{LQL}(D) = exp(-\alpha D - \beta G(\mu T + \delta D)D^2) \qquad (12.34)$$

The survival fraction for the LQL model is always higher than the survival for the standard LQ model.

While the introduction of δ in the manner described above may seem arbitrary at first, Carlone *et al.* [514] showed that the same expression can be obtained by the full solution of Dale's original equations [515] used to derive the LQ model. In Dale's derivation, it was assumed that sub-lethal lesion repair was the dominant process in reducing the number of sub-lethal lesions, neglecting the contribution from lethal exchange processes. By taking into account the contribution from lethal exchange processes in the depletion of sub-lethal lesions, the LQL equations can be derived in a mechanistic manner [515]. In Dale's [515] formulation, the parameter β can be factored as $\beta = p^2 \cdot \epsilon$, where p is the rate per unit dose at which sub-lethal lesions are produced and ϵ is the probability of interaction between lesions. Carlone et al. [514] showed that similarly the parameter δ can be expressed as $\delta = p \cdot \epsilon$. This derivation sets the LQL apart from other proposed extensions of the LQ model since it gives it a mechanistic formulation similar to the original LQ and allows the possibility of using the LQL in general cases other than single acute doses, like split dose experiments and decaying sources [327]. It also gives information about the interplay of radiobiological parameters. For example, Figure 12.3 shows that given the same values of α and δ, late effect tissues with lower α/β ratio exhibit a much larger difference in cell survival between the LQL and LQ models than early responding tissues with higher α/β ratio. Therefore, the effect of the additional parameter δ can increase the therapeutic ratio when large doses per fraction are used. Table 12.1 shows the differences in BED for different values of parameters and illustrates the interplay among α/β ratio and δ, the BED differences are more pronounced for low α/β ratios. The values of δ selected for these tables come from earlier estimations using the correspondence of parameters of the LQL and the LPL, the relationship with the final slope D_0 and the fitting of plots of inverse dose versus dose per fraction [328]. All these estimates suggest a value of the parameter δ of less than 0.2 Gy^{-1}.

12.3.4 Regrowth model

The regrowth model can be expressed [516, 517]:

$$BED^{Regrowth} = D\left(1 + \frac{d}{\alpha/\beta}\right) - \gamma\frac{T}{\alpha} - a(\tau - T)^\delta \quad (\tau > T) \tag{12.35}$$

where the terms of $\gamma\frac{T}{\alpha}$ and $a(\tau - T)^\delta$ describe tumor regrowth during and after radiation treatment course, respectively. T is elapsed treatment duration; $\gamma = \frac{ln(2)}{T_d}$, T_d is potential effective tumor doubling time; τ the follow-up time after the end of the treatment; a and δ are fitting parameters, and δ is a parameter characterizing the speed of tumor cell regrowth after radiation treatment.

For this model, TCP has the form:

$$TCP = 1 - \frac{1}{\sqrt{2\pi}} \int_{-\infty}^{t} e^{\frac{-x^2}{2}} dx \tag{12.36}$$

with

$$t = \frac{e^{-\left[\alpha \times BED - \left(\frac{ln(2)}{T_d}\tau\right)^\delta\right]} - K_{50}/K_0}{\sigma_k/K_0} \tag{12.37}$$

where K_{50} is the critical tumor colonogen number of 50% non-control rate; K_0 is the initial number of tumor colonogens; and σ_k is the Gaussian width for the distribution of critical tumor cell numbers.

12.3.5 Dose limits for SBRT treatments

The journal "Seminars of Radiation Oncology" published an entire issue in April 2016 (Volume 26, Issue 2, 2016) dedicated to dose limits for SBRT treatments (large dose per fraction) with a series of articles corresponding to 10 particular organs at risk (OARs)

TABLE 12.1 Differences in BED in several SBRT fractionated schemes for different values of radio-biological parameters.

α/β(Gy)	δ(Gy^{-1})	d(Gy)	D(Gy)	BED$_{LQ}$	BED$_{LQL}$	BED$_{LQL}$/BED$_{LQ}$	%diff
10.0	0.05	26	26	93.6	71.8	76.7	4-23%
		12	36	79.2	71.7	90.5	
		10	30	60.0	55.6	92.6	
		6	48	76.8	74.1	96.5	
		7	28	47.6	45.5	95.6	
10.0	0.10	26	26	93.6	59.5	63.6	8-34%
		12	36	79.2	66.1	83.4	
		10	30	60.0	52.1	86.8	
		6	48	76.8	71.8	93.5	
		7	28	47.6	43.7	91.9	
10.0	0.20	26	26	93.6	47.0	50.2	14-50%
		12	36	79.2	58.4	73.7	
		10	30	60.0	47.0	78.4	
		6	48	76.8	68.0	88.6	
		7.0	28.0	47.6	40.9	86.0	

α/β(Gy)	δ(Gy^{-1})	d(Gy)	D(Gy)	BED$_{LQ}$	BED$_{LQL}$	BED$_{LQL}$/BED$_{LQ}$	%diff
3.0	0.05	26	26	251.3	178.7	71.1	7-29%
		12	36	180.0	155.0	86.1	
		10	30	130.0	115.2	88.6	
		6	48	144.0	135.1	93.8	
		7	28	93.3	86.3	92.5	
3.0	0.10	26	26	251.3	137.6	54.8	14-45%
		12	36	180.0	136.2	75.7	
		10	30	130.0	103.6	79.7	
		6	48	144.0	127.4	88.4	
		7	28	93.3	80.4	86.2	
3.0	0.20	26	26	251.3	96.1	38.2	24-62%
		12	36	180.0	110.5	61.4	
		10	30	130.0	86.8	66.7	
		6	48	144.0	114.8	79.7	
		7	28	93.3	71.1	76.2	

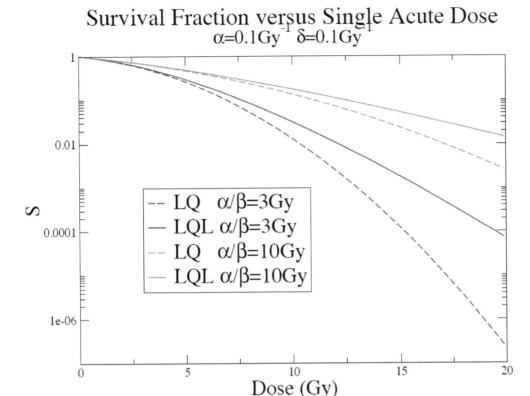

Figure 12.3 Comparison of surviving fraction versus dose for the LQ and LQL models for different α/β ratios. The difference between the models is more pronounced when the α/β is lower.

including thorough reviews of the available literature. In this section we present a summary of their findings.

Each article in the issue is dedicated to one organ at risk (OAR). For each OAR, the authors analyzed one or two of the best available data sets reporting complications outcome for SBRT patients and used it to derive a probit NTCP model derived using the maximum likelihood estimation of the parameters. Based on these models and a systematic review of the literature for each OAR, the authors presented a low risk dose limit and high-risk dose limit. The low and high-risk levels varied depending on the particular structure and are based on the systematic review of the literature. For example, the spinal cord low risk limit is 1% and high risk limit is 3% while the chest wall low risk limit is 10% and the high risk limit is 50%. The NTCP risk estimate was calculated using the derived model.

The results are presented in terms of DVH risk maps. For each D_x, where x is a specific volume, a low risk and high-risk curve is plotted with total dose in the y-axis versus number of fractions in the x-axis. The DVH risk maps are typically developed based on the best available data set and points from additional data sets from the literature that are included for visualization. Tables with the low and high-risk limits are also presented for each structure. The reported studies (one per critical structure) investigate the following organs: spinal cord, chest wall, aorta and major vessels, esophagus, duodenum for pancreatic cancer, cochlea, main bronchi in central lung tumors, toxicities associated with recurrent head and neck cancer treated with SBRT, small bowel and visual pathway. Detail of the limits for each structure and specific studies can be found in references [518, 519, 520, 521, 61, 63, 522, 523, 34, 524, 525, 526, 527, 528, 529, 530].

12.4 BIOLOGICAL MODELS IN TREATMENT PLANNING

Treatment planning tools that use biological models for plan optimization and/or evaluation are being introduced for clinical use. A variety of dose response models and quantities along with a series of organ-specific model parameters are included in these tools. In general, biological models can be broadly characterized as either mechanistic or phenomenological/empirical [152]. The mechanistic models, attempting to account for underlying biological processes for the tumor and normal tissue, are often mathematically complicated and clinically desirable to predict possible endpoints. The empirical biological models, however, have advantages of mathematical simplicity but are not designed to account for detailed biological phenomena, their applications are limited to the situations when the model and model parameters were derived and validated. Since the biological processes in micro- and macro- tumor environment are so profound, assumptions are often made to simplify both types of models for clinical purposes.

Currently, empirical models are mostly used in commercially available treatment planning systems due to their easy adoption and the fact that fewer model parameters are involved. For example, the EUD is the most commonly empirical model that is implemented in several major treatment plan systems, such as Philips and Eclipse.

12.4.1 Plan evaluation

Traditional plan evaluation utilizes physical dosimetric metrics such as DVH, uniformity, conformity and heterogeneity indices etc. to rank the quality of treatment plans. It can be challenging and subjective since the dose-volume based plan evaluation may or may not be correlated to clinical treatment response, therefore, plan evaluation is generally based on an individual clinician's clinical experience and/or literature to define the dose constraints for the IMRT optimization.

Biological model based plan evaluation, however, enables direct comparison of the biological efficacy, such as EUD/TCP/NTCP, by incorporating plausible dose-response models such as TCP/NTCP models. Rather than dose-volume evaluation, the subjective variability of the clinician's knowledge is hopefully less involved.

Although absolute values of predicted outcome probabilities such as TCP/NTCP may not yet accurately predict actual clinical outcome, biological models provide potentially useful dose-response information to steer the optimization process in the desired direction and guide clinical decision making when rival treatment plans are compared, particularly in cases where dosimetric advantages of one plan over another are not clear-cut.

12.4.2   Plan optimization

Plan optimization is the process of finding an optimal plan following the desired objectives through an inverse process. The quality of a clinically acceptable plan depends on many factors including patient anatomy, complexity of the plan, planner's experience, institution specific planning guidelines and TPS used. The traditional planning process requires selecting dose-volume constraints and various optimization parameters to reach the objectives through a lengthy trial-and-error process. This process, however, may be simplified during biological model based planning.

Biological model based planning is potentially more versatile and associated with treatment outcome than those plans created using traditional dose-volume based optimization. It has been demonstrated that the use of the biological models can result in the improved OAR sparing [152, 497, 498]. In Semenenko et al. [498], biologically based plans resulted in smaller EUD values for 26 out of 33 OARs by an average of 5.6 Gy (range 0.24 to 15 Gy) for 6 cancer patients. Plans with equivalent target coverage obtained using the biologically based TPS demonstrate improved dose distributions for the majority of normal structures. In Qi et al. [497], a five-way comparison employing five sets of different optimization algorithms (including both physical and biological optimizations) from three different TPS systems were made. The biological model based optimizations consistently resulted in improved OAR sparing without sacrificing the target coverage. Employing an overall figure-of-merit index by combining all targets and OARs in question, the biological model based optimization often yielded superior plans.

Because the biological models are designed to capture the dose response, biological based optimization might directly reduce the mean dose for a parallel architecture organ or the maximum dose for a serial architecture organ using a different parameter a in the EUD calculation, using fewer optimization parameters as compared to traditional optimization to achieve the most desirable plan quality. Therefore, the biological model based treatment planning allows us to obtain the desirable plan more efficiently as compared to the physically based planning. Unfortunately, as pointed out in a previous study [152], there is no guarantee that a biologically related model does indeed estimate the consequence of dose distributions if they deviate greatly from the baseline dataset that led to the model parameters. However, for the purpose of dose optimization, the use of the biological models can guide the optimization towards favorable dose distributions.

12.4.3   Dose summation using biological models

It is quite often that different radiation modalities such as initial external beam radiation therapy (EBRT) followed by dose escalation using high-dose-rate brachytherapy (HDR) or low-dose-rate brachytherapy (LDR) and/or different fractionation schemes are administered for a single patient for the initial treatment and dose escalation afterwards. Given different modalities and/or fraction sizes, the physical dose is not a simple additive quantity; the physical dose needs to be converted into fractionation schemes with same fraction size (such as NTD) or biological effective dose (such as BED) to estimate the total biological efficacy.

Dose summation based on biological models becomes absolutely essential for the following situations: 1) to capture possible dose response combining initial treatment and subsequent dose escalation using different modalities or fractionation schemes; 2) to design plausible constraints for the OARs for tumor recurrence (or adaptive therapy planning) based on previous treatment fields; and 3) to compensate for the change of biological effective doses due to unplanned breaks during treatment course or prolonged treatment duration, which simply cannot be accounted for using the dose-volume criteria.

12.4.4 Selection of outcome models and model parameters

Careful selection of biological models and model parameters are essential for biological model based plan evaluation/optimization and plan summation. The implementation of biological based optimization in clinical practice is a process of 1) choice of effective cost functions; 2) choice of correct types of cost functions; 3) choice of appropriate volume effect parameters; and 4) a clear idea of what features make a dose distribution acceptable or unacceptable in your clinic [152].

Certain biological models are derived for a range of fraction sizes. The applicability to the fraction sizes beyond the specified range needs to be evaluated before it is used to predict outcomes of different fractionation schemes.

In addition to fraction sizes, other patient or treatment variables may also influence the clinical outcomes, which were not accounted for in biological models. When using published parameter estimates for plan evaluation/optimization, it has to be carefully verified that they apply to the appropriate endpoints, organ volume definitions and fractionation schemes. Caution should be exercised if clinical and demographic characteristics of the patient population under consideration differ substantially from those in the original patient cohort used to derive published parameter estimates.

The models and model parameters used have been applied retrospectively so they would need to be calibrated against your clinical data to make sure they agree with treatment practice you know to be safe and effective.

12.5 COMMERCIALLY AVAILABLE TREATMENT PLANNING SYSTEMS (TPS) EMPLOYING OUTCOME MODELS

The most commonly used commercial TPS that employ biological models for treatment planning optimization are 1) Monaco V1.0 (Elekta Inc., Maryland Heights, MO), 2) Pinnacle V8.0h (Philips Medical Systems, Andover, MA), 3) Eclipse V10.0 (Varian Medical Systems, Palo Alto, CA) and 4) RaySearch V4.0 (RaySearch Laboratories, Stockholm, Sweden). Each of these systems utilizes different biological models and/or different implementation. Initial experiences using the Monaco, Eclipse and Pinnacle TPSs have been reported in [152, 497, 498]. In addition, as a widely adopted imaging review and plan evaluation platform in radiation oncology departments, the biological model based dose evaluation Suite in MIM (MIM Software Inc., Cleveland, OH) is also briefly discussed. Readers are referred to vendor-provided manuals, white paper or training materials for more detailed descriptions about these systems. Here we discuss a summary of system-specific issues.

12.5.1 Elekta Monaco system (Maryland Heights, MO)

The Monaco system can handle both 3D and IMRT planning. It uses both physical and biological cost functions for inverse treatment planning. There are three biological cost functions, 1) Poisson statistics cell kill model, 2) serial complication model and 3) parallel complication model included to handle dose prescription for targets and OARs exhibiting serial and parallel behavior [531, 532]. The Poisson cell kill model is used to create optimization objectives for target volumes and serial and parallel models are used to create constraints for OARs with different FSU architecture. For each of the three functions, a 3D dose distribution in a structure is reduced to a single index that reflects a putative biological response of the structure to radiation. This index is referred to as isoeffect. For the Poisson cell kill model and serial complication model, the isoeffect is expressed in units of dose. For the parallel complication model, the isoeffect is a percentage of the organ that is damaged. Dose or percentage levels specified by the user as optimization goals are referred to as isoconstraints. Following each iteration, isoeffects are recomputed and compared with isoconstraints to determine whether user-specified criteria have been met.

The physical (DV-based) cost functions included in Monaco are 1) quadratic overdose penalty, 2) quadratic underdosage penalty, 3) overdose DVH constraint, 4) underdose DVH constraint and 5) maximum dose constraint. The Poisson cell kill model is a mandatory cost function for targets. In the case of multiple target volumes, the Poison cell kill model must be specified for at least one target. Because the Poisson cell kill model does not include a mechanism to control hot spots, a physical cost function, either the quadratic overdose penalty or maximum dose constraint, should be added to create optimization goals for target volumes. Although the software does not require secondary constraints for target volumes, the optimization algorithm encounters convergence problems resulting in target dose distributions characterized by clinically unacceptable heterogeneity if such constraints are not used.

Monaco is based on a concept of constrained optimization. That is, the serial and parallel cost functions and all physical cost functions are treated as hard constraints. All optimization goals specified using these cost functions must be met by the TPS. The Poisson cell kill model is only an objective. As a result, there are no weights to specify, i.e., effectively the Poisson cell kill model is assigned a very small weight and all other cost functions are assigned very large weights. Because target coverage is only an objective, achieving this objective may often be compromised by constraints on the exposure of nearby OARs or constraints on hot spots in target volumes. A Sensitivity Analysis tool [533] is provided to help the planner to identify the limiting constraints. Desired target coverage could then be obtained by relaxing (increasing) isoconstraint values for the most restrictive optimization criteria.

In addition to the biological constraints for OARs, Monaco allows specification of secondary optimization objectives with these functions. This is referred to as the multi-criterial option. This option could be used to attempt to further reduce OAR doses when adequate target coverage had already been achieved or in special cases when additional OAR sparing is more important than adequate target coverage such as re-treatments for recurrent tumors.

The plans designed in Monaco using biological cost functions were compared with plans created with a conventional TPS using DV-based optimization criteria in a reported study [498]. The Monaco plans were generally characterized by better OAR sparing while maintaining equivalent target coverage. However, less uniform target dose distributions were obtained in each case.

The sensitivity of Monaco model parameters on optimized dose distributions is studied and demonstrated in previous publications [152, 498].

TABLE 12.2   Biological models used for treatment plan optimization in Pinnacle.

Structure Type	Name	Parameters	Objectives/ Constraints	Comments
Target	Min EUD	Volume parameter ($a < 1$)	EUD (Gy or cGy)	Penalizes for too low EUD
Target	Target EUD	Volume parameter ($a < 1$)	EUD (Gy or cGy)	Penalizes for any deviation from the desired EUD
OAR	Max EUD	Volume parameter ($a \geq 1$)	EUD (Gy or cGy)	Penalizes for too high EUD; can be used with both serial and parallel structures

12.5.2   Philips Pinnacle system (Andover, MA)

Pinnacle[3] system (version V8.0h) offers both physical optimization as well as biological optimization features in its P[3]IMRT inverse treatment-planning module. As opposed to Monaco, Pinnacle is not a designated biologically based optimization system, but rather uses biological cost function to enhance the traditional, dose-volume optimization approach [152]. In addition to the physical dose cost functions such as minimum dose, maximum dose, minimum DVH, maximum DVH, uniform dose, and uniformity. The biological objective functions have been developed by RaySearch Laboratories AB (Stockholm, Sweden) [534] including EUD, minimum EUD, and maximum EUD for target(s), and maximum EUD for OARs. Table 12.2 lists biological models used for treatment plan optimization in Pinnacle. The biological based optimization based on the gEUD formula allows for considering tissue-specific properties into the planning process that cannot be done with dose-volume based optimization. Since the gEUD model is parameterized by the single biological parameter a, which is inversely rated to the Lyman model parameter $a = 1/n$, the chosen parameter a should reflect the intended biological properties for the given tumor or organ. For example, large negative values of parameter a [497] are usually taken for tumors to penalize the effect of cold spots for malignant tumor targets, while parameter $a \geq 1$ is generally adopted to penalize too high EUD for both serial and parallel OARs.

Pinnacle[3] inverse planning optimization employs a traditional unconstrained optimization approach where cost functions of target and OAR contribute to the overall cost function in proportion to user-specified weights. Figure 12.4 [497] illustrates a 5-way plan comparison for a representative head and neck (H&N) cancer case using biological and conventional dose-volume based optimization including 1) Elekta XiO physical cost functions, 2) Elekta Monaco biological cost functions, 3) Pinnacle biological and 4) Pinnacle physical based optimization and 5) Tomotherapy dose-volume based optimization criteria. The Pinnacle biological plans were generally characterized by better OAR sparing while maintaining equivalent target coverage.

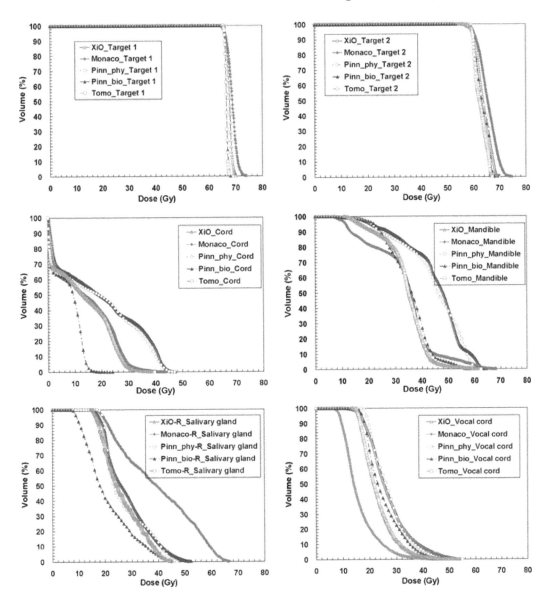

Figure 12.4 Five-way plan comparison for a representative H&N case using biological and conventional dose-volume based optimization, including (1) Elekta XiO physical cost functions, (2) Monaco biological cost function, (3) Pinnacle biological and (4) Pinnacle dose-volume based optimization and (5)Tomotherapy dose-volume based optimization criteria.

12.5.2.1 Sensitivity of model parameters

The usefulness of the biological cost function/optimization depends on the selection of biological models and/or model parameters. The selection of biological models and model parameters in biological optimization supposedly reflect the intended biological properties.

Target coverage and normal tissue sparing can be affected by the selection of parameter for the organ in question. The impact of parameter dependence and optimized dose distributions is demonstrated in Figure 12.5 [152, 497]. When a different parameter a was chosen for a single organ (such as spinal cord) in Pinnacle, the resulting max cord dose was reduced dramatically. When a plan was generated with $a = 5.0$ to a plan with $a = 20$ for the same head and neck case discussed above, the mean dose for a parallel organ, *i.e.*, right parotid gland changed by 24.7% when parameter a varied from 5 to 1. It was also observed that target coverage and normal tissue sparing for other organs can be affected by the selection of parameter a for the organ in question. Caution should be exercised to select appropriate tissue specific parameters in biological based optimization and evaluation.

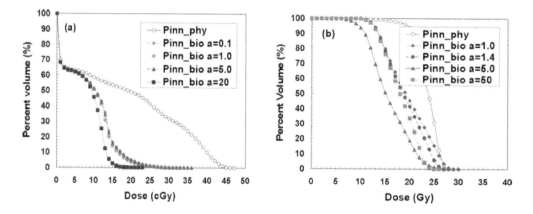

Figure 12.5 Sensitivity of dose-volume histogram (a) for the cord; and (b) the right parotid. The plan was obtained with different parameter a using Philips Pinnacle TPS for a head-and-neck case.

12.5.3   Varian Eclipse system (Palo Alto, CA)

Eclipse version 10.0 and later versions provide the biological optimization and biological evaluation application as a "plug-in" module developed by RaySearch Laboratories AB (Stockholm, Sweden). The biological response to radiation is assessed using the concept of TCP and NTCP that are based on the Poisson-LQ model together with parameter values that have been 'calibrated' against clinical data [535]. The so-called "volume dependence" in normal tissues is handled by the relative seriality (RS) model through the parameter s. Other models, such as the Lyman (LKB) NTCP model, the repairable conditionally repairable (RCR) model and the TCP Marsden model are also implemented in Varian Eclipse [535, 534, 536, 537]. Table 12.3 summarizes the biological objectives functions and model parameters in Eclipse.

12.5.3.1   Objective functions in plan optimization

Biological models are applied to create biologically optimized treatment plans. In Eclipse, the following objective functions are supported:

TABLE 12.3   The radiobiological objective functions in Varian Eclipse.

Biological objectives	Parameters
TCP Poisson-LQ	D_{50} *(cGy)*: The dose where the probability of response is 50%. γ: Maximum relative slope of the response curve. α/β *(Gy)*: Fractionation sensitivity. T_{pot} *(Days)*: Potential cell doubling time. T_k *(Days)*: Kick-off time when accelerated proliferation begins. *Weight*: Relative importance factor.
NTCP Poisson-LQ	D_{50} *(cGy)*: The dose where the probability of response is 50 %. γ : Maximum relative slope of the response curve. α/β *(Gy)*: Fractionation sensitivity. s: Seriality parameter. *Weight*: Relative importance factor.
NTCP Lyman	TD_{50} *(cGy)*: The dose where the probability of response is 50%. m: Defines the slope of the response curve. α/β *(Gy)*: Fractionation sensitivity Fraction size – The dose per fraction used when the parameters were estimated (usually 2 Gy). n: Volume dependence parameter *Weight*: Relative importance factor

TCP maximization: a composite (overall) function that contains one or more constituent TCP functions for different target ROIs. The probability for each ROI is accounted for using specific parameter values.

NTCP minimization: a composite (overall) function that contains one or more constituent NTCP functions for different OARs, which models the probability of complications in various organs.

PPLUS (P+) maximization: P+ is defined as the probability of complication-free tumor control. The function is proposed to express the treatment plan quality in a single value combining a composite TCP (for target ROIs) and a composite NTCP function (for OARs). By maximizing P+ the dose distribution that gives optimal probability of complication-free tumor control according to the used biological models is obtained. The trade-off between tumor control probability and normal tissue complication is controlled by the biological models (and the model parameters), instead of using penalty functions in the dose domain.

12.5.3.2   Plan evaluation

Biological models can be applied to existing treatments plans to assess the radiobiological effect. Biological evaluation in Varian Eclipse can be used to evaluate the biological responses of a single optimized plan, to compare the expected radiobiological effect of two treatment plans with different dose distribution and fractionation schemes, and/or to adjust the prescription dose to meet expected radiobiological responses based on model predictions. In addition, biological response due to involuntary fractionation gap (during treatment course) can be estimated using biological models and potential compensation dose could be accounted for by LQ model, which would complement the radiation oncologist's experience to achieve better outcome.

12.5.3.3 Sensitivity of model parameters

The Eclipse biological module offers default parameters for the selected tumor and OAR type and the biological parameters are also editable if users prefer different values of those parameters.

The sensitivity of dose distribution to model parameters, e.g., gEUD volume parameter, or to threshold levels for the gEUD specified in the optimizer was tested in Eclipse and the results are illustrated in the AAPM Task Group Report 166 [152]. Only one parameter was changed at a time.

12.5.4 RaySearch RayStation (Stockholm, Sweden)

RayStation has an independent module called "RayBiology" that consists of a database of TCP-NTCP models for different sites/structures. For tumor sites, the default model is the Poisson-LQ, which specifies the cancer site and stage, as well as the reference where the model and/or parameters come from. The parameters are those of a standard Poisson TCP function:

D50, γ50 and α/β ratio as well as re-population parameters T_{pot} and T_{start}.

For OARs the default models are the Poisson-LQ or the LKB, both of which have parameters from references in the literature. All the parameters can be edited and customized models can be added for both cancer sites and OARs.

12.5.4.1 Plan evaluation tools

In the plan evaluation workspace, for a given DVH, TCP and NTCP functions can be evaluated for the current dose levels and are plotted as a function of dose providing information of changes in TCP-NTCP if the dose were to be changed.

12.5.4.2 Plan optimization tools

RayStation offers two separate options for plan optimization, one using standard optimization techniques, direct machine parameter optimization, (DMPO) and the other, a more novel option, is multi-criteria optimization (MCO). In the DMPO workspace, all functions from the RayBiology database are available for optimization including P+, the combination of TCP-NTCP. EUD is also available as an optimization function. For MCO, the only biological function available is EUD, which is the recommended function for OARs in all MCO optimizations. However, TCP-NTCP functions are not eligible for objective or constraints in the current versions of MCO.

12.5.5 MIM (MIM Software Inc., Cleveland, OH)

MIM is becoming a practical imaging review and plan evaluation platform widely used in radiation oncology departments. MIM includes tools for multi-modality image fusion, automatic deformable contouring, quantitative functional analysis, remote DICOM review, PACS, and a biological based dose evaluation suite.

The biological based dose evaluation suite [538] offered by MIM is used to estimate tumor and normal tissue response across different treatment modalities and fractionation schemes. Specifically, MIM provides a tool based on the concept of BED to 1) convert different fractionation regimens to standard 2 Gy equivalent dose fractions; and 2) perform dose summation for multi-modality radiotherapy and adaptive radiation therapy using the BED-scaling and TCP modeling tools using common dose response models based on the LQ formalism.

12.5.5.1 Plan summation

When different treatment modalities and fractionation schemes are considered, MIM first converts the physical dose into a biological dose and then performs dose summation. While the BED model discussed in Sec 12.2.3 is used to describe biological effect of conventional fractionation schemes, the linear-quadratic-linear (LQ-L) model (note that this model is not the same as the one described in Section 12.3.3) is used to evaluate biological effectiveness of SBRT treatment in MIM [539]. MIM follows the bipartite method proposed in [332], in which below a transition dose per fraction, d_t, LQ behavior is observed and above which a final linear portion to the survival curve is observable.

$$
\begin{cases}
BED = nd\left(1 + \frac{d}{\alpha/\beta}\right), & (d < dt) \\
BED = nd_t\left(1 + \frac{d}{\alpha/\beta}\right) + n\frac{\gamma}{\alpha}(d - d_t), & (d > dt)
\end{cases}
\tag{12.38}
$$

where, γ is the linear coefficient for the final linear portion of the survival curve; d_t is the transition dose at which LQ-L behavior begins. The ratio $\frac{\gamma}{\alpha}$ can be estimated by using the tangent of the cell survival curve at the point d_t:

$$
\frac{\gamma}{\alpha} = 1 + \left(\frac{2d_t}{\alpha/\beta}\right)
\tag{12.39}
$$

When another treatment modality such as low-dose-rate (LDR) brachytherapy is administrated in addition to EBRT, the following BED formula is used specifically for permanent-seed brachytherapy [540] to calculate BED and then dose summation:

$$
BED_{LDR} = D\left(1 + 2(d_0{\cdot}\lambda)(\beta/\alpha) \cdot \frac{\kappa}{\mu - \lambda}\right) - \frac{\ln(2) \cdot T}{\alpha T_{pot}}
\tag{12.40}
$$

Where

$$
\kappa = \frac{1}{1 - \epsilon}\left\{\left(\frac{1 - \epsilon^2}{2\lambda}\right) - \left(\frac{1 - \epsilon \cdot e^{-\mu T_{eff}}}{\mu + \lambda}\right)\right\}
\tag{12.41}
$$

$$
T_{eff} = \frac{1}{\lambda}\ln\left((1.44{\cdot}d_0) \cdot (\alpha \times T_{pot})\right)
\tag{12.42}
$$

where $\lambda = 0.693t_{1/2}$, $t_{1/2}$ is the half-life of the radioisotope; $\mu \ln(2)/t_{1/2}$, $t_{1/2}$ is the cell repair halftime constant; α and β are the linear and quadratic coefficients; T_{pot} is the potential tumor doubling time; T_{eff} is the effective treatment time $\epsilon = e^{-\lambda T_{eff}}$.

12.5.5.2 Plan evaluation

The Poisson TCP model in conjunction with the LQ formalism can be used to calculate target TCP predictions. Assuming a uniform clonogenic density per cm^3 (ρ) throughout the volume of the tumor V, the formula for TCP is given by:

$$
TCP = e^{-k{\cdot}S} = e^{-\rho \cdot V \cdot e^{-(\alpha \cdot BED)}}
\tag{12.43}
$$

Similarly as the other systems, MIM provides a library with default parameter values for the supplied models that were derived from the literatures [332, 540, 541]. New organ specific user-defined parameters are allowed to add in MIM to meet specific needs.

12.6 CONCLUSIONS

Inverse treatment planning has extended its utilization of dose-volume criteria alone to incorporate information of radiobiological effect of radiation dose. A variety of biological models and organ-specific model parameters are developed to assess radiobiological effects. Such dose-response models are being introduced for radiotherapy treatment plan optimization and/or evaluation and are currently available in major commercially available treatment planning systems for clinical use. The integration of biological models in treatment planning optimization and evaluation demonstrates improved plan quality as compared to the conventional physical based optimization and evaluation. Caution should be exercised in choosing appropriate models and/or model parameters when using the biologically based treatment planning system.

CHAPTER 13

A utility based approach to individualized and adaptive radiation therapy

Matthew J Schipper, PhD

CONTENTS

ABSTRACT

Radiation Therapy treatment requires selection of a dose that balances the increased efficacy (tumor control) with the increased toxicity typically associated with higher dose. Historically, a single dose has been selected for a population of patients. However, this dose is not universally optimal as evidenced by patients who have tumor progression without toxicity (potentially underdosed) or toxicity with no tumor progression (potentially overdosed). Biomarkers offer the possibility of individualized dose selection. We propose a utility based approach to identifying an optimal dose for an individual patient based on biomarkers and clinical factors. First, the numeric utility or "desirability" of each possible bivariate (toxicity and efficacy) outcome is elicited. The optimal dose is defined as the dose which maximizes the expected utility for an individual patient. For binary outcomes, the expected utility can be expressed as the probability of efficacy minus a weighted sum of toxicity probabilities which can each be estimated from separate statistical models as a function of dose, clinical factors and biomarkers. This approach is shown to provide optimal efficacy for any fixed rate of toxicity. Intuitively, this is because it "spends" its toxicity risk (from higher dose) in patients who derive the largest benefit in efficacy. We illustrate this approach using an example from radiation therapy treatment of lung cancer and also present results for a virtual "in-silico" trial comparing standard to utility based treatment.

Keywords: Utilities, biomarkers, plan optimization, radiation therapy.

13.1 INTRODUCTION

Identification of biological markers (biomarkers) for patient outcomes following treatment for cancer (*e.g.*, radiation therapy (RT)) is an area of active research. Despite the widely recognized difficulties in validating biomarker findings, it is likely that in the not too distant future, there will be validated biomarker models for efficacy and toxicity outcomes in a range of cancer treatments. Beyond the obvious benefit of improved prognostic ability, these markers have great potential to inform treatment management and planning decisions. For example, which lung cancer patients might benefit from a higher tumor radiation dose? In this chapter, we describe a novel utility-based framework to treatment planning that naturally accommodates biomarkers for multiple endpoints and can be used to individualize RT plans at baseline or to adapt RT plans based on mid-treatment biomarkers.

13.2 BACKGROUND

13.2.1 Treatment planning in radiation therapy

Many aspects of treating a cancer patient with Radiation Therapy require a balance between the efficacy and toxicity associated with the treatment. This includes what dose to give the tumor, what volume to treat and how to deliver that dose. Current RT plans are typically based on delivering a fixed tumor dose over the course of therapy while also meeting constraints on normal tissue doses each designed to limit a particular toxicity to an acceptable rate. These possibly competing goals are often prioritized. For example, a priority 1 objective in locally advanced lung cancer might be to keep the mean heart dose below 30 Gy while a priority 2 objective would be to deliver 60 Gy to the Planning Target Volume (PTV). While few clinicians would feel that the priority 1 objectives are infinitely more important than the priority 2 or 3 objectives, a strict application of such prioritized optimization requires meeting the priority 1 objectives regardless of the price paid in terms of priority 2 objectives.

It is difficult to compare two treatment plans if one is associated with both greater efficacy and greater toxicity. This is the case when analyzing outcomes data from patients previously treated with one of the two approaches or plans, and also when considering two potential treatment plans for an individual patient and looking at expected outcomes under each plan. We focus on the latter but the discussion is equally relevant to the former setting. In either case, an assessment of which is preferable, requires making some tradeoff to determine whether the increased efficacy of tumor control is worth the added toxicity. One approach to making this tradeoff explicit is to combine the multiple outcomes into a single outcome encompassing both the good and bad things that could happen to a patient. Treatment planning could then proceed by selecting the plan that maximized the expected value of this outcome (assuming higher values were preferable). An early example of this approach is the concept of uncomplicated control, defined as the absence of local progression or toxicity (meeting certain criteria) within a defined time period following treatment [542]. Other examples include Quality Adjusted Life Years (QALYs) or Quality

Time WIthout Symptoms or Toxicity (QTWIST) [543] . Assuming higher values are better, treatment planning could seek to identify the plan (or dose) which maximizes the expected value of QALYs of QTWIST. The expected values here of course require data on similar patients previously treated with similar treatment plans. One drawback to this approach is that many clinical factors and biomarkers are likely specific to a particular clinical outcome (*e.g.*, pneumonitis in lung cancer patients) and not to a composite endpoint like QALYs. Additionally, there may be limited or no overlap between the available biomarkers for different outcomes. In this case, it may make more sense to model each outcome separately and to make the tradeoff after, and separate from, statistical outcome modeling.

13.2.2 Biomarkers in RT

In 1998, the National Institutes of Health (NIH) Biomarkers Definitions Working Group defined a biomarker as "a characteristic that is objectively measured and evaluated as an indicator of normal biological processes, pathogenic processes, or pharmacologic responses to a therapeutic intervention" [544]. Many hundreds and perhaps thousands of biomarkers have been proposed for various clinical endpoints following RT treatment (*e.g.*, [545, 546, 547]). There are several overlapping methods for classifying biomarkers. One useful distinction is between prognostic and predictive biomarkers [202]. A *prognostic marker* informs about a likely cancer outcome (*e.g.*, disease recurrence, disease progression, death) independent of treatment received while a *predictive marker* is generally understood to involve an interaction between treatment (typically standard vs experimental or new).

Whether there is an interaction depends on the scale of measurement. For example, in a logistic regression model, there may be an interaction on the probability scale but no interaction on the $logit(log(p/(1-p))$ scale (used in logistic regression modeling). Letting Y denote the binary outcome, B the biomarker value and T the treatment group, the treatment effect on the probability scale is given by $P(Y = 1|T = t_1) - P(Y = 1|T = t_2)$. An interaction on the probability scale exists if $[P(Y = 1|T = t_1, B = b_1) - P(Y = 1|T = t_2, B = b_1)] \neq [P(Y = 1|T = t_1, B = b_2) - P(Y = 1|T = t_2, B = b_2)]$, which is not equivalent to specifying interactions in terms of treatment effects on the logit scale; $logit(P(Y = 1) = log(P(Y = 1)/(1 - P(Y = 1)))$.

Reports on new biomarkers typically report odds ratios or hazard ratios and measures of discriminatory ability such as area under the receiver operating characteristics curve (AUC) for binary outcomes or concordance index for possibly censored time-to-event outcomes, with associated p-values from tests of no discriminatory ability or no increase in AUC relative to a dose only model. These metrics are useful and relevant if the new biomarker is to be used in a purely prognostic fashion, (*e.g.*, counsel patients regarding risk of toxicity or the likelihood of recurrence). Often however, the goal is to use the new biomarker to select a dose in such a way that overall outcomes are improved. More precisely, there is often an expectation that biomarker based dose selection will result in either 1) increased efficacy at the same toxicity or 2) decreased toxicity at the same efficacy. However, neither of these may occur, even when using a biomarker that substantially increases AUC over the AUC for dose alone. As an illustration, consider the hypothetical scenario depicted in Figure 13.1.

Suppose a biomarker is discovered which enables us to identify two groups of patients, one with a uniformly higher and the other with a lower risk of toxicity. So rather than predicting toxicity using only mean lung dose (*MLD*, purple line) we can now more accu-

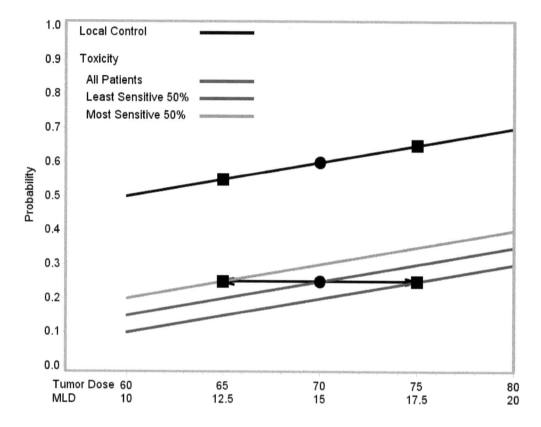

Figure 13.1 Example of a prognostic biomarker. The purple and black lines give the overall probability of toxicity and tumor control for a population of patients while the biomarker identifies more sensitive (red line) and less sensitive (blue line) patients in terms of toxicity.

rately predict toxicity using the new biomarker (blue and red lines). For simplicity of visual presentation, assume that there are 100 patients to be treated, all with the same tumor size and location (so that the ratio of mean lung dose to tumor dose is the same for all patients). Using the dose only toxicity model, we could treat all patients at a tumor dose of 70 Gy (equivalent to $MLD= 15$ Gy) and the toxicity and efficacy rates would be 25% and 60%. With the new biomarker, we could instead treat each patient at their 'personalized' dose corresponding to a toxicity probability of 25% (65 or 75 Gy for more or less sensitive patients). With this approach the overall toxicity rate is still 25%, but he overall efficacy rate is now the average of 55% (for patients treated at 65 Gy) and 65% (for patients treated at 75 Gy), equal to 60%, which is the same as that achieved by ignoring the biomarker. Schipper *et al.* [548] demonstrate that this result also holds in the more realistic setting where the ratio of mean lung dose to tumor dose varies across patients. Thus, the biomarker based dosing approach results in the same marginal probability of efficacy as strategy 1 which is based only on dose. This result depends on the linearity of the dose-outcome relationships and the additive effect of the marker. Another crucial assumption is that there is no interaction between dose and marker. We note that this result does not hold in the presence of a dose×biomarker interaction or when the true relation between the probability of toxicity and dose is not linear.

13.3 UTILITY APPROACH TO PLAN OPTIMIZATION

13.3.1 In phase I trials

In some simplified cases, there is only a single binary outcome following treatment. In these cases, the probability of this outcome is sufficient to use in comparing alternative treatments or dose levels. This setting however rarely holds in oncology where outcomes for patients treated with RT are multi-dimensional. Typically, phase I trials are conducted to identify a maximally tolerated dose based only on toxicity because of the assumption that both toxicity and efficacy would increase monotonically with dose. More recently several authors have proposed phase I designs in which dose assignments are made based on efficacy and toxicity considerations. Several of these designs have been based on what is termed a Utility approach [549, 550]. The general framework is to first elicit, typically from the clinician(s), the numerical utilities (desirability) of each possibly bivariate (efficacy and toxicity) outcome. In the simplest case where efficacy and toxicity are both binary, then the utility can be represented in a 2 x 2 matrix where, without loss of generality, the worst and best possible outcomes are assigned value of 0 and 100, respectively, and a and b are the numerical utility values associated with the intermediate outcomes.

TABLE 13.1 Example utility matrix for binary efficacy and toxicity.

	No Toxicity	Toxicity
Efficacy	100	b
No Efficacy	a	0

The goal of the trial is to identify the dose that maximizes the expected utility. To conduct the trial the first patient is assigned to the first dose level. Subsequent patients are iteratively assigned to the dose currently estimated to have the highest expected utility, subject to constraints on the probability of toxicity and other standard safety constraints such as no dose-skipping. The expected utility for a given dose level is calculated by summing across the cells of the utility matrix, the product of the utility value and the probability of observing that bivariate outcome conditional on dose. Generally, these designs will also include a definition of acceptable or admissible dose levels to ensure that the selected dose level meets some minimal efficacy threshold and is not overly toxic. For additional work in this area, see the recent book by [551].

13.3.2 In RT treatment planning

In Table 13.1, the decreased utility associated with moving from no toxicity to toxicity, is equal to (100-b) in the presence of efficacy and equal to (a) in the absence of efficacy. In some settings it will be natural to constrain these two values to be equal. Let θ denote this shared decline in utility associated with toxicity. Then if we rescale the utilities to range from $-\theta$ to 1, Table 13.2 can be written as:

TABLE 13.2   Example utility matrix for binary efficacy and toxicity with 1 degree of freedom.

	No Toxicity	Toxicity
Efficacy	1	$1 - \theta$
No Efficacy	0	$-\theta$

If we further assume that efficacy (E) and toxicity (T) outcomes are independent conditional on covariates including dose and biomarkers, a little algebra can show that the expected utility is given by:

$$E(U) = P(E = 1) - \theta \times P(T = 1) \qquad (13.1)$$

The probabilities in Equation 13.1 can be estimated from statistical models for E and T. For example, we might have a logistic regression model which estimates the probability of a particular binary toxicity as a function of a normal tissue dose term and, if available, clinical factors and biomarkers. Similarly, we might for example use a Cox regression model to estimate the probability of being alive and progression free 2 years after treatment as a function of tumor dose and, if available, with relevant clinical factors and biomarkers.

Notation: Let D_i denote tumor dose, d_i the normal tissue dose, and T_i and E_i the binary toxicity and efficacy outcomes with $\pi_i = P(T_i = 1)$ and $\pi_i = P(E_i = 1)$, for patient i. We assume availability of models $\pi_i = f^{-1}(d_i, M_i, \beta)$ and $\pi_i = g^{-1}(D_i, G_i, \alpha)$, where M and G denote scalars or vectors of marker values, β and α are parameter vectors and $f()$ and $g()$ are known link functions. This is a key point of distinction from the phase I trial setting where the reason for conducting a phase I dose finding trial is that no data or at most very limited data is available to estimate the parameters in these models. In the RT setting, however, there are often existing datasets from patients treated with a range of doses, particularly for normal tissues but also for tumors. Then, we write the expected utility in equation 13.1 as:

$$E[U(d_i, D_i, M_i, G_i)] = P(E_i = 1|D_i, G_i) - \theta \times P(T_i = 1|d_i, M_i) \qquad (13.2)$$

An isotoxic approach to treatment planning would ignore the efficacy model and simply solve $\pi_i = p^* = f^{-1}(d_i, M_i, \beta)$ for d_i, where p^* is a target probability of toxicity. In many instances, there are multiple possible toxicities that a patient may experience, differing qualitatively and in terms of severity. For example, in lung cancer, esophageal and heart toxicity are also of concern and are commonly considered in treatment planning. The expression in equation 13.2 is then easily generalized to:

$$E(U(d_{ij}, D_i, M_i, G_i)) = P(E_i = 1|D_i, G_i) - \sum_j \theta_j \times P(T_{ij} = 1|d_{ij}, M_i), \qquad (13.3)$$

where normal tissue or organ is indexed by j. We note that the last term in (3) is similar to several recently proposed metrics for phase I trials, including 'total toxicity burden' [552]. Treatment planning is accomplished by maximizing the expected Utility given in Equation 13.3. In addition to controlling the average rate of toxicity in the utility function approach, it may be desirable to also limit an individual patient's probability of toxicity. This can be accomplished by incorporating an additional individual level constraint in the maximization of $U(d_i)$ in Equation 13.3.

13.3.3 Choice of the tradeoff parameter

In practice, use of the utility approach requires one to choose a value for θ. We note that this feature is not unique to this approach since in every treatment plan, there exists a tradeoff, whether implicit or explicit. In equation 13.3, θ_j can be interpreted as the relative undesirability of toxicity j relative to lack of efficacy (*e.g.*, grade 3 lung toxicity relative to local progression within 2 years). This value could thus be elicited from clinicians or it could even be selected by individual patients based on their preferences. A second approach to selection of θ comes from the observation that, as θ increases, Equation 13.1 will be maximized by smaller d. Thus θ could also be viewed as a tuning parameter, with its value chosen to result in an acceptable marginal rate of toxicity for a population of patients, calculated as:

$$P(T = 1) = 1/n \sum_i P(T_i = 1|, \widehat{d}_i(\theta, r_i, M_i)), \tag{13.4}$$

where $\widehat{d}_i$ maximizes equation 13.1 and is thus a function of θ, r_i and M_i. We note that maximizing $\sum_i U_i(d_i, r_i, M_i)$ over d_i and choosing the value of θ such that $P(T) = 1) = p^*$ is analogous to a constrained optimization with a Lagrange multiplier approach. The utility function approach is more general than the isotoxic approach, however, because the isotoxic approach forces $P(T_i = 1) = p^*$ for all i, whereas the utility function method only forces the average of $P(T_i = 1)$ to equal p^*. In both cases, the overall proportion of patients experiencing toxicity will, on average, equal p^*. In the utility approach, patients are exposed to different levels of risk. There is a strong ethical argument for exposing patients to risk of toxicity in proportion to benefit or probability of efficacy (*e.g.*, tumor control).

Figure 13.2 provides an example schematic of plan optimization via the expected utility. From any plan, the relevant dose metrics for each endpoint (one efficacy and three toxicity in this case) are calculated. These dose metrics are used to calculate the probabilities of the various outcomes which form the basis of the expected utility function.

There is a range of potential implementations of the utility approach. In Figure 13.2, only the number and orientation of beams are assumed to be fixed. Then, the optimization of the expected utility is over the set of all possible beamlet intensities. A far simpler approach is to define a reference plan for a given patient and then optimize over all possible scalings of this reference plan. For example, in the case of one toxicity outcome, let d_i denote the normal tissue dose metric and r_i^{ref} denote the ratio D_i^{ref}/d_i^{ref} for the reference plan. Then, we could write Equation 13.2 as a function of only d_i by writing $D_i = r_i^{ref} \times d_i$. In this case the optimization has only a single degree of freedom, thus substantially reducing the computational burden. On the other hand, such an approach is unlikely to be optimal.

13.3.4 Virtual clinical trial

The most convincing evidence of the ability of the utility based approach to treatment planning would come from a clinical trial in which patients were randomized to standard or utility based treatment arms. Of course this evidence would be specific to the type of patients enrolled, including disease site, as well as the particular biomarkers used. Prior to actually conducting such a study it is possible to conduct an "in-silico" or "virtual" clinical trial using simulation methods. Two key ingredients are necessary to do this. The first is data on a population of patients similar to those in whom the real trial would be conducted. This would include the necessary imaging to plan hypothetical treatments as well as values

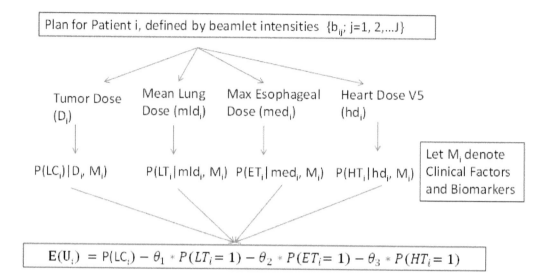

Figure 13.2 A schematic of utility based treatment planning in case of lung cancer.

for clinical factors and biomarkers included in the models. The second ingredient is fully specified approaches to standard and utility based treatment. The former might for example include a prioritized list of planning objectives. The latter would include the precise form of the expected utility function, including the specific biomarker based models and tradeoff parameters (θ in equation 13.3. Both would also include any hard constraints on normal tissue doses. Given these ingredients, the study would follow the following steps:

1. Plan each patient in the dataset using the standard and utility planning techniques (P_S and P_U).

2. Calculate expected outcomes for each patient under each potential plan.

3. Average over patients to estimate the overall expected outcomes if all patients had been treated with a particular approach (P_S and P_U).

An illustration of this approach is given in [548]. Based on a database of 82 locally advanced non-small cell lung cancer patients, several statistical models were fit to predict lung toxicity as a function of mean lung dose (d) and various cytokines measured at baseline or mid-treatment. These are given as rows in Table 13.3. Using baseline miRNA biomarkers, a model for local progression free survival (LPFS) at 2 years was also fit using a Cox proportional hazard model to account for patients who were censored during the first 2 years. Our goal was to use the biomarkers to identify patients who would (or would not) benefit from higher RT dose. We used a stepwise procedure in which dose and dose×biomarker interactions were included as potential covariates. For biomarkers, we used expression values of 84 miRNAs which are detectable in serum. In addition to dose, 23 dose×miRNA interactions were jointly significant at the 10 level. From the fitted Cox model we calculated the dose coefficient for patient i as $\alpha_0 + \sum_j \alpha_j \times X_{ij}$ where α_0 and α_j denote the parameter coefficients for dose and the interaction between dose and the jth miRNA term and X_{ij} denotes the value of the jth miRNA for the ith patient. Although this biomarker could be used in a continuous fashion, we chose to group patients into 2 groups for illustration purposes in the simulation study. Patients whose dose coefficient was in the lowest 1/3 had significantly worse LPFS and no dose effect compared to patients with higher dose coefficient so we used these 2 groups. LPFS at 2 years is a commonly utilized endpoint in RT lung trials and we use it here as our binary efficacy endpoint and estimate it from the fitted Cox model as:

$$\widehat{S}\left(2|D,G\right) = \widehat{S}_0\left(2\right)^{exp(\alpha_1 D \times I(G=1) + \alpha_2 D \times I(G=2))}, \tag{13.5}$$

where $\widehat{S}_0\left(2\right)$ is the Nelson-Aalen estimate of baseline survival at 2 years. All methods were calibrated so that they resulted in an overall rate of lung toxicity equal to 15%. Thus they can all be directly compared in terms of expected efficacy outcomes. Figure 13.3 and the rightmost 3 columns in Table 13.3 give the overall expected values of LPFS at 2 years under a variety of treatment planning approaches. For additional details see [548]. From table 13.3, it can be seen that using biomarkers for toxicity and efficacy via the utility approach has the potential to improve efficacy outcomes without increasing toxicity. In this particular example, expected LPFS increased from 0.40 to 0.52. The intuition behind this result is that the utility approach achieves this benefit by "spending" its toxicity wisely. Specifically, the utility approach preferentially exposes patients to increased risk of toxicity in proportion to their benefit quantified by improvement in expected LPFS.

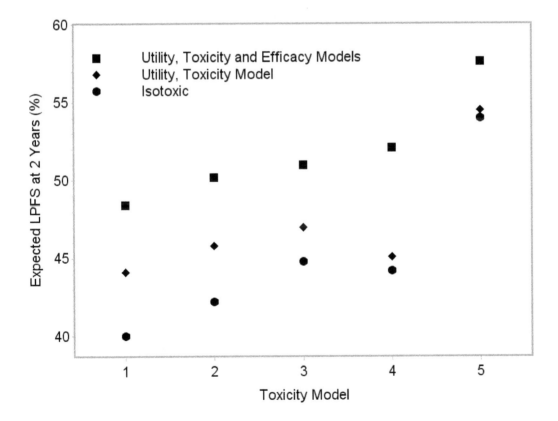

Figure 13.3 Toxicity and efficacy results when dose is selected according to the given model.

The other interesting result from Table 13.3, is that a model with lower AUC (*e.g.*, model 4) may be more useful for individualization of RT than a model with higher AUC (*e.g.*, model 2 or 3) if the former model includes biomarkers with a larger degree of dose effect modification (*i.e.*, stronger interaction with dose).

TABLE 13.3   Toxicity and efficacy results when dose is selected according to the given model. Toxicity rate is equal to 15% in every case. (* Hypothetical model).

Model	Toxicity Equation (on logit scale)	AUC	Local Progression Free Survival at 2 Years (%)		
			Isotoxic	Utility Without Efficacy Markers	Utility With Efficacy Markers
1	-5.16+0.2*d	0.72	40.0	44.1	48.4
2	-3.66+.22*d-.85*log(IL8)	0.78	42.2	45.8	50.2
3	-3.72+.19*d-.79*log(IL8)+.03*TGFβ*d	0.83	44.8	47.0	51.0
4	-5.34+0.17*d+0.07*d*I(TGFα 3)	0.76	44.2	45.1	52.1
5*	-5.6+0.12*d+0.16*d*I(TGFα 3)	NA	54.0	54.5	57.6

13.4   CONCLUSIONS

The utility approach to treatment planning has several appealing features. First, it makes the efficacy/toxicity tradeoff explicit and quantitative. This will be increasingly important as treatment planning moves towards incorporation of biomarkers. Second, it is a very flexible framework for treatment planning and can accommodate multiple toxicity endpoints of varying types and severities. It is also flexible with regard to timing of use. It could be used at baseline to individualize and again mid-treatment (*e.g.*, using change in cytokine values) to adapt a course of RT. Third, and most importantly, it is optimal in the sense of maximizing the measure of efficacy (e.g. local control) for any fixed toxicity level.

There is a number of research questions related to utility based treatment planning that merit further study. The first involves the computational and optimization issues surrounding full plan optimization using the utility function which is the sum of a number of functions, some of which may be non-convex. How to combine the utility based approach to treatment planning with mid-treatment image-based re-planning (*e.g.*, providing boost dose to PET avid region as in RTOG 1106 trial) is also an area requiring additional study. A third question is whether, conditional on treatment factors, a patient's efficacy and toxicity outcomes are correlated. In calculating the expected value of the utility matrix in X, we assumed conditional independence. Finally, as noted previously, utilities are inherently subjective so it may be possible to elicit these from individual patients. For example, some patients may prefer a more aggressive treatment and be willing to accept additional toxicity for the payoff of higher probability of local control. Other patients are inherently more conservative and would be more concerned about treatment toxicity.

CHAPTER 14

Outcome modeling in Particle therapy

J. Schuemann

A.L. McNamara

C. Grassberger

CONTENTS

ABSTRACT

The use of particle therapy for cancer treatment has become increasingly widespread and each year new particle (mostly proton) therapy facilities are built around the world. The interaction of particles with matter is fundamentally different from photons, resulting in distinct energy deposition patterns on the microscopic scale. Increasing the mass and charge of the irradiating particles results in a more concentrated distribution of energy deposition events in space. Accordingly, the extent of biological damage and response of the irradiated tissue may differ. For a given physical dose, particle irradiations are typically more effective than photon irradiation, and the concept of the relative biological effectiveness (RBE) was introduced to capture this effect. RBE is defined as the ratio of doses required to produce the same biological effect for reference (photon) and particle irradiations. In addition to the underlying difference in biological response, particle beams stop within the patient, producing highly conformal dose distributions. Most notably, particle beams have (nearly) no dose distal to the target. However, most of our clinical data is based on photon treatments. Relating experiences from photon treatments to particle therapy outcome modeling is thus not straight forward. In this chapter, we highlight the main differences of particle to photon therapy and how this impacts outcome modeling, we discuss important concepts for particle therapy and their potential applications.

Keywords: outcome models, particle therapy, relative biological effectiveness.

14.1 HOW ARE PARTICLES DIFFERENT FROM PHOTONS?

The physical, biological and clinical properties of ion beams for radiotherapy have been studied extensively, starting in the early 1950s at the Lawrence Berkeley Laboratory using helium ions and protons [553]. Due to their charged nature, the interaction of ions with matter is fundamentally different compared to photons. This was first described by Bethe and later complemented by Bloch [554, 555]. Contrary to photons, charged particles interact continuously with the surrounding orbital electrons and atomic nuclei of the medium through the Coulomb force, resulting in a continuous loss of energy over the entire particle path. This continuous loss of small amounts of energy results in the main clinical characteristic of charged particles: their finite range.

From a physics point of view, the microscopic energy deposition and the macroscopic finite range of charged particles are intimately connected, however, they are usually studied in very different contexts. The distribution of energy deposition on the micrometer or nanometer level is relevant for investigating cell survival via DNA damage or radiation induced damage in electronic components for space radiation research. Commonly track structure Monte Carlo simulations are used in these types of studies [556, 557, 558, 559].

The macroscopic differences between photons and particles are applicable for clinical contexts, such as dose-volume effect models for normal tissue or relative biological effectiveness (RBE) models. Figure 14.1 shows these two aspects of charged particles, the left panel shows the microscopic energy distribution (track structure) of photons compared to carbon ions of different energies. The right panel presents a comparison of photon and proton dose distributions within a liver patient.

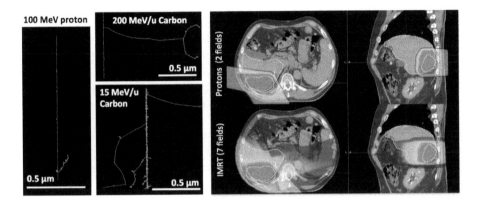

Figure 14.1 The energy deposition of a particle track seen on a microscopic (left) and macroscopic (right) level. The left panel demonstrates the microscopic energy distribution (track structure) of protons and carbon ions for different incident particle energies. The right compares the proton to photon dose distribution in a liver patient, demonstrating the finite range of protons. As a result, the proton integral dose to the patient is significantly reduced compared to the photon case.

Clinically, the main advantage for the utilization of particles is the superior dose distribution which leads to a reduced integral dose ($\sim 30\%$ lower than for photon treatments), as well as the sharp distal fall-off that enables sparing of critical structures close to the target. Protons and carbon ions have been investigated for a range of indications, from lung, gastrointestinal (GI) to prostate, though randomized clinical trials are exceedingly rare [560, 561]. Despite the absence of such Level I evidence, proton therapy is considered superior in pediatric central nervous system (CNS) malignancies, ocular melanomas and chordomas [562]. Many compacting factors contribute to this discrepancy between the superior dose distribution of particles and the short list of indications where there is clear clinical evidence for superiority. The pace of technological developments, larger uncertainties and the longer life cycle of particle therapy centers certainly all play a role [563], but there are indications where the superior dose distribution translates into little clinical benefit. For this reason outcome research plays a crucial role in particle therapy, in this chapter we highlight the main differences of particle to photon therapy and how this impacts outcome modeling:

1. The concept of Linear Energy Transfer and its limitations

2. Models for relative biological effectiveness

3. Role of Monte Carlo in particle therapy

4. Additional implications of particle therapy for outcome models

14.2 LINEAR ENERGY TRANSFER (LET)

The determination and explanation of the damage that various types of radiation inflict on biological systems is a long standing and recurrent problem in radiobiology [564, 565, 566]. However, all types of ionizing radiation share a common trait, their immediate physical actions are principally mediated by the formation of ionization and excitation events. Therefore a common approach when investigating the differences among charged particles compared to photons is to use the variations in the microscopic absorbed dose distributions. To this end the concept of the linear energy transfer (LET) can be useful. LET is a macroscopic average of the microscopic single-event lineal energy distribution, y, and was introduced in order to evaluate the energy deposition patterns of charged particles in microscopic regions.

All LET definitions are based on the stopping power S, which is often divided into electronic, nuclear collision and radiative terms. For ions in the energy range of medical applications, electronic excitation and ionization processes dominate, reducing the stopping power to

$$S = \frac{dE_{el}}{dx} \quad .$$
(14.1)

Here dE_{el} is the energy loss of the charged particle due to electronic collisions while traversing a distance dx. This matches the ICRU definition of unrestricted LET (L_∞) as *unrestricted linear electronic stopping power* [567]. ICRU also defines a restricted LET (L_Δ), which includes only the local collisions with energy transfers less than a threshold value Δ, usually denoting an energy in electron volts (eV) and not a range. The purpose of the restricted LET was to allow the exclusion of very energetic delta rays, which deposit most of their energy far from the primary particle track and are thus not contributing to the local biologic effect of that track. In particle therapy, LET is usually denoted in units of keV/μm or MeV/cm.

For patient dose calculations, the use of the *mass stopping power* S_ρ, which normalizes the stopping power to unit density and accounts for macroscopic density variations, is essential. From here on, the mass stopping power will be used in all formulas.

As mentioned above, LET describes a statistical mean of the stopping power over measurements within a microscopic region or for Monte Carlo simulations, the two most common approaches are the *dose averaged* or *track averaged* LET.

14.2.1 Dose averaging, track averaging and limitations

Consider a point in a radiation field at which the LET is calculated. The stopping power at that specific point is only well-defined for mono-energetic beams, since it varies with energy. For all practical purposes, *e.g.*, a treatment field of multiple contributing beams treating a target in a patient, the above definition is not valid and thus an average stopping power has to be defined [568, 569, 570]. The *track-averaged LET* (LET_t), or fluence averaged LET, weighs the stopping powers of all contributing particles by their fluence ψ_E:

$$LET_t(x) = \frac{\int_0^\infty \psi_E(x) S_\rho(E) dE}{\int_0^\infty \psi_E(x) dE} \qquad (14.2)$$

For the *dose averaged* LET (LET_d) on the other hand, the contributing particle stopping powers are weighed by the dose contributed at that point as

$$LET_d(x) = \frac{\int_0^\infty \psi_E(x) S_\rho^2(E) dE}{\int_0^\infty \psi_E(x) S_\rho(E) dE} \qquad (14.3)$$

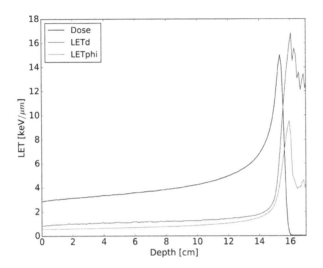

Figure 14.2 Energy deposition of a 160 MeV proton beam (black) together with dose averaged (green) and track averaged (blue) LET, simulated using TOPAS v3.0.1 [571]. The values behind the Bragg peak stem from secondary protons created by neutrons that traveled past the maximum proton range. Figure courtesy of David C. Hall (Massachusetts General Hospital).

Both of these LET definitions depend heavily on the energy spectrum of the contributing particles ψ_E; the broader the energy spectrum, the more they differ (Figure 14.2). $LET_d \geq LET_t$ always holds, the equality proving true only for mono-energetic beams [570]. Generally proton therapy applications utilize a dose averaged LET [569, 572]. Dose is the quantity of interest in clinical cases and since LET is often used in models of relative biological effectiveness (RBE) that modify physical dose, the dose averaged LET is a natural choice. This also keeps the averaging within fields and between different treatment fields consistent. In case one does not use dose averaged LET, RBE-weighted doses from different treatment fields can not simply be added together.

One complicating factor is that ^{4}He and heavier ions undergo fragmentation, the particles can either undergo fragmentation themselves or cause the target nuclei to fragment. This leads to a host of secondary particles and therefore mixed radiation fields. In that case the LET is usually restricted to the "primary" particle, and does not include other particle types present in the beam. However, especially for protons, the inclusion of secondary protons coming from nuclear collisions should be considered as it can contribute up to 10% of the dose in the plateau region [573].

14.3 RELATIVE BIOLOGICAL EFFECTIVENESS

The effect of radiation on living cells depends on a complex sequence of physical, biochemical and biological processes and a simple relationship between dose and biological effect does not seem to exist. The difficulty in predicting biological outcome is further complicated by the dependency of the dose-response relationship on the type of treatment modality. Due to a lack in understanding the biological effects, treatment planning in the clinic is still mostly based on dosimetric indices, such as the dose prescription to the target and dose constraints to healthy tissue. Thus when treating patients with different modalities, the potential difference in the biological effectiveness, or radiation quality, needs to be taken into account. The dose prescription of the modality is generally scaled by a factor to achieve the same biological effect as a reference radiation (*e.g.*, photons). This factor is captured in the concept of the relative biological effectiveness (RBE). The RBE is defined as the ratio of doses of two different radiation modalities, required to reach the same level of effect. This definition of the RBE assumes that the dose is homogeneous and macroscopic within a specific volume (*i.e.*, organ, cell culture sample or CT voxel).

Experimentally, RBE has been shown to be dependent on many factors such as the LET [569], dose [574], particle nuclear charge (Z), track structure [575], depth and volume of the target as well as the reference radiation used. Additionally, RBE is dependent on biological factors that influence cell radiosensitivity, such as cell type and the chosen biological endpoint [576].

Investigating and developing RBE models that accurately predict the dose delivered to the patient is a very active area of research. However, any biological model used routinely in radiotherapy would need to be capable of predicting the radiation induced biological response with simplicity and robustness. Currently RBE models are either phenomenological, fitting a radiobiological model to existing data, or mechanistic. A true mechanistic approach would accurately predict the physical interactions of the ionizing particle within the cell and the resulting level of DNA damage from both the direct interaction of particles as well as free radical production, leading to the prediction of possible biological responses such as repair or cell death. True mechanistic approaches are highly challenging since so few of the biological pathways involved are well understood. Current mechanistic models exist but vastly approximate the overall process [310].

With the increasing popularity of proton radiotherapy and availability of proton experimental data, several phenomenological models for proton RBE predictions have been proposed, *e.g.*, Wilkens & Oelfke [569], Tilly et al. [577], Carabe et al. [578], Wedenberg et al. [579], [580] and McNamara et al. [581]. The majority of these models are based on the linear-quadratic (LQ) model and are dependent on LET. Ion RBE is generally modeled using a semi-mechanistic approach.

Biophysical models that have been developed for both protons and heavy ions include the microdosimetric-kinetic-model (MKM [306]), the local effect model (LEM [291]), the repair-misrepair-fixation model (RMF [310]) and the track structure (δ-ray) model [582, 583]. We discuss some of these models in more detail below.

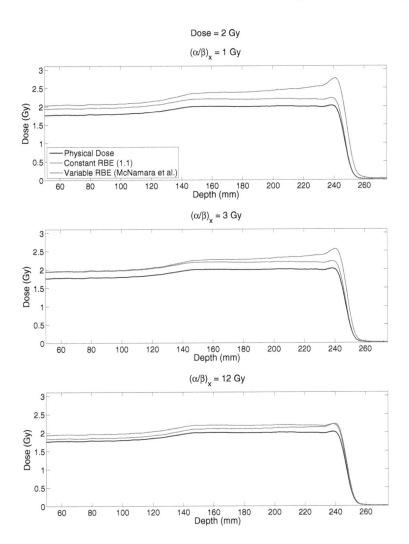

Figure 14.3 Comparison of the predicted biological dose for three different $(\alpha/\beta)_x$ values and a dose of 2 Gy for a simulated SOBP with a modulation width of 100 mm and range of 250 mm. The physical dose is shown by the black curve while the red curve shows the biological dose calculated by scaling the physical dose by a constant value of 1.1. The blue curve shows predicted biological dose for a variable RBE model [581].

14.3.1 The 1.1 conundrum in proton therapy

Somewhat controversially, dose in proton radiotherapy is generally prescribed by scaling the physical proton dose by 1.1, *i.e.*, protons are assumed to be 10% more efficient than photons [584]. This assumes a spatially invariant RBE within all treatment fields but experimental evidence (mostly based on cell survival) strongly suggests that the RBE varies across a treatment field and with cell line.

Clinically however there is no significant evidence that suggests the RBE deviates significantly from 1.1 in proton therapy. Furthermore, experimental data suggests that the constant RBE, especially as a conservative estimate for the target, may be a valid approximation [584, 585]. RBE depends on many factors such as dose, LET, tissue specific parameters (α and β values of the linear quadratic dose response relationship) and other physical beam characteristics. All these values can vary substantially for individual treatment sites and the uncertainties in these parameters may potentially be larger than the effect that needs correcting.

Caution in a constant RBE should, however, be considered for some specific sites or beam arrangements. These include sites with vital organs coinciding with the distal falloff of the spread out Bragg peak (SOPB) or for specific tissues, those with low (α/β) values (≤ 3 Gy). Figure 14.3 shows the difference in the predicted biological dose for a simulated SOBP of range 25 cm for a constant RBE of 1.1 compared to a variable RBE model. Continued investigation into clinical toxicities associated with the use of a constant RBE is essential to improve treatment planning [586, 587, 588, 589].

14.3.2   LET based RBE models

Biologically, in higher LET regions there is an increase in the clustering of DNA damage resulting in more complex lesions which are less likely to be repaired by the cell. This potential increase in radiation effectiveness requires a modification in dose to achieve the same biological effect as a given reference radiation. In proton therapy, the LET in a uniform SOBP increases from the proximal to the distal end, which may lead to a small increase in the RBE across the SOBP (Figure 14.3). The region of highest LET occurs downstream of the Bragg peak. These properties are similar for ion beams, *e.g.*, carbon, but the lateral fall-off is steeper for ions than protons.

The majority of LET based RBE models are for protons. A large dataset of *in vitro* cell survival experimental data exists for protons [585]. LET-based models are generally phenomenological using parts of these data and rely on the linear quadratic (LQ) model, which is extensively used in radiation biology. The LQ model, based on the tissue specific quantities α and β, expresses cell survival as a function of dose.

$$\left(\frac{N}{N_0}\right)_{\text{surv}} = e^{-\alpha D - \beta D^2} \quad . \tag{14.4}$$

The LQ model is however limited to doses between $\sim 1 - 10$ Gy and is potentially dependent on the sub-populations of the cell culture. The values of α and β are obtained from radiobiological experiments and the ratio α/β is used to quantify the radiosensitivity of a particular cell type.

Using the LQ model, the RBE between photons and protons can be calculated as

$$\text{RBE} = \left(\frac{D_x}{D}\right)_{\text{same biological effect}}$$

$$= -\frac{1}{2D}\left(\frac{\alpha}{\beta}\right)_x + \frac{1}{D}\sqrt{\frac{1}{4}\left(\frac{\alpha}{\beta}\right)_x^2 + \frac{\alpha}{\alpha_x}\left(\frac{\alpha}{\beta}\right)_x D + \frac{\beta}{\beta_x}D^2} \quad . \tag{14.5}$$

Here D and D_x are the proton and reference (photon) radiation dose, respectively, and x indicates the reference radiation. To relate RBE to macroscopic dosimetric parameters, the dose-averaged LET (LET_d) has been shown to be a reasonable approximation [590].

Below we discuss four empirical models, based on the LQ model, that have been developed to determine proton RBE assuming an LET dependence.

Model by Carabe et al. 2012

Dale and Jones [591] were the first to include the concept of a maximum RBE in the LQ model. The model by Carabe-Fernandez et al. [592, 593] introduced the minimum RBE to account for possible changes in β with LET. The maximum value of the RBE is the asymptotic RBE value at zero dose and can be determined as α/α_x, while the minimum RBE is the asymptotic RBE value at infinite dose and corresponds to $\sqrt{\beta/\beta_x}$ [592]. In the Carabe et al. model [578], RBE_{min} and RBE_{max} are determined using linear regression of experimental data obtained from V79 cells. The model assumes that the RBE is proportional to LET_d and inversely proportional to $(\alpha/\beta)_x$, except for those tissues with $(\alpha/\beta) = 2.686\,\text{Gy}$ (the average value of the experimental data used),

$$\text{RBE}_{\text{max}} = 0.843 + 0.154\frac{2.686\,\text{Gy}}{(\alpha/\beta)_x}\text{LET}_d \tag{14.6}$$

$$\text{RBE}_{\text{min}} = 1.09 + 0.006\frac{2.686\,\text{Gy}}{(\alpha/\beta)_x}\text{LET}_d. \tag{14.7}$$

Model by Wedenberg et al. 2013

The Wedenberg et al. model [579] assumes that a linear relationship exists between α and LET and that the quadratic parameter β of the LQ model is independent of LET i.e., $\beta/\beta_x = 1$. The model additionally assumes that the factor α is inversely proportional to $(\alpha/\beta)_x$. A single parameter q is fit to the experimental data,

$$\alpha/\alpha_x = 1 + \frac{q}{(\alpha/\beta)_x}\text{LET}_d \tag{14.8}$$

where $q = 0.434\,\text{Gy}\,\mu\text{m}\,\text{keV}^{-1}$ (95% confidence limit (0.366,0.513)).

Model by Wilkens & Oelfke 2004

The model by Wilkens & Oelfke 2004 [569] also assumes that the parameter β is constant and that a linear relationship exists between α and LET,

$$\alpha = \alpha_0 + \lambda\text{LET}_d \quad . \tag{14.9}$$

The parameters are determined from experimental data for V79 cells and find $\alpha_0 = 0.1 \pm 0.02\,\text{Gy}^{-1}$ and $\lambda = 0.02 \pm 0.002\,\mu\text{m}\,\text{keV}^{-1}\text{Gy}^{-1}$.

Model by McNamara et al. 2015

The McNamara et al. [581] model makes a similar assumption regarding $\text{RBE}_{\max}$ to the Carabe et al. [578] model; *i.e.*, that it has a linear relationship with respect to LET_d as well as a dependence on $(\alpha/\beta)_x$. However, it assumes that $\text{RBE}_{\min}$ has a dependence on $\sqrt{(\alpha/\beta)_x}$ with,

$$\text{RBE}_{\max} = 0.99064 + \frac{0.35605}{(\alpha/\beta)_{\text{x}}}\text{LET}_{\text{d}} \tag{14.10}$$

$$\text{RBE}_{\min} = 1.1012 - 0.0038703\sqrt{(\alpha/\beta)_{\text{x}}}\text{LET}_{\text{d}} \quad . \tag{14.11}$$

Despite the different assumptions made by the above four LET based proton RBE models, all agree relatively well with each other, within the error of the model, for most LET, dose and α/β values (within the clinically relevant range).

14.3.3 Non-LET based

The non-LET based RBE models discussed here represent the effectiveness of radiation fields without the direct use of LET. Generally this is achieved within the framework of microdosimetry or nanodosimetry.

The Repair-Misrepair-Fixation (RMF) model

The repair-misrepair-fixation (RMF) model [309, 310] is based on a combination of the repair-misrepair (RMR) model [594] and the lethal-potentially-lethal model [283]. The DNA lesion of importance in the RMF model is the double strand break (DSB), single strand breaks (SSBs) and base damages are neglected. The model accounts for the interaction of DSBs, through intertrack and intratrack binary misrepair of lesions. DSBs generated by a single particle track at effectively the same time and within close proximity are more likely to interact (intratrack) than DSBs created by different tracks (intertrack).

In the RMF model, the formation of DSBs of a certain complexity (a minimum of lesions that are unrejoinable) are linked to the α and β parameters.

$$\alpha = \theta\Sigma + \kappa\bar{Z}_F\Sigma^2 \tag{14.12}$$

$$\beta = \kappa\frac{\Sigma^2}{2} \tag{14.13}$$

$$\alpha/\beta = \frac{2(\theta/\kappa)}{\Sigma + 2\bar{Z}_F} \tag{14.14}$$

Here Σ is the number of DSBs per Gy per gigabase pair (or per cell) and $\bar{Z}_F$ is the frequency mean specific energy in a cell nucleus, with $\bar{Z}_F \cong 0.206\,\text{LET}_d/d^2$ as a first approximation. The DSB induction parameter, Σ and $\bar{Z}_F$ are dependent on the particle type and energy. Θ and κ are cell specific parameters that relate to the biological processing of DSBs into lethal lesions (*e.g.*, chromosome aberrations) and are independent of proton LET in the energy range relevant to clinical proton beams (*i.e.*, up to $\sim 75\,\text{keV}\mu\text{m}^{-1}$). For low LET reference radiations (*e.g.*, MV x-rays), the RMF model for the endpoint of reproductive cell death predicts that the minimum and maximum RBE are

$$\mathrm{RBE_{max}} \equiv \frac{\alpha}{\alpha_x} \cong \mathrm{RBE_{DSB}} \left(1 + \frac{2\bar{Z}_F \mathrm{RBE_{DSB}}}{(\alpha/\beta)_x} \right) \tag{14.15}$$

$$\mathrm{RBE_{min}} \equiv \sqrt{\frac{\beta}{\beta_x}} = \mathrm{RBE_{DSB}} \quad . \tag{14.16}$$

where $\mathrm{RBE_{DSB}}$ is the ratio of Σ for the particle to that of the photon (^{60}Co) reference radiation. Values for $\mathrm{RBE_{DSB}}$ have been estimated with Monte Carlo Damage Simulation (MCDS) [313]. MCDS in combination with the RMF model has been used to predict spatial variations in RBE for protons and carbon ions in the SOBP [310].

The Microdosimetric-Kinetic Model (MKM)

The microdosimetric-kinetic model (MKM) was first introduced by Hawkins 1996 [305] and combines a microdosimetric approach with a kinetic description of lesion creation and repair based on the repair-misrepair (RMR) model [594], the lethal-potentially-lethal model [283] and the theory of dual radiation action [282].

The MKM uses a microdosimetric approach to characterize any radiation field, by subdividing the nucleus into microscopic compartments called domains. Since the energy deposition within the whole cell nucleus is macroscopically inhomogeneous, each domain within the nucleus is chosen with a homogeneous dose distribution. The energy absorbed by the domains can thus vary randomly, and a domain can be thought of as a homogenous reaction vessel in which lesions can randomly diffuse. Lesions are affected by repair mechanisms as well as biological barriers (e.g., membranes), which limit the distance lesions can travel to encounter and interact with other lesions. This is accounted for by the finite size of the domain. For simplicity, domains are modeled as uniformly dense spheres. The diameter and mass of each sphere is dependent on the type of cell and determined experimentally with tissue equivalent spheres. The diameter of a domain has been shown to be between 0.5 and 1 μm [305].

The MKM assumes that ionizing radiation induces potentially lethal lesions in the DNA of the cell and that the lesions can undergo transformations to either a lethal and unrepairable lesion, a repaired lesion or can remain unchanged for a period of time before being converted to a lethal lesion [305, 595]. The quantity of interest is the lineal energy $y = \epsilon/\bar{l}$, where ϵ is the energy deposited within the domain and $\bar{l}$ is the mean chord length of the domain. Due to the large fluctuations in the energy deposition in domains, the ionizing radiation is characterized by the probability distribution of lineal energy events, $f(y)$, the frequency-mean lineal energy, $\bar{y}_F$, and the dose-mean lineal energy, $\bar{y}_D$. In terms of the LQ parameters, the MKM assumes that β is constant ($\beta = \beta_x$) and that α can be calculated from $\bar{y}_D$,

$$\alpha = \alpha_0 + \frac{1}{\rho \pi r_d^2} \bar{y}_D \beta \tag{14.17}$$

Here ρ and r_d are the density and the radius of the spherical domain. The parameters α_0 and β are dependent on the cell line and are equal to the LQ parameters in the limit of zero LET. The cell-dependent parameters r_d and α_0 are tabulated in the literature [306, 596, 597, 169].

14.3.3.1 Track structure (δ-ray) model

Katz et al. [582, 598, 583] introduced the δ-ray theory of track structure based on the multi-target theory. The multi-target theory assumes that the cell consists of sub-targets, modeled as infinitely thin cylinders orientated coaxial to the ion path with a defined cross-section [582]. The biological effect depends on four experimental parameters m, E_0, σ_0 and κ as well as two modes of cell kill. The parameters m and E_0 describe the biological response to photons and are derived from cell survival experiments to γ-radiation. The parameter m accounts for the accumulation and repair of sublethal damage and relates to the subtargets that need to be inactivated to reach a certain endpoint and E_0 describes the dose necessary to cause on average one lethal event per target (intrinsic radiosensitivity). Parameters σ_0 and κ describe cell response to ion irradiation, where σ_0 is the saturation cross-section of the radiation sensitive part of the nucleus and κ characterizes the sensitive subtarget radius, a_0 [598]:

$$\kappa = \frac{E_0 a_0^2}{2 \times 10^{-11}\,\mathrm{Gy\,cm^2}} \tag{14.18}$$

The biological response (e.g., the survival fraction S) is calculated from the radiosensitivity parameters, the effective charge Z^*, the relative velocity β and the fluence F. The δ-ray model differentiates between two modes of cell killing depending on the value of $Z^*/\kappa\beta^2$. The γ-kill or multi-target/single-hit mode dominates at high fluence and low LET where $Z^*/\kappa\beta^2 \ll 1$; here cells are killed due to an accumulation of sublethal damage (intertrack damage). In this mode a set of subtargets needs to be inactivated to cause cell death. The ion-kill or single-hit/single-target mode occurs when $Z^*/\kappa\beta^2 \gg 1$, here a single ion is sufficient to induce cell death (intratrack damage) and dominates at low fluence and high LET. RBE reaches a maximum for approximately $Z^*/\kappa\beta^2 = 2$. Two regimes are assumed for the ion-kill mode. The grain-count regime assumes a cell is inactivated by a single primary ion that passes through the nucleus and the track-width regime assumes that cell kill results from overlapping δ-rays that transverse the nucleus, without the primary ion passing through the nucleus. Protons have low LET and are limited to the grain-count regime [575].

The RBE can be calculated as the ratio of the isoeffective x-ray dose and ion dose D_x/D [598] with

$$D_x = E_0 \ln[1 - (1 - S)]^{1/m}. \tag{14.19}$$

Here S is the total survival fraction, usually consisting of a mixture of both killing modes.

The Local Effect Model (LEM)

The local effect model is based on the assumption that equal local doses lead to equal local effects, independent of the radiation quality. The particle effectiveness is determined from the microscopic dose distribution within the sensitive target, the nucleus. This local dose is calculated from an amorphous track structure representation of the energy deposition as a function of the radial distance to the particle path. The LEM has several implementations (LEM I - LEM IV) based on this general concept.

In the first implementation (LEM I) [304], the local biological effect is derived from the photon dose response curve, $S(D)$, represented by the LQ parameters for a particular biological endpoint. The biological effectiveness of the ion radiation field is derived by first calculating the mean number of lethal events $N_{l,ion}$ per cell,

$$\bar{N}_{l,ion} = \int \frac{-\ln[S_x(d(x,y,z))]}{V_{\text{nucleus}}} dV_{\text{nucleus}} \qquad (14.20)$$

Here $S_x(D)$ represents the survival for photon radiation as a function of dose D and $d(x,y,z)$ is the distribution of the local dose within the cell nucleus (target) of volume V_{nucleus}. LEM I assumes a Poisson distribution of lethal events around the mean value and the surviving fraction is simply determined from $S_{ion} = e^{\bar{N}_{l,ion}}$. The RBE is then calculated from the ratio of photon dose to ion dose leading to the same biological effect ($S_x = S_{ion}$).

LEM I performs reasonably well in predicting data for carbon ions in therapy relevant conditions, however, predictions have large deviations for high-energy, low-LET ions (e.g. protons or He ions). The model has thus continuously been improved resulting in different LEM versions.

LEM II [599] made improvements for very high doses by accounting for the DSB yield resulting from SSBs occurring in close proximity to each other ($< 25\,\text{bp}$). LEM II also accounts for indirect effects such as free radical diffusion near the target. LEM III [600] introduced a more detailed track structure description with an energy dependent extension of the track core.

LEM IV [601] extends on LEM I by assuming that the final biological response of the irradiated cell is dependent on the initial DNA damage spatial distribution rather than the local dose distribution. This is based on the premise that the cell reacts to the spatial DSB distribution (clustering of damage) rather than the energy deposition, and that this is a better measure of the effectiveness of different radiation qualities. In LEM IV, the structure of DNA within the cell nucleus (chromatin fiber arrangement) needs to be considered as well as the different types of radiation damage (*e.g.*, isolated DSB, clustered DSBs).

14.3.4 Uncertainties

The majority of experimental data used for the basis of RBE models, whether for purely empirical or mechanistic purposes, are based mostly on clonogenic cell survival data *in vitro*. The question arises if this data is relevant for clinical purposes. Furthermore, the complex inter-dependencies of the RBE, with RBE dependent on so many different factors such as dose, LET, endpoint, beam characteristics and tissue specific parameters, each with their own potentially large uncertainty, add to the difficulty in defining the error in RBE measurements.

Generally most experimental data suggests that the RBE increases with decreasing dose, however, some proton experiments have shown a reverse effect. For most clinical fractionation schemes, doses of 2 Gy or less are of interest, a regime that is challenging to collect experimental data especially *in vivo*. Another challenge for RBE modeling is finding the correct α and β parameters for the tissue of interest. The biological endpoint has to be very well defined. A small shift in the dose response curve can have large effects on the measured α/β value. Furthermore, we still don't fully understand the fundamental biological processes induced in the cell from different particles. Different particles may induce specific effects in the cell such as activation of signaling pathways, gene expression, angiogenesis and other cell cycle effects. None of these effects are considered in current RBE models.

14.4 THE ROLE OF MONTE CARLO

Monte Carlo (MC) simulations are considered the gold standard in determining particle transportation in media and offer multiple advantages over analytical dose calculation engines. Due to the concept of following each particle along the track for each interaction step, MC simulations give users access to a multitude of physical properties not available in analytical (pencil beam) dose calculation algorithms (ADCs). One area where MC simulations are frequently used is for calculating LET distributions and investigating the effects of RBE models.

14.4.1 Understanding dose and LET distributions

Clinical dose calculations need to fulfill two main criteria, they need to be sufficiently accurate and sufficiently fast. While MC is the most accurate method to determine dose distributions, typical MC approaches are too slow for routine treatment planning in clinical practice. Thus, ADCs are (still) predominantly used for treatment planning. ADCs offer fast dose calculations with adequate accuracy. ADCs calculate dose by deconvolving the delivered treatment field (beam) into pencils that are propagated along their central axis, thus this method is often also referred to as a pencil beam algorithm. To avoid confusion with pencil beam treatments we use the term ADC here. The lateral spread of ADCs is typically determined by the material along the central pencil axis and expressed by a combination of one or two Gaussians. This approximation explains the main shortcoming of ADCs; they are typically blind to changes in material compositions lateral to the central axis of a pencil; *i.e.*, ADCs are insufficient describing the effects of multiple Coulomb scattering (MCS) in particle therapy.

MC simulations have been essential in understanding the approximations made in clinically used ADCs, which lead to potentially inaccurate predictions of the range of the particle field, the distribution of dose and the distribution of LET, *i.e.*, of RBE; across the field. Treatment margins are typically added around target volumes to account for uncertainties, including those from dose calculations of ADCs. MC can thus be used to reduce treatment margins and improve our understanding of radiation effects.

14.4.1.1 Range uncertainties

MCS results in a re-distribution of dose, in particular in the presence of density gradients in beam direction, i.e. parallel to and inside the field. Charged particles are preferentially scattered out of high density material into the lower density material via MCS. This results in a reduction (increase) in dose behind high (low) density material relative to the prediction of ADCs. Treatment fields passing through the cavities in the head as illustrated in figure 14.4 are particularly tricky for ADCs and thus ideal to illustrate this effect. The resulting dose dips or peaks relative to ADC predominantly affect the dose at the distal end of the treatment field, i.e. the predicted range of the field, and to a minor extent, the dose in the target (see next section).

It is well known that the approximations used by ADCs result in uncertainties of the predicted dose distributions. In order to ensure target coverage despite these uncertainties, the range of treatment fields is extended by the so-called range margin, which is (hospital-dependent) in the order of 3.5% of the prescribed range of the treatment field plus 1 mm. Originally, this margin was established to cover the uncertainties in dose cal-

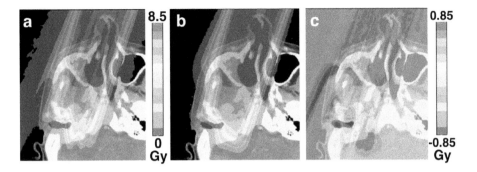

Figure 14.4 Illustration of the effects of approximations made in ADC calculations. Example treatment field to illustrate MCS effects, shown are MC dose (a), ADC dose (b) and the difference (MC-ADC) (c). The redistribution of dose can clearly be observed in (c). d: The resulting uncertainties in the prediction of range for single treatment fields and various treatment sites.

culation and uncertainties related to the conversion of CT to proton stopping powers or material compositions. The constant term (1 mm) covered the setup uncertainty [602]. The range uncertainties were re-evaluated by Paganetti in 2012 [603]. Paganetti concluded that current range uncertainties are between 2.7% and 4.6% of the prescribed range (plus 1.2 mm) when using ADCs depending on the heterogeneity of the patient geometry. However, these margins could be reduced to 2.4% if MC simulations were used in treatment planning, independent of the patient heterogeneity. The amount of various contributions to the range uncertainties were estimated by multiple groups, for example Schuemann et al. [604] used MC simulations to investigate the effect of dose calculation uncertainties in ADCs. They found that for some sites (liver, prostate, whole brain) range uncertainties could be reduced to 2.7% (+1.2 mm) but for more heterogeneous sites (breast, lung, head and neck) range margins would need to be increased to 6.3% to guarantee single field coverage. While these results suggest that range margins should be nearly doubled for heterogeneous patient geometries, the authors pointed out that this is only the case if treatment fields are blindly applied and only single fields are delivered.

Most treatments consist of multiple fields from various angles. This greatly mitigates the effects of range uncertainties on the dose distribution in the target. For example, if a patient receives 4 treatment fields from different angles (for simplicity assuming equal beam weights), the effects of range uncertainties occur in 4 distinct locations around the target. Thus even in an extreme case, *e.g.*, where one beam stops too early and delivers absolutely no dose to the distal end, that region would still receive 3/4 of the prescribed dose from the other 3 fields. Due to the applied range margins, target doses were found to be typically not compromised (see next section). Additionally, experienced treatment planners and clinicians know about the shortcomings of ADCs and adjust the direction and combination of treatment fields accordingly and try to avoid passing through regions with a high density gradient.

MC simulations remove the dose calculation uncertainties, thus if treatment plans were generated using MC simulations, the range margins could be reduced to 2.4% + 1.2 mm [603]. Accordingly, many vendors of treatment planning systems are working on including

MC in their optimization process. While typical MC systems require a prohibitively long calculation time for treatment planning purposes, GPU-based MC systems rival ADCs in speed and standard MC codes in accuracy [605, 606, 607]. Such GPU MC codes are likely to replace ADCs as standard dose calculation engines in the near future, which would result in a reduction of range margins of $\sim 1\%$ of the prescribed range, i.e. sparing around 2-3 mm of healthy tissue for deep seated tumors.

14.4.1.2 Considerations for dose and DVH

Typically the main goal for radiation therapy is to cover the target with a homogeneous dose. The target coverage is generally measured by parameters derived from dose-volume histograms (DVHs). There have been multiple studies investigating the effects of uncertainties in patient setup, CT conversion and dose calculation on target coverage. However, only the latter can be addressed by MC simulations. Figure 14.5a shows the effect of dose calculation uncertainties (including the above mentioned range uncertainties) on DVH-based properties and the resulting effect on tumor control probabilities (TCP) for five common treatment sites as reported in [608]. MC doses were found to be around 1-2% lower than predicted by ADCs. If MC simulations were used to correct the ADC results accordingly, all treatment parameters would be within the recommended uncertainties of $\pm 2.5\%$. Even without corrections, the effects on TCP were small and typically within 5%, except for lung patients where prescribed doses are often limited by the surrounding lung. Thus for most treatment sites, the combination of clinically applied margins and ADC dose calculations can be considered a safe (but not optimal) approach.

A more obvious case for the use of the MC method in treatment planning is shown in figure 14.5b. Again, the difference between ADC and MC is illustrated, but this time for five treatment sites of small, stereotactic treatments in the head [609]. The left side compares pure ADC calculations to MC predictions. Differences in DVH parameters of up to 15% can be seen. Small fields are particularly difficult for ADCs as the loss of lateral equilibrium due to in-field multiple Coulomb scattering is typically not correctly described and the effects can even be observed in the central part of the treatment field. It should be noted, that such fields are never delivered purely based on the ADC calculations. Treatment planners know about the shortcomings and during commissioning of stereotactic treatments, potential correction factors are experimentally determined. However, if instead the treatment plans were designed using the MC method (right side of figure 14.5b), the target coverage can be greatly improved without the need of additional corrections.

One of the main reasons that ADCs are used in clinical treatment planning is their fast calculation time, albeit their extensive approximations in dose calculations. Inawa et al. proposed a pencil beam redefinition algorithm (PBRA) [610], which improves the ADC accuracy by introducing additional calculation steps. However, the PBRA and similar approaches also escalate the calculation time, resulting in computation times similar to that of MC simulations. As previously mentioned, the solution is likely going to be GPU-based MC simulations as the new standard for treatment planning.

14.4.1.3 LET

One of the main advantages of MC simulations, is that particle histories are followed step-by-step through the world, simulating each interaction individually (except for the grouped

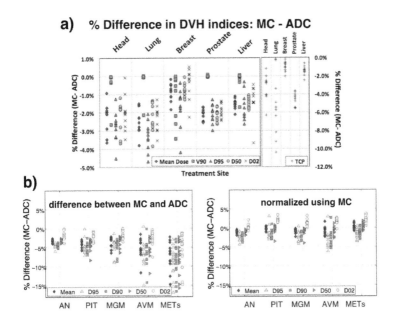

Figure 14.5 a: The effect on target doses and other DVH based parameters when considering full (multi-field) proton treatment plans. The percentage difference for five sites and five dosimetric parameters. DXX is maximum dose covering XX% of the target volume, and V90 is the percentage of the target volume covered by 90% of the prescribed dose. TCP is the tumor control probability calculated based on mean dose and treatment plan parameters. Reprinted from [608] with permission from Elsevier. b: The difference between analytical dose calculations and the Monte Carlo method (left) and the potential of correcting small field dosimetry using the Monte Carlo method (right), reprinted from [611] copyright Institute of Physics and Engineering in Medicine. Reproduced by permission of IOP Publishing. All rights reserved.

MCS and continuously slowing down approximations). This provides access to properties that are not typically assessed for treatment planning, such as the neutron background dose or the LET. Such properties are useful to estimate secondary effects of radiation treatment such as secondary cancer induction rates from neutrons or the increase in RBE due to the increasing LET at the distal end of a Bragg peak.

While the MC method has been extensively used to investigate LET distributions in patients [572, 612], the typically used dose-averaged LET (LET_d) itself is not well suited for MC simulations. MC codes follow histories along their tracks in finite steps. The step sizes and the initiated physics processes are typically determined by the interaction probabilities or geometric boundaries, *e.g.*, when entering a new voxel of a CT. At any given step, the initiated physics process determines the energy deposited at that position. For LET_d calculations, this will result in some unphysical spikes caused by artificial calculation steps, i.e. by steps that happen to be very short but undergo a discrete interaction depositing a relatively large amount of energy. To avoid these spikes, one can instead calculate track-averaged LET (LET_t); however, for RBE calculations in a clinical scenario the dose averaged LET is thought to be more meaningful. The two LET concepts and their limitations were recently discussed by Ganville and Sawakuchi [613].

14.4.2 RBE modeling

Besides providing a more accurate dose and LET distribution than analytical calculations, MC simulations are particularly useful to investigate biological effects of radiation. MC codes have long been the only method to determine LET distributions in a patient. Due to the increased interest in the field, some treatment planning platforms have recently included LET calculations and implemented LET-based RBE models. However, some RBE models require additional information that is typically only directly available in the MC particle tracking.

Thus the role of MC simulations is not only to calculate RBE distributions for various models that have been developed (see Section 14.3), but to provide a platform to further develop and test RBE models. Monte Carlo codes are ideal to test new hypotheses since simulations can be re-run for a variety of parameters, changing material compositions, biological constants or radiation types without incurring large costs. One can even develop more sophisticated models that do not depend on LET but the microscopic structure of particle tracks and the target cells. Efforts to understand the biological effects from the (sub-) cellular scale up are being developed using track structure MC codes. Physics interactions with DNA are simulated explicitly and mechanistic models try to translate DNA damages to biologically observed effects. The hope is, that such multi-scale modeling may soon provide a better understanding of why different types of radiation have different biological effectiveness. These models could potentially even predict a cell-dependent RBE and provide a means to determine which radiation modality is optimal for a given tumor.

14.4.3 Example MC simulations using TOPAS

Monte Carlo tools typically require a high level of programming knowledge and understanding of the MC method. The TOPAS (TOol for PArticle Simulations) application [571] is a MC tool aimed at medical physicists in proton therapy. TOPAS removes the need to write and compile code, instead simulation settings are controlled by text based parameter files. In addition to providing multiple scoring options such as dose and LET, TOPAS includes multiple RBE models. TOPAS is available at `www.topasmc.org`. Here we present an example of patient simulation parameter files to illustrate the basic principles of TOPAS. For simplicity the simulation uses only two pencil spots as the beam source. The design of a full treatment field either consists of hundreds to thousands of pencils (for pencil beam scanning, PBS) or a complex treatment head design (double scattering). The two spots are used to illustrate the concept of how a PBS treatment plan can be designed with multiple spots. Additional and extended examples can be found at `http://topas.readthedocs.io/`.

14.4.3.1   2-spot pencil setup

The left side of Figure 14.6 shows an example parameter file consisting of two pencil beams shot into an empty world. The two single pencils can be replaced by a map of pencils with varying energy and intensity. An alternative approach is to use a phase space file generated by a planning system as a beam source.

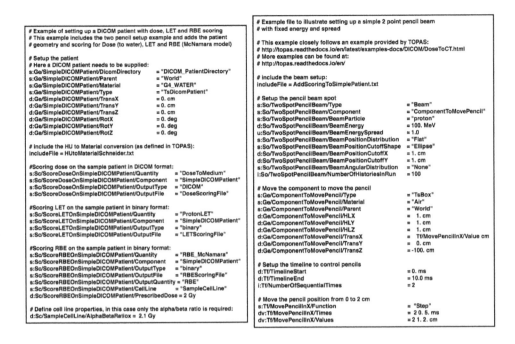

Figure 14.6 Left: Setup of two pencils moving along the X direction in an empty world. Right: The additional lines to include a DICOM patient setup and dose, dose averaged LET scorer and an RBE scorer based on the McNamara model.

14.4.3.2 Expansion to include patient setup, dose, LET and one RBE scorer

The right side of Figure 14.6 shows the setup of a DICOM patient with scorers for dose, LET_d and the McNamara RBE model [581] in the patient. The dose is scored in DICOM format, LET and RBE are scored in a binary file. The patient can be replaced by changing the DICOMDirectory parameter name, the positioning is automatically assigned as defined in the DICOM files. In case of multiple treatment fields, both LET_d and RBE need to be combined using a dose-averaged sum instead of a simple addition. The methodology is described for example in [614].

14.5   IMPLICATIONS OF PARTICLE THERAPY FOR OUTCOME MODELS

Up to now we have described how the different energy deposition patterns of particles impact cell survival, *i.e.*, DNA damage induction. For many years particle radiobiology has focused on the RBE models presented above, and has been restricted to a few endpoints, mainly cell survival and carcinogenesis. Only lately it has been recognized that irradiation with charged particles activates signaling pathways distinctly different compared to photons, and has a differential effect on the microenvironment, impacting tumor as well as normal tissue response [565, 615, 566].

Only few models describing these differential effects exist, therefore the purpose of this section is to discuss important differences between charged particles and photons that are not covered by the models described above and can lead to differences in outcome. Generally caution is advised when outcome models based on photons are extended to charged particle irradiation.

14.5.1 Target effects

Over the last decade it has become increasingly clear that tumor response relies, to a large extent, on the interaction between cancer cells and their microenvironment, represented by endothelial cells, lymphocytes and others. In parallel to this development, particle radiobiology has shifted its research focus from cellular endpoints to expression factor and microenvironment studies to investigate how the tumor-host dynamics are differentially impacted by particle irradiation. The main differences we will delineate here concern angiogenesis, invasion and immunological response, which have been reviewed in great detail [565, 615, 566].

Angiogenesis, the development of new blood vessels, is one of the hallmarks of cancer and crucial to the continued growth and progression of the tumor. It is well known that photon irradiation promotes angiogenesis via increased expression of various factors like VEGF, HIF-1A and others [616, 617]. Protons on the other hand have been shown to inhibit blood vessel formation in animal models as well as in resected tissue of patients [618, 619, 615, 620]. Similar results have been reported for carbon ions [621], providing evidence that particle irradiation might be superior in inhibiting angiogenesis compared to photon irradiation, most likely by modulating the VEGF level in the tumor microenvironment [622].

Similar to angiogenesis, tumor cell migration and invasion are essential to disease progression and greatly affect patient outcome. It has been shown that particle irradiation decreases cell migration and the invasive potential of cancer cells *in vitro* in fibrosarcoma and lung cancer cell lines [623, 620]. These results have been confirmed *in vivo*, where it has been demonstrated that single doses of protons inhibit the metastatic potential significantly in a hamster melanoma model [624]. Similarly, protons were shown to reduce the lung colonization potential of irradiated A459 cells compared to photons [615].

There are studies indicating that proton irradiation also leads to a markedly different inflammatory response of the irradiated tissue [625, 565]. Carbon ions have also been shown to confer some resistance to re-exposure to tumor cells [626]. The effect seems to be mediated among others by CD8+ T-cells, possible synergizing with certain immunotherapeutic approaches. Whether particle therapy can offer an advantage combined with immunotherapy compared to photons remains to be seen; though the topic is gaining popularity and clinical trials are being conducted [627, 566, 628].

Another interesting effect reported after irradiation with carbon ions is the reduced effect of intra-tumor cell-population heterogeneity on the outcome. Glowa et al. [629] irradiated different grades of syngeneic prostate cancer cell lines in rats with either photons or carbon ions. They showed that carbon ion irradiation not only led to a steeper dose response curve, but also that the RBE increased for higher-grade, aggressive tumors compared to more differentiated, low-grade tumors.

14.5.2 Normal Tissue effects

In addition to the factors outlined above, normal tissue effect models are further affected by the pronounced differences in dose distribution in the normal tissue surrounding the tumor. Figure 14.7 shows two treatments plans for photons and protons for a target in the liver, together with the DVH. The low dose bath is greatly reduced in proton therapy, while the high-dose area is slightly larger, owing to extra margins for additional uncertainties.

Normal tissue complication probability (NTCP) models that connect these dose distributions to an outcome usually consist of two separate steps. First, the dose distribution is converted using the biologically effective dose (BED) transformation to account for the fractionation differences among points receiving different doses per fraction [630]. Then a volume-effect model, such as the Lyman-Kutcher-Burman (LKB) [503] or Equivalent Uniform Dose (EUD) [631] model, is applied to account for the effect of organ architecture on the outcome. Both of these steps are impacted by the significantly different dose distributions exhibited by particles.

In the BED transformation, due to its linear-quadratic nature, high dose areas are weighted more heavily, *i.e.*, for the same physical mean dose the particle dose distribution has a higher BED [632]. Since the α and β parameters necessary for the transformation are not well known, generic values between 2 and 3 are usually used, adding additional uncertainties. For hypofractionated regimens, this uncertainty translates into large additional errors in the high dose region, which is typically larger in proton therapy.

A range of Dose-Volume models has been developed to describe the relationship between uniform and partial volume irradiations [633]. Especially in liver [634], lung [635] and brain [636] these models have been extensively fit to clinical patient data. However, since the dose distributions possible with particle therapy can differ significantly from photon distributions, it is unclear how well these parameterizations derived from photon therapy extend to particle therapy. The very popular LKB and EUD approaches do not take volume thresholds into account; *i.e.*, even small volumes irradiated to very high doses continually increase the complication probability, which is not always in line with clinical observations [637]. Thus more data is clearly necessary to better understand the subtleties of organ effects for particle therapy.

14.6 APPLICATION IN TREATMENT PLANNING

In order to provide accurate biological models for particle therapy, treatment plans have to be designed using MC simulations. The main strength of MC simulations is their superior accuracy. The main limitation is the computation time to obtain sufficient statistical accuracy. GPU based MC simulations offer a new approach to MC simulations with computational speed comparable to ADCs while maintaining accuracy similar to full MC simulations [638, 639, 605, 607, 606]. Such GPU based MC codes take seconds to minutes to calculate a treatment plan and are currently in use at the Mayo clinic as a secondary dose calculation check or to obtain LET and a simplified RBE [639, 606]. Full RBE and other effect models can also be included with the MC method.

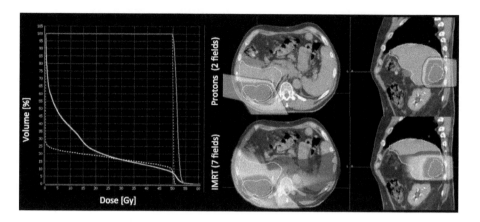

Figure 14.7 Dose-Volume Histograms and dose distribution for proton (top) and photon (bottom) treatment plans. Dotted lines represent protons, full lines photons for liver (green) and target volume (red) in DVH.

14.6.1   Vision for the future

GPU-accelerated MC simulations in the clinic would help achieve the goal of fully biologically optimized, adaptive and image guided particle therapy. The possibility of the following scenario could become standard practice. A patient would be planned using GPU-MC. For each treatment day, online imaging data from (cone-beam) CT or MRI would be used to re-optimize the treatment plan within seconds after the initial positioning of the patient on the table. The adapted treatment plan could be based on deformable image registration to the planning CT, or, depending on the imaging quality and computation speed, directly re-optimized on the new imaging data.

However, in order to move to a fully adaptive treatment system, the level of confidence in the accuracy of the re-planning system must be high. Careful and highly accurate validations for each component are required. In double scattered particle beams, for example, a minimal set of experimental measurements was described in [640]. For PBS, the range and shape of each pencil beam has to reproduce the experimental measurements with high accuracy, the golden beam data necessary is described in Clasie et al. [641]. Additional measurements can be added to increase confidence in the calculation algorithm, such as evaluating physics accuracy [642] or accuracy in dose distributions for heterogeneous patient geometries [643]. The deformable image registration has to be similarly validated. Safety limits for the adaptation would likely be imposed, defining the size of adaptations allowed without a re-evaluation by the clinical team.

Once sufficiently validated, such an adaptive treatment system could allow daily optimized treatments, adapted for daily inter-fraction changes in the patient geometry. Such daily re-optimization would allow a daily assessment of organ effects. TCP and NTCP models would have to include single fraction effects. If the computational speed is further increased, one can imagine ultra-fast adaptation of each pencil in the lateral direction and energy (depth) to follow a tumor guided, for example, by MRI imaging data. Together with online range verification measurements using prompt-gamma detectors that are currently being developed, such ultra-fast adaptation capabilities would remove the need for margins imposed due to dose-calculation uncertainties, range uncertainties, setup uncertainties, motion uncertainties and intra-fractional patient changes, practically allowing near marginless treatment of tumors. Margins induced by physics limitations could be reduced to around 1 mm. However, the biological margins or the extent of the tumor still remain and are often difficult to estimate, current safety margins may in fact also cover uncertainties of the tumor extent. Nevertheless, such minimal-margin treatments would allow the design of new treatment options, fully utilizing the sharp distal end of the particle beam to obtain highly conformal treatment fields, which would significantly reduce treatment side effects.

CHAPTER 15

Modeling response to oncological surgery

J. Jesus Naveja

Leonardo Zapata-Fonseca

Flavio F. Contreras-Torres

CONTENTS

ABSTRACT

In this chapter, the authors discuss computational modeling regarding response to oncological surgery from a conceptual perspective. In particular, emphasis is placed on biomedical mechanistic models for predicting the clinical outcomes in the context of oncological surgery. The medical background of oncological surgery is briefly addressed; the understanding of variables involved in oncological surgery is highlighted; the fundamentals of different approaches for modeling oncological surgery are explained, and results to be expected from each model are discussed. Finally, a computational model of oncological surgery, which considers tumor growth, nutrient diffusion and immune system variables, is explained using the R programming language.

Keywords: Cancer, multiscale modeling, oncological surgery, R programming.

15.1 INTRODUCTION TO ONCOLOGICAL SURGERY

ONcological surgery is fundamental for both curative and palliative treatment of diverse types of cancer. When tumor detection is timely made, surgery typically represents the best therapeutic choice regarding locoregional control of the disease. Moreover, the prognosis can be significantly improved if the management comprises either adjuvant systemic therapy or radiotherapy [644]. Unfortunately, the capabilities of the treatment for being curative diminish as the tumor grows and metastasizes. In such late stages, palliative surgeries are usually performed with the aim of relieving obstructive symptoms, pain and infections. Crucially, they are only attempted if the expected benefits in quality of life or survival overcome the surgery risks [645].

Studies conducted with patients undergoing resection of recurrent or metastatic pancreatic cancer demonstrated an improvement in both overall and disease-free survival, more significant when recurrence of the disease was locoregional or metastasis was in the lung [646, 647, 646]. Therefore, oncological surgeries performed at later stages are not only suitable for palliating symptoms, but they also comprise therapeutic options with a potential benefit on survival. For instance, studies carried out in patients with pancreatic cancer showed that resection of local or metastatic recurrence is correlated with a longer survival even when curative surgery is not indicated [648]. Similar results have been found after resection of liver metastases of colorectal carcinoma [649, 650], as well as neuroendocrine tumors [651]. However, the rationale for deciding whether a patient should undergo a therapeutic oncological surgery has remained an elusive one.

In addition, these surgical procedures are associated with medical complications in up to 16% of the cases, and it has been estimated that the mortality rate for oncological surgeries is about 1.2% [652, 653]. Under such circumstances, the expected benefit of surgery must be weighed against the risks on a personalized basis. Table 15.1 describes the different approaches for oncological surgery. Remarkably, a universal tool for personalized assessment of the expected benefit/risk ratio of oncological surgery is still missing.

15.1.1 Clinical and surgical factors modifying patients' outcomes

Prognosis of patients undergoing oncological surgery highly relies on the body region being affected. And according to such distribution, each type of cancer has its own dependent variables which are in turn related to lower survival rates and higher tumor aggressiveness. Here some examples are listed:

Colorectal cancer: poor nutritional state and an older age [663].

Lung cancer: the histological type and positive smoking status [664].

Bladder cancer: older age, advanced pathological stage and surgical factors (*e.g.*, positive margins during surgery and less than 10 lymph nodes resected) [665].

Gliomas: advanced grade, older age, lower general performance status, performing biopsy instead of resection [666].

Breast, kidney and lung cancers: high neutrophil/lymphocyte ratio and the lack of intraoperative use of nonsteroidal anti-inflammatory drugs [667].

A particular case is pancreatic cancer, in which locally advanced disease can lead to tumors that invade or compromise very relevant anatomic structures, such as a major vein or artery. Thus, although technically removable, such tumors are unresectable in clinical practice [668]. Nonetheless, it has been suggested that intensifying neoadjuvant chemotherapy could be considered in the cases of unresectable masses [669, 670]. Interestingly, in borderline pancreatic cancer the use of neoadjuvant agents plus the correct arterial and venous resection increase the likelihood of negative margins during the surgery, hence improving the outcomes [671, 672, 673]. Therefore, in this type of cancer local progression and the absence of distant metastasis can be considered as the main factors in survival rate [674].

These examples depict the complexity and variability underlying oncological surgery outcomes. Consequently, it is mandatory to develop models devoted to a particular cancer type taking into account its specific prognostic variables.

15.1.2 Complementary therapies to oncological surgery

It is important to keep in mind that patients undergoing oncological surgery typically have the chance of receiving secondary neo- or adjuvant therapies, which can be of the utmost importance for a successful treatment [675, 676]. In general terms, neoadjuvant therapies are administered before the surgical procedure, whereas adjuvant therapies take place after the surgery is performed [676]. As different therapies may be administered on the same patient, the interactions among them should be considered for a better modeling of oncological surgery outcomes.

TABLE 15.1   List of references according to different types of oncological surgeries.

Surgery	Description	References
Curative	Surgery intended to eliminate cancer from the patient.	[654, 655, 656, 657]
Therapeutic	Surgical interventions and re-interventions aiming to increase survival.	[648, 647, 658, 659]
Palliative	Surgeries intended to reduce symptoms in advanced disease with low chances of long term survival.	[660, 661, 645, 662]

Significant scientific advances have been done in modeling response to chemo- [677] and radiotherapy [678] in a wide variety of cancer types. The same applies to models for immunotherapy [679], oncolytic virus [680], as discussed in Chapter 11 of this book. These pioneer studies can be taken as a basic framework on cancer response to various therapies. Additionally, the burgeoning field of systems biology has been recently contributing to cancer modeling, yielding a promising opportunity to integrate the available data and finally develop personalized treatments [681, 682]. In the next section we attempt to provide a holistic standpoint of tumor modeling by presenting relevant variables, with emphasis on those useful for oncological surgery modeling.

15.2 MODELING OF ONCOLOGICAL SURGERY

Two main mathematical approaches for predicting cancer behavior can be applied, namely, computational oncology and physical oncology [683]. The former is related to the detection of patterns and retrieval of information by means of *omics* data mining that could be neglected otherwise; whereas the latter aims to develop mechanistic models that recapitulate cancer behavior and therefore, to capture the trajectories that different types of cancers might follow. These theoretical approaches are discussed focusing on the main variables considered by each model. Remarkably, different classifications of oncological surgery models exist (*e.g.*, [684]), though we consider the aforementioned the most convenient.

15.2.1 Computational oncology models

Analysis of metabolic and gene expression patterns provide valuable information about tumors structure and dynamics, and it has been showed that they are tightly linked to clinical outcomes [685]. For instance, an unsupervised machine learning approach to modeling pancreatic cancer was able to cluster apart metastatic cells from those of primary tumors solely by looking into the molecular features of the neoplasm, which is important for defining the best therapeutic options for a particular patient [686]. Such classification can be even more thorough leading to new subdivisions that make more exhaustive predictions on response to treatment [686]. A transcription analysis of pancreatic tumors also reported that the phenotype (clustered into classic, quasimesenchymal and exocrin-like) is related to clinical outcomes and differential response to therapies [687].

The advances in computational oncology have integrated genomic, transcriptomic, proteomic and microenvironmental data [688, 682], but its application on prediction of oncological surgery outcomes started to emerge later than those on diagnostics and drug discovery. In the next paragraph we describe some examples of computational oncology models that have proven to enhance biologically-based predictions using quantitative models.

Smith *et al.* applied a Support Vector Machine model (see Chapter 9) to predict whether a patient with colorectal cancer will respond to neoadjuvant therapy with data from serum protein spectroscopy (SELDI-TOF); the outcomes from this non-invasive proteomics analysis achieved acceptable specificity and sensibility (ca. 80%) [689]. A similar model using these data was tested in the context of patients' follow-up after surgery, being able to detect advanced stages of the disease [690]. Another application of this analysis focused on liver metastasis of colorectal cancer [691], obtaining specific signatures of the disease. Accordingly, another study showed that gene expression patterns can predict response to neoadjuvant therapy using a 54-genes signature, with similar sensibility and specificity [692].

In the same line, gene expression has also been used as a predictive biomarker for recurrence after resection [693, 694]. Importantly, studies including routinely made histopathological assessments had highlighted the need of more powerful biomarkers in order to select candidates for adjuvant therapy [695].

Interestingly, computational oncology approaches are versatile in terms of the type of outcomes that can be predicted, and with an adequate data set, virtually everything could be subject to predictive modeling. *E.g.*, Hosseini *et al.* used clinical information, namely the preoperative assessment related to different types of surgical interventions (oncological included), for generating a linear regression model that predicted the duration of surgeries, with an R^2 of 0.65 [696]. Predicting and controlling more variables related to the patients would lead to a better and integral health care and allocation of resources [696]. Another study applied artificial neural networks and logistic regression models to the epidemiological data openly provided by the Surveillance, Epidemiology, and End Results Program (SEER) from the NCI (http://seer.cancer.gov/). In this case the prediction was high in sensitivity and specificity of 1-, 3- and 5-years survival for thyroid cancer, in which surgery is a substantial part of the treatment [697]. As a third and final example, Tanaka *et al.* developed a decision tree model for identifying candidates with better prognosis after locoregional surgical treatment for hepatocellular carcinoma. They found that a total tumor diameter < 4 cm and an alfa-fetoprotein level lower than 73 ng/dL are factors that predict a positive prognosis with a p<0.0001 [698].

It is clear now that computational oncology exploits the vast data sources to optimize cancer treatment and prognosis calculation. Complementary types of data that can be used by such approaches include non-coding RNA data [699], processed natural language from clinical files [700, 701], clinical imaging analysis [702], among others.

Case study: Gene expression and surgery prognosis in HCC

In a recent publication, Nault *et al.* [685] validated a transcription data-driven prognosis score for hepatocellular carcinoma (HCC). Surgery is a potentially curative treatment for HCC, despite the high recurrence rates. These authors take advantage of the fact that recurrence is associated with specific events such as local invasion and *de novo* tumor formation, which in turn correlate with both tumor and host phenotypic traits. Previous clinical assessment tools included only macroscopic and liver function variables. To add more details, and given the phenotypic diversity observed within HCC, this research studied the expression pattern of 103 genes in resection samples of more than 400 patients undergoing a potentially curative resection, as determined by clinical evaluation. After a 60 months follow up period the authors applied a multivariate Cox model and found 5 genes that predicted survival in a better way than previous prognostic models; thus, defining a new biomarker for prediction of outcomes in HCC.

15.2.2 Mechanistic models from physical oncology

Mechanistic models attempt to reproduce the behavior of tumors step by step, in as much detail as possible. They are able to generate a dynamic picture of the tumor as it develops and interacts with its microenvironment. These models also allow definition of the bio-physical laws of cancer in a mathematical framework, which can afterwards be used for predicting the tumor's growth and response to therapies [683]. Physical models of cancer, as a general attempt to recapitulate the tumor's behavior, are capable of including aspects related to therapeutics (*e.g.*, chemo- and radiotherapy), immunotherapy, oncolytic viruses, surgery, and also combinations of them (see [703] for a detailed review). It is also possible to include as many physiological variables as desired even before the addition of therapeutic interventions. In this way, a much more complex and realistic framework can be created. Nevertheless, the core of every model relies on the basic assumptions regarding the way in which the tumor both develops and reacts to its environment. In the following sections, some of the most important variables are discussed; afterwards, some representative models are illustrated.

15.2.2.1 *Relevant variables*

Previous to the implementation of mechanistic models, it is fundamental to revisit the variables for modeling the behavior of a tumor within a general framework. Tracqui [703] and Chauviere *et al.* [683] have already identified many topics and variables appropriate for the initial conception of mathematical/physical models. The following list aims towards the update and convenient reorganization of the generic concepts considering the current state-of-the-art in computational oncological surgery.

1. *Intrinsic factors*

   (a) **Tumor size**. This central variable influences diffusion barriers, indirectly defining hypoxic regions and resistance to therapies [704]. In some models (*e.g.*, *gliomas*), it might play a more crucial role, since mass effect in these tumors has a determinant effect on survival [705].

   (b) **Cancer stem cells**. Typically a small fraction of the tumors' tissue consists of cancer stem cells which hold the potential of self-renewal, differentiation and tumorigenicity (*i.e.*, the capability of generating new tumors) [706]. Significant efforts have been made for targeting this important subpopulation [707].

   (c) **Tumor invasion and aggressiveness**. In medicine, *invasion* might be regarded as the capacity and velocity at which a tumor spreads out of its own limits, directly affecting the functioning of the surrounding healthy tissue. On the other hand, *aggressiveness* refers to the rate at which it develops, grows or disseminates. For example, Tanase and Waliszewski [708] proposed a quantitative measure of prostate carcinoma aggressiveness. They defined a low destructive behavior in terms of a low potential for metastasis formation in conjunction with the absence of significant influence on survival. Importantly, the degrees of invasion and the aggressiveness can coexist and even be correlated [709]. For instance, positive surgical margins due to higher invasiveness result in poorer outcomes [710]. A third term, *progression* —a predominantly clinical term— is somewhat ambiguous and difficult to model computationally, as it is related to the course of a disease, specifically when it becomes worse or spreads throughout the body [711].

(d) **Hypoxia-induced phenomena**. Low oxygen availability results in ulterior angiogenesis, necrosis and metabolic changes, which promote cancer progression and metastasis [712, 713].

(e) **Diffusion phenomena**. Nutrients and oxygen are required by the neoplastic cells. Given that the need for oxygen grows relative to availability, hypoxic regions lead to angiogenesis providing neovasculature to the tumor and supporting its growth [714]. Additionally, deprived irrigation causes a reduction in the transportation of oxygen, nutrients and also the drugs that could be administered, hindering their potential benefit [715].

(f) **Angiogenesis**. This phenomenon is acknowledged as crucial for the tumoral development and is among the hallmarks of cancer [716]. As the tumor growths and develops hypoxic regions, it secretes angiogenic factors, aiming to support its metabolic activities. Therefore, angiogenesis is often related to tumor progression and sometimes it correlates with a more advanced disease [717]. Also, antiangiogenic drugs can be effectively combined with other treatments [718].

(g) **Mechanisms of immune evasion and resistance to therapies.**
Indeed, many of the aforementioned factors provide the tumor with resistance not only to the host immune response, but to many of the therapies available. Some factors also involved in tumor response to perturbations are: the diffusion barrier; mutations of susceptibility genes, enhanced adaptation to DNA damage, expression of multidrug resistance proteins; and population heterogeneity, [719, 720, 721, 722].

(h) **Metastasis**. Given the poor prognosis associated with this phenomenon, several models of metastasis have been developed, and some of them consider oncological surgery [723, 724]. Metastasis is a quite stochastic process that varies according to histological type, as well as to epidemiological and biological variables such as nutrients availability and angiogenesis [725, 723, 724]. Metastasis occurs when tumor cells invade the host stroma, penetrate blood vessels, and enter the systemic circulation to produce new colonies in distant organs [723]. Given the critical impact of this process in patients' survival, Bezencry *et al.* validated a mathematical model capable of predicting the postsurgical metastatic potential, and then validated it clinically [724].

Some insights should be noted from this first set of tumor intrinsic variables. If no constraints were to be exerted on the tumor, only the growth rate would determine its size or cellularity at any given time t, consequently producing an exponential growth function; however, it is known since many decades that this is not the case [726]. Of note, the tumor is embedded from the very beginning within numerous factors that limit its growth. Hence, tumor cells deploy a variety of strategies to overcome these limitations.

So far, we have discussed some tumor intrinsic factors that pursue adaptation to a medium that is hostile for the development and survival of such type of cells. In the following section, we mention some of the adaptive exchange that result from the interaction between a tumor and its medium.

2. *Factors related to the niche of tumor and microenvironment*

(a) **Host tissue**. Tissue surrounding the tumor and its microenvironment plays an important role by exerting selective pressure on cells, which favors those with the highest malignant potential [727]. Besides, the microenvironment is crucial for fulfilling the necessities of the tumor and provides resistance mechanisms against therapies and immune response, as well as a hypoxic medium [728, 729, 712].

(b) **Mechanical forces**. It has been shown that pressure exerted as a consequence of the tumor development can be a limiting factor for its own growth [730]. When this factor is neglected, models cannot predict mass effect and deformation of the healthy tissue [731]. Also, tumor invasiveness is influenced by the malleable properties of the extracellular matrix [732].

3. *Systemic or external*

(a) **Immune response and inflammation**. In patients with metastatic colorrectal cancer, the resection of the primary tumor is associated with a reduced systemic inflammation and a prolonged survival [733]. This could be the case also in other types of cancer.

(b) **Treatment schemata**. Depending on the treatments available for each type of cancer, different settings of treatment could be modeled. For example, in pancreatic cancer, besides neo- and adjuvancy [734, 735], resection of recurrences of local disease [648] improves survival of patients.

(c) **Multiscale modeling**. Different scales within the models should be considered for cancer modeling and therapy proposals instead of being overlooked [736, 737]. Compatible with this is the systems biology paradigm that recognizes many explanatory levels in cancer development, progression, and its interaction with the host. Accordingly, the biological process is regarded as an integrated whole from punctual mutations, to gene networks, to metabolites, to cell populations, to evolution, to systemic and multiorganic responses.

15.2.2.2 Implemented models

A variety of approaches have been applied to the simulation of tumor growth and resection. Eikenberry *et al.* proposed a model integrating the development of melanoma, providing plausible hypotheses for explaining its recurrence and proclivity to metastasize after surgery [738]. Their approach included modeling coupled boundary limit ordinary differential equations (ODEs) that recapitulated tumor growth, invasion, metastasis, surgery, and immune response. They concluded that after surgery, the growth rate and aggressiveness of micro-metastases rises, leading to recurrence of the disease. Indeed, ODEs and partial differential equations (PDEs) have been extensively used for modeling these events, both complementing experimental and clinical instances, and investigating the mathematical validity of current paradigms [739].

Bezencry *et al.* [724] adapted a previous metastasis framework to incorporate surgery as a variable for modeling breast cancer behavior after surgery. They considered growth and dissemination of metastasis, as well as of the primary tumor. They estimated the parameters of the ODEs through nonlinear mixed effects model, and validated their results through mice experiments and clinical data. Although the immune system was not explicitly included in the model, they observed mice that displayed a reduced growth rate or eradication of the metastasis after surgery, and this was attributed to the immune system heterogeneity within the population. Finally, they analyzed clinical data regarding more than 2,000 patients who underwent surgery for breast cancer treatment, and concluded that a strong nonlinear relation exists between presurgical tumor size and postsurgical metastasis growth and survival. They developed a prognosis tool that considered the metastatic potential and aggressiveness of the lesion to predict the survival after surgery and therefore assess whether surgery should be performed. This elegant study demonstrates that even simple theoretical models are able to make strong predictions.

In another study by Eikenberry *et al.* [738], the growth of glioblastoma was simulated with mechanistic assumptions and a set of stochastic partial differential equations describing tumoral cells behavior and, to some extent, their microenvironment. They retrieved the parameters of the model from the literature and considered tridimensional growth within the brain. The tumors were subject to surgery and chemo- and radiotherapy, emulating clinical settings. They were able to predict outcomes from simulations of different combinations of therapies. Finally, the model was applied in a personalized manner to recapitulate the clinical history of a patient who underwent three resections, as well as cycles of radio- and chemotherapy. After tuning the parameters, and considering the Magnetic Resonance Imaging (MRI) studies for the initial values of the tumor size and location, the simulated clinical history was similar to the actual one. However, probably because the model did not consider mechanical forces, the mass effect exerted by the tumor was not reproduced. Major advances in modeling glioblastoma treatment have been done, such as the use of MRI for estimating hypoxic regions prone to higher migration rates and less survival [740, 741, 742], as well as targeting angiogenesis and predicting its effects after resection [743], and even the generation of a personalized clinical outcomes calculator inputting MRI data for estimating diffusive capacity and growth rate and then simulating the net effect of the treatment in terms of survival [744].

A major problem in the assessment of these models is related to the lack of public repositories where sufficient clinical data is made available for benchmarking against the theoretical results. Consequently, researchers cannot evaluate the models, or they obtain very small samples. The US National Cancer Data Base offers plenty of clinical data regarding oncological surgery, though it is not publicly available [745]. Promisingly, the Comprehensive Neuro-Oncology Data Repository (CONDR) will be openly released after June 2017, providing a vast amount of follow-up neuroimaging, as well as histological and clinical data [745]. The development of theoretical models extends our knowledge of cancer behavior, and should be clinically validated for its application to predicting treatment outcomes of patients.

15.3 EXAMPLE: A BIDIMENSIONAL ONCOLOGICAL SURGERY SIMULATION MODEL

In this section, a simulated toy example is applied to study the modeling of a tumor resection using an adapted multiscale approach [746, 747, 748, 749]. This model takes into account a bidimensional space (x, y), diffusion of essential (M) and division-related (N) nutrients, interactions among cells, and the response of the immune system. Natural killer (NK) and cytotoxic T lymphocytes (CTL) are considered in order to include the innate and adaptive immune cell response against the tumor. The framework is a partial differential equation (PDE) boundary-value problem at the level of diffusion (of nutrients) and a cellular automaton (CA) at the corresponding cell level. The following pseudocode was modified from [748]:

1. Define parameters, as well as the bidimensional spatial domain.

2. Determine CA initial conditions.

3. Solve steady-state PDEs for calculating the availability of nutrients at every point in the CA grid.

4. Randomly assign an event to every tumoral cell.

5. Calculate the probabilities of the relevant events happening according to the defined CA rules; then compare the values obtained against random numbers to determine whether the events will happen or not.

6. Evaluate the rules for determining the fate of the immune cells (CTL and NK).

7. Update the CA.

8. Repeat steps 3-6 until the assigned final step, or if tumor reaches domain edge or it is finally eradicated.

15.3.1   Step 1: diffusion of nutrients

We can start by describing the behavior of tumoral cells restricted by the availability of nutrients. The non-dimensionalized reaction-diffusion equations can be expressed as [746, 748]:

$$\frac{\partial N}{\partial t} = \nabla^2 N - \alpha^2(H+I)N - \lambda_N \alpha^2 TN, \tag{15.1}$$

$$\frac{\partial M}{\partial t} = \nabla^2 M - \alpha^2(H+I)M - \lambda_M \alpha^2 TM, \tag{15.2}$$

where t is the time measured in relative adimensional units, the Laplacian operator represents the diffusion of nutrients, whereas the substracting terms are the consumption rates of the nutrients by cells. Table 15.2 summarizes the description of the rest of the variables and parameters. This model assumes that tumoral cells consume nutrients at a higher rate that than healthy host cells (due to the formers' increased metabolism). For the sake of simplicity, these will hereby be assumed as static parameters, and thus we use the default values provided within the references cited above.

Using the package texttttReacTran in texttttR programming language [750, 361], Equations 15.1 and 15.2 can be considered as diffusion-reaction functions (see Appendix 15.6 for the code). For the integration of the PDEs, boundary values were set at $(x,0) = (x,L) = 1$. Periodic boundary conditions were considered in the ranges $(0,y)$ and (L,y). Thus, PDEs were solved as steady-state since it is assumed that diffusion occurs almost instantly as compared to other cellular processes [746]. Figure 15.1 shows the progression of diffusion scalar fields as the tumoral population grows.

TABLE 15.2   Description of the variables and parameters considered in the modeling of nutrients' diffusion. Adapted from [748].

Variable/parameter	Description
$N(x, y, t)$	Non-essential (related to cell division) nutrients concentration
$M(x, y, t)$	Essential (related to survival) nutrients concentration
$H(x, y, t)$	Proportion of normal host cells
$I(x, y, t)$	Proportion of live immune cells (NK+CTL)
$T(x, y, t)$	Proportion of live tumor cells
$NK(x, y, t)$	Proportion of live NK cells
$CTL(x, y, t)$	Proportion of live CTL cells
λ_N	Excess consumption factor of tumoral over non-tumoral cells on nutrient N
λ_M	Excess consumption factor of tumoral over non-tumoral cells on nutrient M
α	Rate of nutrients consumption by host cells
L	Grid side size

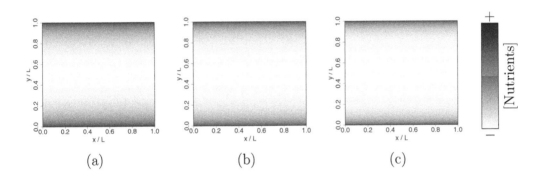

(a)   (b)   (c)

Figure 15.1 Diffusion scalar fields for nutrient N in a grid of $L = 250$, at $\alpha = 4/L; \lambda_N = 100; a)$ $T = 0.01, b)$ $T = 0.04, c)$ $T = 0.08$, x, y axis are divided by L

15.3.2   Step 2: CA rules and tumor growth constrained by the nutrients concentration and immune system response

The rules for the CA take into account the following assumptions:

1. Simulations start at $t = 0$ and $T = 1/L^2$.

2. Tumoral cells are randomized at every step to undergo death, division or migration events. Afterwards, the probability of each event is compared against a random number to add variability and stochasticity to the model.

3. Tumoral cells can die either due to the lack of nutrients (P_{nec}) or the immune system activity around the cell (P_{imdth}).

4. Tumoral cells may divide (P_{div}) with a probability proportional to nutrient N availability and inversely proportional to the actual number of tumoral cells.

5. Tumoral cells may migrate (P_{mig}) with a probability proportional to the concentration of nutrient M and the number of tumoral cells. Thus, the tumor attempts to avoid overcrowding.

6. Immune cell population condition is a parameter fixed at $I_0 = 0.01$. To hold the immune cell population, P_{NK} is calculated at any given t and for every grid cell in the CA. Thus NK cells can increase to meet the initial condition I_0.

7. NK cells die whenever they kill a tumoral cell. Afterwards, this event induces the recruitment of CTL.

8. CTL cells are able to kill a fixed number k of tumoral cells. At any time CTL cells kill a tumoral cell, the former can be further induced/recruited with a probability P_L proportional to the number of tumor cell neighbors. After NK cells die for killing a tumoral cell, more of them can also be recruited. Therefore, CTL cells die with a probability P_{LD} inversely proportional to the number of tumoral cells around them.

Therefore, the following rules are defined for CA conditions considering the parameters of shape θ_i.

$$P_{nec} = exp\left[-\left(\frac{M}{T\theta_{nec}}\right)^2\right], \tag{15.3}$$

$$P_{imdth} = 1 - exp\left[-\left(\sum_{j\in\eta} I_j\right)^2\right], \tag{15.4}$$

$$P_{div} = 1 - exp\left[-\left(\frac{N}{T\theta_{div}}\right)^2\right], \tag{15.5}$$

$$P_{mig} = 1 - exp\left[-T\left(\frac{M}{\theta_{mig}}\right)^2\right], \tag{15.6}$$

$$P_{NK} = I_0 - \frac{1}{L^2}\sum_{j=1}^{n^2} I_j, \tag{15.7}$$

$$P_L = exp\left[-\left(\frac{\theta_L}{\sum_{j\in\eta} T_j}\right)^2\right], \tag{15.8}$$

$$P_{LD} = 1 - exp\left[-\left(\frac{\theta_{LD}}{\sum_{j\in\eta} T_j}\right)^2\right]. \tag{15.9}$$

There, the summations account for the cell neighbors adjacent to any given CA grid cell. For more details on these rules and their derivation, please see the original references in [746, 747, 748, 749].

This model approaches the tumor growth as a bulky mass with a necrotic core, or with a papillary architecture [746]. A papillary pattern can be reproduced depending on the diffusion parameters regardless of the immune system [746, 751] and agrees with pathology observations, in which the size of tumors can be determined based on the shape and mitotic activities at the edges [752]. Figure 15.2 shows a semi-log growth curve and the image of a simulated papillary and a bulky tumor, neglecting immune system. Figure 15.3 depicts the tumor and the immune cells distribution at the end of the simulation considering the immune system with $I_0 = 1\%$. Although the morphology is similar to that depicted in Figure 15.2b, a greater proportion of necrotic cells is observed. Interestingly, highly populated tumor zones show lower immune cells density.

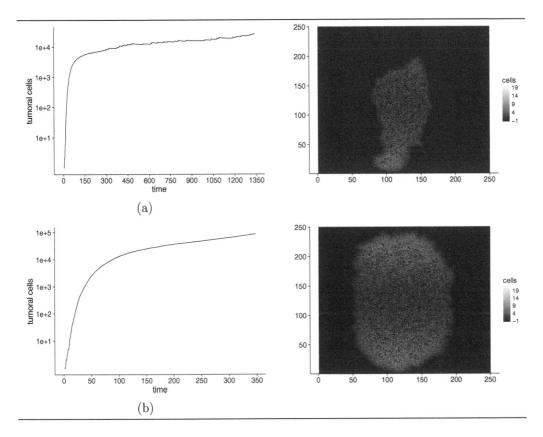

Figure 15.2 Semi-log growth curve and image of a papillary (a) and a bulky (b) tumor at the end of the simulation. The number of cells is rendered by color, being higher density of cells being brighter, and necrotic cells represented by negative values. Parameters in a) $L = 250; t = 1340; \alpha = 2/L; \lambda_N = 100; \lambda_M = 10; I_0 = 0; \theta_{nec} = 0.03; \theta_{div} = 0.3; \theta_{mig} = 1000$. Parameters in b) $L = 250; t = 348; \alpha = 1/L; \lambda_N = 50; \lambda_M = 25; I_0 = 0; \theta_{nec} = 0.03; \theta_{div} = 0.3; \theta_{mig} = 1000$

15.3.3 Step 3: surgery

An incomplete resection could be assumed such that it leaves behind tumoral cells in the margins of the tumor. Therefore, in the simulation we can remove an inner grid of tumoral cells containing a given percentage of them. Afterwards, we can track the development

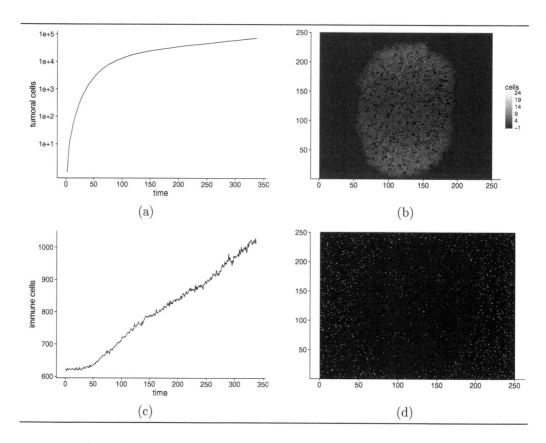

Figure 15.3 Semi-log growth curve and image at the end of the simulation of a bulky tumor (a and b, respectively) and its associated immune cell response (c and d, respectively). The number of cells is rendered by color, being higher density of cells brighter, and necrotic cells represented by negative values. $L = 250; t = 337; \alpha = 1/L; \lambda_N = 50; \lambda_M = 25; I_0 = 1\%; \theta_{nec} = 0.03; \theta_{div} = 0.3; \theta_{mig} = 1000; k = 20; \theta_L = 0.3; \theta_{LD} = 0.9$

and behavior of these margins. Figure 15.4 shows the behavior of a tumor with the same parameters and immune response as in Figure 15.3, but incorporating a 95% resection at time $t = 150$. Not surprisingly, after resection margins resume exponential growth, as the competition for the nutrients is dramatically decreased, and these cells are also nearer to the nutrients sources (Figure 15.4a)). Interestingly, immune cell response does not experience the same exponential growth (Figure 15.4c)). This tumor behavior is in agreement with the higher rates of metastasis seen after surgery in melanoma and breast cancer [738, 753, 754], and provides a plausible explanation for the usefulness of adjuvant therapy for eradicating the remaining cells, which are able to drive recurrency [755]. It has also been considered that the parameters of the immune cell response may determine the final eradication of a tumor after surgery [756], so the parameters in this model could be adapted to represent real patient data. It is of the utmost importance to understand tumor margin behavior and its interactions.

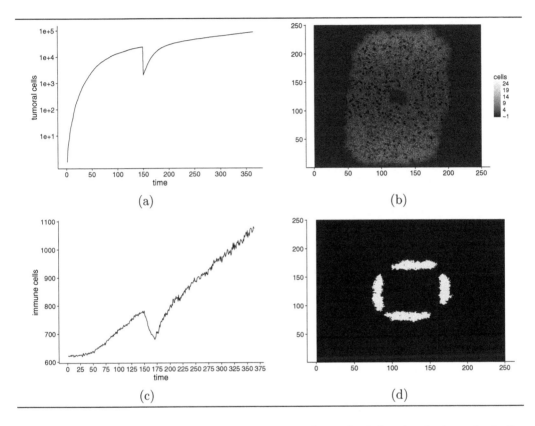

Figure 15.4 Semi-log growth curve and image at the end of the simulation of a bulky tumor (a and b, respectively). In b), the number of cells is rendered by color, being higher density of cells being brighter, and necrotic cells represented by negative values. The associated immune cell response is shown in c). A 95% resection was performed at $t = 150$, and during the immediately following time lapse, exponential growth was resumed. Margins after resection are shown, depicting only presence (white) or ausence (black) of tumoral cells (d). $L = 250; t = 337; \alpha = 1/L; \lambda_N = 50; \lambda_M = 25; I_0 = 1\%; \theta_{nec} = 0.03; \theta_{div} = 0.3; \theta_{mig} = 1000; k = 20; \theta_L = 0.3; \theta_{LD} = 0.9$

15.4 DISCUSSION

The proposed model is studied as a toy example, in which the implementation of a multiscale modeling and the considerations of the immune system and cell interactions were addressed to recapitulate the determinant impact they have on tumor growth. However, the model proposed shows some disadvantages: (1) even though the nutrients concentration influences the cells' outcomes to some extent, in this model they do not exert a direct influence on the selection of the event (*i.e.*, it is randomly assigned); (2) it is assumed that cell death, migration and division events occur in the same time scale, which is an oversimplification of these rather complicated processes [757]; (3) this model has not been validated in clinical settings [748]; (4) originally, its main purpose was to model the behavior of early tumor stages, so it neglects angiogenesis [748], though it considers that metastasis could occur when the tumor reaches the outer limit of the spatial domain [749]. In principle, this model could be extended by including angiogenesis, for example with the avascular to vascular transition modeled by Santagiuliana *et al.* [758].

Similar models have been proposed, some of them enriched with clinical evidence. Haeno *et al.* simulated pancreatic cancer progression from a single cell, which reproduces, mutates its genes, and finally acquires metastatic potential [759]. Remarkably, this framework allowed the study of the behavior of tumor sub-populations and their response to chemo- and radiotherapy, as well as surgery. They concluded, upon clinical validation of the model, that aggressive systemic therapy would have a strong impact on survival, regardless of the stage of the disease [759]. Kim used a hybrid model intended to predict whether glioblastoma cells would migrate or proliferate after surgery. It comprised a tumoral molecular regulating mechanism that is well-known for this type of cancer, as well as spatial modeling, chemo-attractants reaction-diffusion equations and mechanical forces . The conclusion was that local administration of chemoattractants after surgery could prevent dissemination and accelerate relapse detection by imaging studies [760].

It cannot be assumed that tumors undergoing surgery are in initial stages; however, this model can be considered as a simulation of the tumor margins, which in turn play a strikingly important role in patients' outcomes [761]). Under such circumstances, the model proposed constitutes a to-scale simulation of the cellular interactions between the tumor and the immune system.

For comparison, a simplified mathematical model based on a system of ordinary differential equations recapitulating tumor growth and immune system response, was clinically validated by De Pillis *et al.* with patients suffering from melanoma [762]. Within this framework, in a recent study we proposed a model for predicting the clinical outcomes of patients undergoing tumor resection [756]. However, the approach presented in this chapter has the advantage of incorporating spatial data and diffusion of nutrients, and as a multiscale approach for modeling, it can incorporate many different variables, treatment included.

15.5 CONCLUSIONS AND PERSPECTIVES

Computational medicine is a novel field which focuses on developing predictive models. Oncological surgery models are still under development since biological and clinical variables represent a challenge to modeling research. In contrast to other therapies, clinical oncological surgery possesses relatively few data available to validate the predictions. Newer physical models have now incorporated imaging studies as initial conditions for the simulations, with good results predicting outcomes in small samples of patients [763, 764, 765, 731, 766]. Remarkably, imaging techniques can provide data not only regarding the size and growth of tumors, but also about their physical conditions, architecture and vascularization [767, 763].

Other types of data should be integrated as well, namely the physical cell interactions, which can be effectively measured *in vitro* [768]. As it has been recently highlighted, newer clinical studies should provide better resources for posterior data mining [769]. This approach is named "data farming" and its application entails better availability of data for validating *in silico* models. Hence, it is plausible to expect the emergence of new tools from physical models, as well as hybrid models incorporating both approaches. Given that oncological surgery has potentially severe complications, we consider that theoretical approaches can provide better strategies in clinical settings for choosing the adequate surgical treatment.

Acknowledgments. Financial support from the National Council of Science and Technology of Mexico (CONACYT, grant 269399) is greatly appreciated. J. Jesus Naveja and Leonardo Zapata-Fonseca thank PECEM from the Faculty of Medicine, UNAM, for the support and the organization of a combined MD PhD program. Leonardo Zapata-Fonseca acknowledges funding from PAPIIT-DGAPA-UNAM (grants IN113013, IN106215 and IV100116), CONACYT (grants 221341 and 167441) and the Newton Advanced Fellowship from the British Academy of Medical Sciences.

15.6   APPENDIX 1: R CODE

R implementation of the simulated toy example as described in the chapter, thoroughly annotated. Note that results may differ slightly as it is a probabilistic model. Please, cite this chapter when using the code provided below.

```
#Initial conditions
L=250 #length of the side
celltype=matrix(data=rep("host", L**2), nrow=L, ncol=L) #matrix of cell
Tum=matrix(data=rep(0, L**2),nrow=L, ncol=L) #empty tumor matrix
I0-0.01 #initial proportion of immune cells
NK=trunc(L**2 * I0) #initial number of NK cells
celltype[sample(1:L**2, NK, replace=T)]="NK"
Tum[round(L/2),round(L/2)]=1 #initial tumoral cell
celltype[Tum > 0]="tumor"

#Surgery conditions
tsurg=150 #vector determining timing(s) of the surgery
rct=0.95 #vector determining percentage(s) of resection

#Function for modeling nutrient diffusion
require(ReacTran)
require(rootSolve)
pde2D=function (t, y, parms) {
CONC=matrix(nr=L, nc=L, y) #vector to 2D matrix
Tran=tran.2D(CONC, D.x=Dx, D.y=Dy, dx=dx, dy=dy #transport phenomenon
C.y.up=BNDy, C.y.down=BNDy) #boundary conditions
```

```
dCONC=Tran\$dC - r * CONC #change in concentration by consumption
return(list(as.vector(dCONC)))}

#Diffusion parameters
L=L #same length of the side
dy=dx=1 #grid
Dy=Dx=1 #diffusion constant
BNDy=rep(1,L) #boundaries
a=1/L #alpha papillary: 2
ln=50 #lambda N papillary: 100
lm=25 #lambda M papillary: 10

#Cellular automata parameters
L=L #same length of matrix
tnec=0.03 #theta necrosis
tdiv=0.3 #theta division
tmig=1000 #theta migration
tout=800 #number of steps
I0=I0 #initial number of immune cells
k=20 #CTL number of possible deaths
tl=0.3 #theta L
tld=0.9 #theta LD

#Functions
#function for adding all neighbors
#based on The game.of.life() function by Petr Keil. 2012
#https://www.r-bloggers.com/fast-conways-game-of-life-in-r/
shift=function(X){
col=ncol(X)
row=nrow(X)
allW=cbind(rep(0,row), X[,-col]) #west
allNW=rbind(rep(0,col), cbind(rep(0,row-1), X[-row,-col])) #northwest
allN=rbind(rep(0,col), X[-row,]) #north
allNE=rbind(rep(0,col), cbind(X[-row,-1], rep(0,row-1))) #northeast
allE=cbind(X[,-1], rep(0,row)) #east
allSE=rbind(cbind(X[-1,-1], rep(0,row-1)), rep(0,col)) #southeast
allS=rbind(X[-1,], rep(0,col)) #south
allSW=rbind(cbind(rep(0,row-1), X[-1,-col]), rep(0,col)) #southwest
X2=allW + allNW + allN + allNE + allE + allSE + allS + allSW #adding all
return(X2)}

#Function stopcrit()
#for determining if limits are reached by the tumor, or it is eradicated
stopcrit=function(X){
lims=sum(X[1,], X[nrow(X),], X[,1], X[,ncol(X)]) #cells within limits
total=sum(X) #total cells
return(lims > 0 | total == 0)}

#The simulation
```

```
t=1
nmessages=100 #100 progress messages
nimages=10
storage=array(0, c(L, L, nimages))

#Initializing variables that will be measured
NI=NH=NNTum=vector(length = tout)

print(system.time(while (t <= tout & ! stopcrit(Tum){
#guess of base nutrients concentration
if(! exists("N")) N=list(); N$y=matrix(nr=L, nc=L, 1)
if(! exists("M")) M=list(); M$y=matrix(nr=L, nc=L, 1)

#Populations total
NI[t]=sum(as.numeric(celltype=="NK" | celltype=="CTL"))
NH[t=sum(as.numeric(celltype=="host"))
NNTum[t]=sum(Tum)
Tot=NI[t] + NH[t] + NNTum[t]

#Populations proportions
I=sum(as.numeric(celltype=="NK" | celltype=="CTL"))/Tot #immune cells
H=sum(as.numeric(celltype=="host"))/Tot #host cells
NTum=sum(Tum)/Tot #tumoral cells

#Optimization
#PDEs are only updated if the number of tumoral cells changes
if(t==1){
#calculating diffusion of N
r=a**2 * (H+I) + a**2 * NTum * ln #consumption rate
N=steady.2D(y=N$y,func=pde2D,parms=NULL,
dimens=c(L,L,lrw=1e+8,cyclicBnd=1)
#cyclic boundaries in x
#image(N, main="2D steady-state diffusion of Nutrients N")
#calculating diffusion of M
r=a**2 * (H+I) + a**2 * NTum * lm #consumption rate
M=steady.2D(y=M$y,func=pde2D,parms=NULL,
dimens=c(L,L,lrw=1e+8,cyclicBnd=1)
#cyclic boundaries in x
#image(M, main="2D steady-state diffusion of Nutrients M")}else{
if(NNTum[t] != NNTum[t-1]) {
r=a**2 * (H+I) + a**2 * NTum * ln #consumption rate
N=steady.2D(y=N$y,func=pde2D,parms=NULL,
dimens=c(L,L,lrw=1e+8,cyclicBnd=1)
#cyclic boundaries in x
r=a**2 * (H+I) + a**2 * NTum * lm #consumption rate
M=steady.2D(y=M$y,func=pde2D,parms=NULL,
dimens=c(L,L,lrw=1e+8,cyclicBnd=1)
#cyclic boundaries in x}}
```

```
#Evaluating the cellular automaton rules
#determining events to happen
#tumoral cells
#assign event and random number per cell
result=celltype
result[result=="tumor"]=sample(c("die", "divide", "migrate"),
size=sum(result=="tumor"), replace=T)

tmpTum=Tum #copy of Tum for adding modifications

#Death by lack of nutrients
die1=matrix(F, nrow=L, ncol=L)
die1[result=="die"]=(exp(-(M$y[result=="die"]/(Tum[result=="die"]*
tnec))**2))=runif(sum(result=="die"))

#Updating
result[die1]="necrotic"
tmpTum[die1]=0
#death by immune response
die2 <- matrix(F, nrow=L, ncol=L)
die2[result=="die"]=(1-exp(-shift(matrix(as.numeric(celltype %in%
c("CTL", "NK")),ncol=L))[result=="die"]**2)) > runif(sum(result=="die"))

result[die2]="necrotic"
tmpTum[die2]=0
result[result=="die"]="tumor" #keep tumor cells that did not die

#Division of tumoral cells
div=matrix(F, nrow=L, ncol=L)
div[result=="divide"]=(1-exp(-(N$y[result=="divide"]/
(Tum[result=="divide"]*tdiv))**2)) >  runif(sum(result=="divide"))

result[result=="divide" & ! div ]="tumor" #cells that will not divide
div=which(result=="divide", arr.ind=T)
tumorcrowd=shift(Tum)

if (sum(result=="divide") > 0) {for (i in 1:nrow(div)){
cell=div[i,] #select the cell
grid=data.frame(row=rep((cell[1]-1):(cell[1]+1), 3),
col=rep((cell[2]-1):(cell[2]+1), each=3)) #cell neighbors
neigh=matrix(celltype[as.matrix(grid)], nrow=3) #celltype neighbors
if(sum(neigh %in% c("host", "necrotic")) > 0){
sel=sample(which(neigh %in% c("host", "necrotic")), 1)
#selecting a suitable space
result[as.matrix(grid[sel,])]="tumor"
#updating results
tmpTum[as.matrix(grid[sel,])]=1}
else { #if there are only tumor spaces in the neighborhood
tosel=as.matrix(grid[neigh == "tumor",])
```

```
#selecting tumoral cells
tosel=tosel[row.names(tosel)!=5,] #avoiding the dividing cell
sel=sample(which(tumorcrowd[tosel] == min(tumorcrowd[tosel])),1)
#sampling least crowded
#updating results
tmpTum[matrix(tosel[sel,], ncol=2)]=tmpTum[matrix(tosel[sel,], ncol=2)]+1 }}}

tmpTum[result=="necrotic"]=0
#if tumoral cell died migration is aborted for consistency
result[result=="divide"]="tumor" #remove divide label

#Migration of tumoral cells
migr=matrix(F, nrow=L, ncol=L)
migr[result == "migrate"]=(1-exp(-(Tum[result=="migrate"]*
M$y[result=="migrate"]/tmig) **2)) > runif(sum(result=="migrate"))
result[result == "migrate" & ! migr ]="tumor"
#tumor cells that will not migrate
migr=which(migr, arr.ind = T)
if (nrow(migr)>0) {for (i in 1:nrow(migr)) {
cell=migr[i,] #select the cell
grid=data.frame(row=rep((cell[1]-1):(cell[1]+1),3),col=rep((cell[2]-1):
(cell[2]+1),each=3))
#cell neighbors
neigh=matrix(celltype[as.matrix(grid)],nrow=3) #celltype neighbors
if(sum(neigh %in% c("host", "necrotic")) > 0){
sel=sample(which(neigh %in% c("host", "necrotic")), 1)
#selecting a suitable space
result[as.matrix(grid[sel,])]="tumor"
#updating results
tmpTum[as.matrix(grid[sel,])]=1}else{
#if there are only tumor spaces in the neighborhood
tosel=as.matrix(grid[neigh=="tumor",])
#selecting tumoral cells
tosel=tosel[row.names(tosel)!=5,]
#avoiding the dividing cell
sel=sample(which(tumorcrowd[tosel] == min(tumorcrowd[tosel])),1)
#sampling least crowded
tmpTum[matrix(tosel[sel,], ncol=2)]=tmpTum[matrix(tosel[sel,], ncol=2)]+1}

tmpTum[matrix(migr[i,], ncol=2)]=tmpTum[matrix(migr[i,], ncol=2)]-1
#migrating the cell
if(tmpTum[matrix(migr[i,], ncol=2)]==0)
result[matrix(migr[i,], ncol=2)]="host"
else
result[matrix(migr[i,], ncol=2)]="tumor"}}
tmpTum[result == "necrotic"]=0 #if cell died migration is aborted

#Immune cells
#replacing NK cells
```

```
random=matrix(runif(L**2), nrow=L) #calculate random number for grid
result[random < (I0 - I)]="NK"
tmpTum[random < (I0 - I)]=0

#CTL flagged induction
if(sum(celltype=="flag")>0) {
if(!exists("kvec")) {
kvec=cbind(matrix(which(celltype=="flag",arr.ind = T),ncol=2),
rep(k,sum(celltype=="flag")))}else{
kvec=rbind(kvec,cbind(matrix(which(celltype=="flag",arr.ind=T),ncol=2),
rep(k,sum(celltype=="flag")))) }
result[celltype=="flag"]="CTL"
#induction if NK cell died due to lysing a tumoral cell}

#Effects after lysing tumoral cell
#determining the causal cell and affecting it
tolook=which(die2 == T, arr.ind = T) #lysed cells
if(length(tolook)>0) {for (i in 1:nrow(tolook)){
cell=tolook[i,] #select the cell
grid=data.frame(row=rep((cell[1]-1):(cell[1]+1),3),
col=rep((cell[2]-1):(cell[2]+1),
each=3))
#cell neighbors coordinates
grid=grid[apply(grid, 1, function(x) all(x<L) & all(x>0)),,drop=F]
#selecting a valid grid
neigh=matrix(celltype[as.matrix(grid)], nrow=3) #celltype neighbors
tosel=as.matrix(grid[neigh %in% c("NK", "CTL"),]) #selecting immune cells
sel=sample(nrow(tosel),1) #sampling for determining the causal cell
#if NK dies and poses and induction flag for CTL
#if CTL, its k is reduced by 1 and it may further induce CTLs

if(celltype[matrix(tosel[sel,], ncol=2)]=="NK")
result[matrix(tosel[sel,], ncol=2)] ="flag"
else{
kvec[,3][kvec[,1]==tosel[sel,][1] &
kvec[,2]==tosel[sel,][2]] = kvec[,3][kvec[,1] == tosel[sel,][1] &
kvec[,2] == tosel[sel,][2]] - 1 #reducing k

#For the neighbors evaluate CTL induction probability
cell=tosel[sel,]
grid=data.frame(row=rep((cell[1]-1):(cell[1]+1), 3),
col=rep((cell[2]-1):(cell[2]+1),each=3))

#Cell neighbors coordinates
grid=grid[apply(grid, 1, function(x) all(x<L) & all(x>0)), drop=F]
#selecting a valid grid
neigh=matrix(celltype[as.matrix(grid)], nrow=3) #celltype neighbors
tosel=as.matrix(grid[neigh %in% c("host", "necrotic"),])
#selecting free cells around
```

```
pind= 1-exp(- (tl/tumorcrowd[tosel])**2) #probability of induction
pind[tumorcrowd[tosel] == 0]=0 #completeness
sel<-tosel[pind > runif(nrow(tosel)),]
#select comparing against random number
result[matrix(sel, ncol=2)] = "CTL" #add CTL
if(length(sel) > 0) kvec=rbind(kvec, cbind(matrix(sel, ncol=2),k))
#consider k}}}


#Death of CTL
#by having too few tumor neighbors
if(sum(celltype=="CTL") > 0) {
dieCTL=(1 - exp(-(tld/tumorcrowd[matrix(kvec[,1:2], ncol=2)])**2)) >
runif(nrow(kvec)) | apply(kvec, 1, function(x) any(x <= 0))
result[kvec[dieCTL,1:2,drop=F]]="necrotic"
kvec=kvec[!dieCTL,,drop=F]
kvec=kvec[!apply(kvec, 1, function(x) any(x <= 0)),,drop=F]
#remove CTLs when out of domain (after migration)}


#Migration of immune cells
#taking result matrix for not considering immune cells that died
#NK with no tumor neighbors will move one space away randomly
migrNK=which(result=="NK" & tumorcrowd == 0, arr.ind = T)
if (nrow(migrNK) > 0) { for (i in 1:nrow(migrNK)){
cell=migrNK[i,] #select the cell
grid=data.frame(row=rep((cell[1]-1):(cell[1]+1), 3),
col=rep((cell[2]-1):(cell[2]+1),
each=3))
#cell neighbors
grid=grid[apply(grid, 1, function(x) all(x<L) & all(x>0)),,drop=F]
#selecting valid grid
neigh=matrix(celltype[as.matrix(grid)], nrow=3) #celltype neighbors
if(sum(neigh %in% c("host", "necrotic"))>0) {
sel=sample(which(neigh %in% c("host", "necrotic")), 1)
#selecting a suitable space
result[as.matrix(grid[sel,])]="NK" #updating results
result[matrix(migrNK[i,], ncol=2)]="host" #migrating the cell}}}


#Performing the surgery
if(t %in% tsurg) {
grid=tmpTum
i=1
while(i<L & sum(grid)/sum(tmpTum)>rct) {
grid=tmpTum[(i+1):(L-i),(i+1):(L-i)]
i=i+1}
result[(i+1):(L-i),(i+1):(L-i)]="host"
tmpTum[(i+1):(L-i),(i+1):(L-i)]=0}
t=t+1 #updating time
tmpTum[result!="tumor"]=0 #reassuring consistency
Tum=tmpTum #updating tumor grid
```

```
celltype=result #updating grid
if(t %in% seq(1,tout,tout/nmessages)) print(paste(100*t/tout, "% done..."))
if(t %in% seq(1,tout,tout/nimages))
image(Tum); storage[,,which(seq(1,tout,tout/nimages)
==t)]=celltype}))
```

CHAPTER 16

Tools for the precision medicine era: developing highly adaptive and personalized treatment recommendations using SMARTs

Elizabeth F. Krakow

Erica E. M. Moodie

CONTENTS

ABSTRACT

In this chapter, we aim to acquaint the reader with a new clinical trial design for developing personalized longitudinal treatment strategies. Sequential multiple assignment randomized trials (SMARTs) offer several benefits over traditional randomized and factorial designs such as, for example, the ability to detect synergistic and delayed effects of sequential treatments. SMARTs are especially suitable for identifying key tailoring characteristics that indicate the superiority of one sequence of treatments compared to alternatives for a given patient as the patient evolves over time. SMARTs are analyzed with statistical techniques rooted in machine learning, which offers methods for handling high-dimensional data (*i.e.*, where the number of predictors relative to the number of patients or events is large) and may detect unexpected interactions between covariates and treatment. We describe a data-driven, evidence-based approach that can be adopted by hematologists and oncologists, who are challenged with assessing many novel, possibly prescriptive biomarkers in the context of burgeoning sequentially-administered treatments. As a motivating example, we consider the prevention of graft-*versus*-host disease throughout the chapter.

Keywords: outcome models, personalized medicine, adaptive treatments, adaptive trials, SMART.

16.1 INTRODUCTION

Here are two approaches to predicting the effect of a treatment. Population-level ("marginal") predictions report the average of the predicted outcomes had everyone in the population been given a treatment, say A, compared to the average of the predicted outcomes had everyone been given some alternative, say B. By contrast, individual-level ("conditional" or "personalized") prediction entails predicting the outcome for an individual exposed to treatment A or B, given his relevant characteristics. Usually the primary goal of a randomized clinical trial (RCT) is to compare treatments by comparing groups of highly similar patients exposed to different treatments (or placebo), *i.e.*, focusing on population-level prediction. In practice, we recognize that the treatment effect might vary in magnitude or even in direction (protective *vs.* harmful) depending on key individual characteristics. In the presence of such effect adaptation or interaction, we are usually careful to report separate estimates for populations restricted to each level of the modifying factor, typically in the form of stratified or subgroup analyses.

Tailoring variables (also called prescriptive variables) are an important subtype of effect modifier for which knowledge of the value of the variable helps to inform optimal treatment choice. That is, tailoring variables actually changes what ought to be written on the prescription pad. For instance, with regard to all patients with acute myeloid leukemia (AML), translocation between chromosomes 15 and 17 resulting in the *PML/RARA* fusion gene identifies who should be prescribed all-*trans* retinoic acid. To provide the most highly personalized treatment recommendations, all relevant tailoring variables must be identified.

The presence of the *PML/RARA* fusion gene is a static baseline characteristic, but many tailoring variables naturally assume different values over time. Sequential multiple-assignment randomized trials were developed specifically to study interventions that adapt treatment to an individual's responses and other changing characteristics over time

[770, 771]. In a SMART, patients are re-randomized to interventions each time a treatment decision must be made; data must be analyzed with suitable statistical techniques. Like traditional RCTs, SMARTs can be used to assess the comparative efficacy of the options available at each isolated decision point, but their advantage lies in providing data essential (1) for comparing the effectiveness of longitudinal strategies and (2) for personalizing treatment over time by revealing useful tailoring variables [772].

16.2   STUDYING TREATMENTS IN SEQUENCE

16.2.1   Adaptive treatment strategies

Adaptive treatment strategies (ATSs), also called dynamic treatment regimes, are sequences of decision rules (one per stage of intervention) for adapting treatments to time-varying states of an individual, where decisions made at one stage may affect those to be made at future stages and where the long-term effect of the current treatment may depend on the performance of future treatment choices [773]. For example, giving prednisone 1 mg/kg to a patient presenting grade II acute graft-*vs.*-host disease (GVHD) is legitimate because if it fails, evidence suggests that she can likely be salvaged with 2 mg/kg or other agents, and starting with the lower dose will not compromise her long-term outcome [774]. Longitudinal decision points in an ATS may occur at regular intervals, such as bimonthly visits, or at defined clinical events, such as remission, relapse or onset of a complication. ATSs are dynamic because the recommended actions for a particular patient can change based on observations made over time (*e.g.*, whether GVHD responds to glucocorticoids). By formalizing the study of ATSs, we hope to improve long-term outcomes. This can be done by attempting to identify an optimal ATS from data or by comparing an ATS (or more than one) with another treatment protocol in terms of their utility.

16.2.2   Decision rules

An ATS is a collection of decision rules. Assume two possible treatments at each interval j are coded as "action" $A_j \in \{0, 1\}$. Consider, for example, the rule "Give induction chemotherapy $A_1 = 1$ if the patient responds, follow by maintenance chemotherapy $A_2 = 1$, otherwise (*i.e.*, if no response) give second-line induction $A_2 = 0$." Under this rule, some patients will experience the trajectory $A_1 = 1 \rightarrow$ response $\rightarrow A_2 = 1$ while others will experience the trajectory $A_1 = 1 \rightarrow$ no response $\rightarrow A_2 = 0$. All treatment-relevant information from the patient's history, *i.e.*, tailoring variables such as responses to past treatments or lack thereof, can be incorporated into a decision rule. The decision rule outputs individualized treatment recommendations such as when to start, stop, or modify treatment.

16.2.3   Tailoring variables are key for personalized recommendations

For a variable to be useful in tailoring treatment, it must discriminate optimal treatment choices for at least some sub-population. Let O represent a putative tailoring variable (and o represents specific levels of that variable), A the treatment type (coded as $a = 0$ or $a = 1$), and Y the primary outcome for which a higher value is preferred such as quality of life, or disease-free survival time. Suppose that Y is described by a linear regression model:

$$Y = \beta_0 + \beta_1 O + (\psi_0 + \psi_1 O)A + \text{error}. \tag{16.1}$$

Then, mathematically, the concept of tailoring can be expressed as follows: If $(\psi_0 + \psi_1 O)$ is positive for some values of O and zero or negative for other values of O, then O is a tailoring variable (Figure 16.1), and the optimal treatment rule would be to treat with option $A = 1$ whenever $(\psi_0 + \psi_1 o) > 0$ since this will lead to a positive increase in the expected outcome for an individual with $O = o$. The magnitude of ψ_1 determines the strength of the "prescriptive" effect of the tailoring variable, $i.e.$, its impact on the choice of therapy.

Other forms of decision rules are possible. For example, Cotton and Heagerty [8] consider ATSs in end-stage renal disease that adjust erythropoietin dose A_j at time j multiplicatively based on the dose in the previous month, A_{j-1}, and the most recent hematocrit, O_j:

$$A_j = \begin{cases} A_{j-1} \times (0, 0.75) & \text{if } O_j > \varphi - 3 \\ A_{j-1} \times (0.75, 1.25) & \text{if } O_j \in (\varphi - 3, \varphi + 3) \\ A_{j-1} \times (1.35, \infty) & \text{if } O_j \geq \varphi + 3 \end{cases} \quad (16.2)$$

where φ is the midpoint of the target hematocrit range. To find the ATS that would maximize survival time, φ was varied from 31% to 40%, and patient hematocrit at a given time point, O_j, took the role of tailoring variable.

Defining good tailoring variables is a common objective in designing an optimal ATS. Categorical and continuous covariates may be tailoring variables. The net effect of multiple covariates can also be used for tailoring treatment. That is, if a regression model for the outcome contains interactions of covariates $O_1, ..., O_p$ with A such that $(\psi_0 + \psi_1 O_1 + \psi_2 O_2 + ... + \psi_p O_p)$ may be positive for some values $(o_1, o_2, ..., o_p)$ and zero or negative for other values of the linear combination of predictors, then this linear combination of variables, $(\psi_0 + \psi_1 O_1 + \psi_2 O_2 + ... + \psi_p O_p)$, can be used to tailor treatment to the individual by assimilating many characteristics simultaneously.

When decisions are taken over time, further considerations are required. For instance, in the erythropoietin example given above, the treatment decision threshold φ does not vary over time; it is the same at each month j. In some situations, it is desirable to use different thresholds at different disease stages. For example, in treating chronic myeloid leukemia (CML), the $rate$ at which patients who initiate a tyrosine kinase inhibitor (TKI) ought to achieve particular reductions in BCR/ABL mRNA transcripts to maximize survival time can be captured by assigning φ decreasing values of transcript burden the further away from the patient is from TKI initiation ($e.g.$, $\leq 10\%$ International Scale by 3 months and $\leq 0.1\%$ IS by 12 months) [775, 776]. An ATS for CML might be represented by the following rule:

$$A_j = \begin{cases} \text{Switch to an alternative TKI} & \text{if } O_j > \varphi_j \\ \text{Continue current TKI} & \text{if } O_j \leq \varphi_j \end{cases} \quad (16.3)$$

where A_j is the action to be taken at time j, O_j is the observed quantification of BCR/ABL transcripts at time j, and φ_j is the time-varying upper acceptable limit for O_j. This approach is of interest in malignancies where quantification of minimal residual disease (MRD) is incorporated into treatment algorithms but "safe" and "dangerous" MRD levels likely depend on the initial disease burden, the time since initiating therapy, toxicity of alternative treatments, and patient-specific covariates. Figure 16.2 illustrates a hypothetical SMART designed to address the questions such as when MRD should be assessed to maximize survival or whether the timing of MRD assessment should depend on baseline characteristics ($e.g.$, initial white cell count) or on evolving characteristics ($e.g.$, a comorbidity index

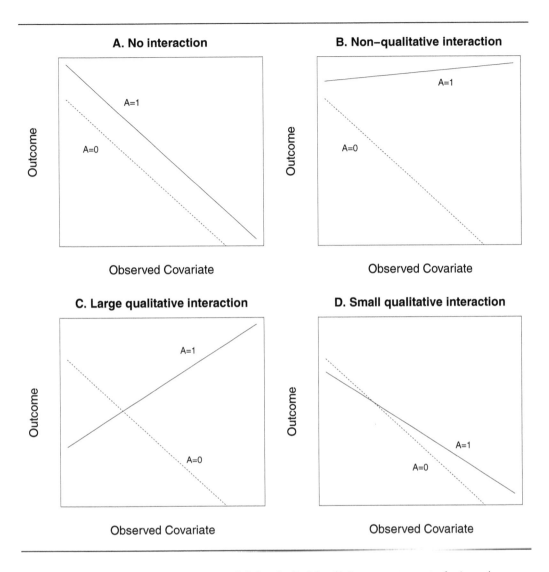

A. No interaction

A=1

A=0

Outcome

Observed Covariate

B. Non–qualitative interaction

A=1

A=0

Outcome

Observed Covariate

C. Large qualitative interaction

A=1

A=0

Outcome

Observed Covariate

D. Small qualitative interaction

A=1

A=0

Outcome

Observed Covariate

Figure 16.1 Some covariates are useful for individualizing treatment choice. Assume the goal is to maximize the outcome Y (*e.g.*, Y is survival time or hospital-free days). Assume the horizontal axis displays concentrations of a serum protein biomarker in a population, increasing from left to right. Panel A shows no interaction: Treatment $A = 1$ is equally and uniformly better than treatment $A = 0$ at all levels of the biomarker. Panel B shows a non-qualitative interaction: the magnitude of the benefit of treatment 1 over treatment 0 differs depending on the level of the biomarker. However, the biomarker is not a tailoring variable because the same treatment choice would apply to all patients. Panel C shows a large qualitative interaction. Treatment $A = 0$ is better at low levels of the biomarker while treatment 1 is preferred at higher levels of the biomarker. The biomarker is useful for tailoring treatment to the individual patient. Panel D shows a qualitative interaction of lesser magnitude, but the same conclusion as in panel C applies.

[777]). Further, researchers may wish to compare competing ATSs, which incorporate different treatment-decision thresholds φ_j, as well as different tailoring variables such as baseline Sokal scores, to see which optimizes survival.

16.2.4 Machine learning "teaches" us the optimal ATS

Machine learning (ML) is a branch of artificial intelligence that allows computers to "learn" from example inputs and then make predictions or decisions. The fields of ML and causal inference have yielded several methods to estimate the optimal ATS, in both frequentist, *e.g.*, [778, 779, 780] and Bayesian, *e.g.*, [781] paradigms. We will briefly outline one ML approach, Q-learning [778], which is appealing because of its simplicity. In fact, there are methods better suited to the SMART design as they leverage knowledge of the randomization mechanism, but these are less intuitive and knowledge of methods beyond regression is required to understand these alternatives.

For simplicity, Q-learning will be explained for the setting of two treatment intervals; however the method extends naturally to any fixed number of intervals. Longitudinal data on a single patient are given by the trajectory $(C_1, O_1, A_1, C_2, O_2, A_2, Y)$, where C_j and O_j ($j = 1, 2$) denote the set of covariates measured prior to treatment beginning at the j-th interval. Y is the outcome at the end of interval 2. A_j ($j = 1, 2$) represents the action (treatments) which is assigned at the j-th interval subsequent to observing O_j; we will assume only two treatment options at each interval, coded $\{0, 1\}$. Two types of covariates are distinguished. Those represented by O_j interact with treatment and are called tailoring or prescriptive variables; they directly impact the optimal choice of A_j. Those represented by C_j do not interact with treatment but are predictive of the outcome. The data consist of a random sample of n patients. The history at each interval is defined as $H_1 = (C_1, O_1)$ and $H_2 = (C_1, O_1, A_1, C_2, O_2)$; these may be further split into main effects terms $H_{j,0} = H_j$ and tailoring variables that interact with treatment, $H_{j,1} = O_j$, $j = 1, 2$. A two-interval ATS consists of two decision rules, (d_1, d_2), with d_j a function of history H_j (or more specifically, O_j) yielding an action, A_j, to be taken. The goal of Q-learning is to build regression models that correspond to a transformation of the outcome so as to estimate treatment rules of the form "treat when $(\psi_{j0} + \psi_{j1}O_{j1} + \ldots + \psi_{jp}O_{jp}) > 0$" for $j = 1, 2$.

The Q-learning algorithm for a continuous outcome proceeds as follows:

Step 1 Posit a model for $E[Y|H_2, A_2]$, denoted $Q_2(H_2, A_2; \beta_2, \psi_2) = \beta_2^\top H_2 + (\psi_2^\top O_2)A_2$. Estimate the interval 2 parameters using ordinary least squares to regress the outcome on all predictive variables, tailoring variables, the treatment of interest, and interactions between the treatment and potential tailoring variables.

Step 2 Estimate the optimal interval 2 rule for each history $H_2 = h_2$:

$$\hat{d}_2^{opt}(h_2) = \arg\max_{a_2} Q_2(h_2, a_2; \hat{\beta}_2, \hat{\psi}_2). \tag{16.4}$$

Step 3 Create a new outcome of interval 1 by assuming that patients had received optimal treatment for interval 2. Patient by patient, determine which A_2 treatment would lead to the best outcome given each patient's observed values of all previous treatments and tailoring variables, using the regression parameter estimates from Step 1. For each patient i, create a new (transformed) outcome that is the predicted outcome for that

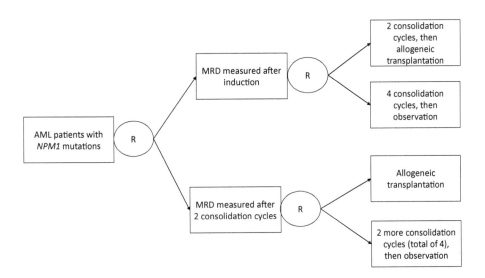

Figure 16.2 Suggested design for constructing a trial based on patient characteristics and *NPM1* minimal residual disease (MRD) responses. The chance of relapse of cytogenetically normal AML is influenced by the presence and persistence of detectable mutations in the gene encoding nucleophosmin (*NPM1*). In general, when the relapse risk is judged to exceed 35%, allogeneic stem cell transplantation during first complete remission may be warranted, instead of simply continuing post-remission ('consolidation') chemotherapy. However, when and how sensitively to measure *NPM1* MRD, and whether the results should influence the decision of whether to proceed to allotransplant in first remission, are areas of active research. Both the ideal timing (*e.g.*, after induction or after 2 cycles of consolidation) and the target depth of MRD response can be learned from the two clinically-relevant ATSs and the non-adaptive strategy embedded in this SMART: (1) Measure MRD only after induction → transplant (after 2 cycles of consolidation) if post-induction MRD exceeds threshold ψ, otherwise complete 2 more cycles of consolidation and then observe. (2) Measure MRD only after 2 cycles of consolidation chemotherapy → transplant if MRD exceeds threshold ψ', otherwise complete 2 more consolidation cycles and then observe. (3) Ignore MRD measurements (compare all patients randomized to 2 cycles of consolidation plus transplant vs. all patients randomized to consolidation alone). Note that the values of the thresholds ψ and ψ' might depend on other salient baseline and evolving patient characteristics, such as the transplant comorbidity index; including such attributes in the longitudinal data collection enables the development of more highly tailored ATSs, providing truly personalized treatment recommendations. The trial design could be further enhanced by measuring MRD at multiple time points in all patients. CR: a complete remission. R: time points at which patients are to be randomized.

patient under his own optimal treatment at the A_2 decision point:

$$\tilde{Y}_{1,i} = \max\{\hat{\beta}_2^\top H_{2,i}, \quad \hat{\beta}_2^\top H_{2,i} + \hat{\psi}_2^\top O_{2,i}\} \tag{16.5}$$

$$= Q_2(H_2 = h_{2,i}, A_2 = \hat{d}_2^{opt}(h_{2,i}); \hat{\beta}_2, \hat{\psi}_2). \tag{16.6}$$

Step 4 Estimate the interval 1 parameters. Using ordinary least squares regression, regress the new outcome from Step 3 on all baseline predictive variables, all baseline tailoring variables, interval 1 treatment, and interactions between the interval 1 treatment and all tailoring variables, fitting the model $E[\tilde{Y}_1|H_1, A_1]$, denoted $Q_1(H_1, A_1; \beta_1, \psi_1) = \beta_1^\top H_1 + (\psi_1^\top O_1)A_1$.

Step 5 Estimate the optimal interval 1 rule for each history $H_1 = h_1$:

$$\hat{d}_1^{opt}(h_1) = \arg\max_{a_1} Q_1(h_1, a_1; \hat{\beta}_1, \hat{\psi}_1). \tag{16.7}$$

Note that in Steps 2 and 5, the optimal ATS is identified for each patient given the tailoring variables included in the models constructed in Steps 1 and 4: treatment is personalized on the variables for which interactions with treatment were included in the model. The estimated optimal ATS is given by $(\hat{d}_1^{opt}, \hat{d}_2^{opt})$. In Step 4, the algorithm uses patient-specific predicted outcomes rather than observed outcomes. In doing so, all individuals are put on an level playing field, so that we can isolate and contrast only the impact of the interval 1 treatment, with the impact of any sub-optimal second interval treatment removed.

Q-learning and other forms of ATS estimation can be adapted to a variety of outcomes including binary response (*e.g.* 2 year disease-free survival) or time-to-event outcomes such as survival time, as well as for treatments that are not binary, such as a continuous-valued dose of a drug. As the focus of this chapter is on the design or trials, rather than the many possible forms of analysis, we leave it to the interested reader to pursue a deeper understanding through references such as the suggested further readings given below.

16.3 COMPARISON TO TRADITIONAL METHODS

16.3.1 Why might RCTs fail to identify good treatment sequences?

Individual patients often respond differently to the same treatments, either in terms of a primary outcome, side effects, or both. Inter-patient variability and intra-patient variability over time motivate the *personalized medicine* paradigm, which emphasizes systematic use of an individual's updated information to optimize care [773]. Caring for a person with cancer is challenging because personalization must be marshalled through multiple stages of intervention, such as induction, consolidation, maintenance, and salvage. The same is true for many chronic diseases such as chronic GVHD, where immunosuppressive regimens must frequently be adjusted based on an individual's symptom burden, organ involvement pattern, experience of side effects, and infections.

The optimal treatment identified by a generic RCT might not be best for a particular patient because (1) the combination of baseline characteristics that is his "signature" might not be captured by coarse strata or coarse subgroups in the RCT; (2) the effect of time-varying characteristics might dwarf the initial effect of the baseline strata or subgroup to which he belonged; and (3) there may be delayed synergism or antagonism between the treatment examined in the trial and further therapies that are required beyond the initial, randomized treatments and yet it is well-known that post-randomization variables cannot be included in an analysis without the risk of introducing spurious associations.

To expand on this last idea about contamination from downstream interventions, one might be tempted to develop a model where first-stage treatment, response status after first-stage treatment and subsequent treatment options are input as predictors of the long-term outcome. However it is well-known that post-randomization variables cannot be included in an analysis without the risk of introducing spurious associations. For that reason, post-randomization variables cannot be incorporated as predictors of the main treatment effect.

16.3.2   Why can't we combine results from separate, single-stage RCTs?

Combining results from separate RCTs to construct a clinical practice guideline poses difficulties for several reasons. First, exchangeability of source populations is difficult to verify. Second, single stage RCTs may not adequately explore treatments that would be of interest for multiple stages of treatment due to the single stage RCT's inability to account for the performance of future treatment options. Consider the example of acute GVHD prophylaxis and treatment. A guideline developer wishes to know the efficacy of anti-thymocyte globulin (ATG) compared to a "standard" intervention, such as cyclosporine and methotrexate without ATG for prophylaxis, or mycophenolate for acute GVHD treatment. The optimal speed of tapering immunosuppressants among patients with no or minimal acute GVHD is also in question. Suppose the incidence of acute GVHD is 40% with standard prophylaxis and that 52% of patients who develop acute GVHD require salvage treatment (defined as additional immunosuppressants beyond glucocorticoids), regardless of the type of prophylaxis received [782]. Further suppose ATG prophylaxis (added to cyclosporine/methotrexate) decreases the incidence of steroid-refractory GVHD by half. However, also suppose that ATG treatment leads to a 1-year overall survival (OS) among those requiring salvage GVHD treatment of 59% *only among* patients who did not receive ATG prophylaxis, and that due to infection and relapse, ATG treatment actually decreases 1-year OS to 25% among those who previously received ATG prophylaxis. Finally, suppose that among those without GVHD or with steroid-responsive GVHD, rapid immunosuppression taper improves 1-year OS compared to slow taper, at 76% *vs.* 66%, in the standard prophylaxis arm, but due to infection and relapse, slow taper decreases survival among those who received ATG prophylaxis to 60%. Each of the eight possible treatment and clinical outcome trajectories, and the associated survival rates, are depicted in Figure 16.3.

Practice guidelines are often constructed by stringing together single-stage trial results. For example, at the A_1 decision point, nine randomized trials compared ATG *vs.* standard prophylaxis, showing reduction in the incidence of grade II-IV acute GVHD but no benefit in overall survival [783, 784, 785]. Another trial compared rapid (A_2-taper = 1) *vs.* slow (A_2-taper = 0) tapers in responders, *i.e.*, patients without GVHD or with steroid-responsive GVHD [786]. Among patients with acute GVHD who did not respond to steroids alone, another trial compared ATG (A_2-treatment = 1) to high-dose methylprednisolone (A_2-treatment = 0) on the basis of the proportion achieving complete or partial responses after 1 month, toxicity, and survival [786]. It is tempting to piece together these (and other) separate trials to suggest a "best" treatment strategy among the possible A_1, A_2 choices by first selecting the apparent-best prophylaxis (A_1) for preventing GVHD, then the best treatment (A_2) for prolonging survival if GVHD were to develop. However, in this (hypothetical) example depicted by Figure 16.3, the treatment that gives the best intermediate

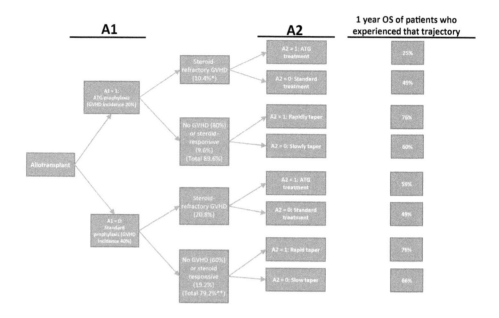

Figure 16.3 Hypothetical trajectories post allotransplant in a population. At the A_1 decision point, action 1 refers to ATG prophylaxis and action 0 refers to "standard" (non-ATG) prophylaxis. At the A_2 decision point, patients with steroid-refractory GVHD may receive ATG treatment (action 1) or standard treatment (action 0). Patients with no or minimal steroid-responsive GVHD may be prescribed a rapid taper (action 1) or a slow taper (action 0) of immunosuppression. Abbreviations: ATG - Antithymocyte globulin, GVHD - graft-vs-host disease, OS - overall survival.

response ($A_1 = 0$ yields a lower proportion of steroid-refractory GVHD cases) also renders intensification at the time GVHD develops less effective and so the probability of 1-year survival is lower even when the optimal second-stage treatment is given.

Similar considerations are pertinent to many areas of oncology. Achieving deep remissions with intensive therapy might only be beneficial in the long-term if early and late treatment-related mortality is low and undesirable effects, such as peripheral neuropathy or drug-induced renal insufficiency, do not preclude effective salvage options. Selecting resistant clones is also a concern. On that note, a re-analysis of a prostate cancer RCT using appropriate ATS methodology suggested that, out of four chemotherapeutic combinations, the front-line treatment favored in a traditional analysis that focuses on intermediate outcomes may in fact be inferior to the alternatives when the entire treatment sequence was considered because, although fewer patients progressed initially, those who did progress responded more poorly to salvage [787]. In summary, because of delayed effects and difficult-to-predict net effects, studying entire strategies is preferable to proposing a sequence of treatments based on studies that examine only one component of the sequence.

16.3.3 What are the advantages of SMARTs?

Two key advantages of a SMART over a series of single-stage RCTs are the abilities (1) to estimate whether long-term outcome differs substantially between two or more competing sequences and (2) for a specific ATS, to improve outcomes for future patients by suggesting how to further tailor treatment according to baseline or evolving characteristics. Note: A SMART is designed to propose sequences of *treatments that are adapted to the individual patient*, however *a SMART is not an adaptive trial design*. In adaptive trials, the probability of being randomized to a particular treatment is updated depending on the outcomes of patients already enrolled to give higher probability to treatments that look more promising. Adaptive trials classically seek to minimize the number of patients exposed to inferior treatments. In contrast, although randomization probabilities might depend on the covariates in a SMART, the randomization probabilities remain fixed throughout the study's duration [773]. There is some research into integrating adaptive trial methodology into SMARTs [788, 789]; however, an adaptive trial design is most useful for studying conditions where outcomes are known quickly relative to the pace of trial enrollment and this is typically not the case when studying sequences of treatments for chronic diseases.

Let us briefly consider how one might actually implement a SMART so as to account for synergism or antagonism between prophylaxis and both taper speed and GVHD treatment. One approach is to randomize patients up-front to ATG or standard prophylaxis. At a well-defined point in clinical care, those who develop no or minimal GVHD would be randomized to rapid or slow immunosuppression taper while those with steroid-refractory GVHD would be randomized to ATG or standard second-line treatment. Allowing for optimal randomization probability at the prophylaxis stage, 80% power and 5% type I error in detecting each salient interaction outlined above, and no losses-to-follow-up, such a trial would require enrolling 4932 patients. Alternatively, the same trial design could be operationalized by a single randomization per patient to one of four possible treatment strategies:

AB: ATG prophylaxis followed by ATG treatment or rapid taper, depending on response;

Ab: ATG prophylaxis followed by standard treatment or slow taper, depending on response;

aB: Standard prophylaxis followed by ATG treatment or rapid taper, depending on response;

ab: Standard prophylaxis followed by standard treatment or slow taper, depending on response.

Both approaches to the randomization will provide data in the same form, and the choice in approach may be guided by practical considerations such as maintaining blinding as well as ethical concerns, such as whether patients providing consent are equally able to understand and be informed for each study design.

Conducting a SMART requires careful consideration of power calculations which is beyond our scope; the interested reader is referred to recent publications [772, 790, 791]. A recent case study comparing a large SMART to all single-stage RCTs conducted in the same time period for the same treatments and condition found that the SMART had lower rates of patient attrition [792], suggesting that not only can SMARTs provide data on longer-term outcomes and sequences of treatments, but that the data may be of a higher quality if patient retention is improved.

16.3.4   Motivating example

Returning to our hypothetical trial aimed at developing an ATS for GVHD, we conducted a simulation in which the survival patterns for a population of 20,000 transplant recipients given their specific prophylaxis, taper speed, GVHD outcomes (none vs. steroid-responsive vs. steroid-refractory), and GVHD treatment regimens (ATG vs. non-ATG) were consistent with the patterns given in Figure 16.3. Then, to simulate a trial, we randomly drew a sample of size n from the population. We 'conducted' the SMART by randomizing the n trial participants first to A_1 and then to A_2, each time using a randomization probability of 0.5. We then employed Q-learning to determine the best ATS. Finally, to simulate clinical practice after the trial is reported, we randomly sampled 300 new patients from the same population and applied the best ATS for each of the new patients. For each patient, we compared the individual patient prediction to what was known from the (pre-programmed) observed trajectories in the population to be the true best choice for that patient at that time. We repeated this procedure of drawing random samples, running the trial, fitting the model, and out-of-sample prediction for 300 random, non-trial participants 1000 times for each trial sample size n, varying n from 40 to 996.

Our metric for comparing the SMART procedure across sample sizes is the proportion of the 1000 trials in which the Q-learning algorithm selected the correct GVHD prophylaxis and (as applicable) taper speed or salvage treatment for all subsequent 300 non-trial patients. Figure 16.4 shows that with approximately 560 trial participants, all treatment-tailoring variable interactions are detected with perfect patient-level accuracy by models developed from at least 80% of trials.

These comparisons are in estimated treatment strategies only, effectively comparing point estimates. This may not reflect real world practice, where institutions or clinical societies are likely to be hesitant to adopt an ATS that has not been shown to be (statistically) significantly better than an alternative treatment strategy or current practice. A much larger sample size would be needed to ensure significant inference about the optimal ATS. However SMARTs are often viewed as developmental studies, used to suggest a candidate ATS that can then be verified in a head-to-head comparison with another treatment strategy in a confirmatory trial.

16.4   VALIDATING A PROPOSED ATS

16.4.1   If we find an optimal ATS with a SMART, do we still need an RCT?

SMARTs are often viewed as a way of constructing evidence-based ATSs when there exists legitimate equipoise about the optimal sequence of treatments or the utility of tailoring variables. Most methodologists advocate that a RCT be performed to confirm the best ATS suggested by a SMART, *e.g.*, by randomizing patients to either ATS-directed care or usual care. Alternatively, the benefit of the proposed ATS might be evaluated non-experimentally in observational studies before its adoption on a wider scale. For instance, the performance of institutions that implement the ATS could be compared to those that do not, or within a given institution, the ATS could be implemented with outcomes compared to

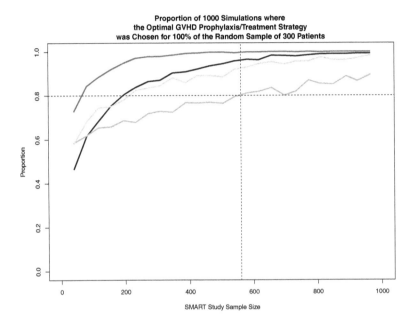

Figure 16.4 Simulation of the results of a SMART for GVHD prevention and treatment. The graph depicts the probability of providing the optimal ATS (correct prophylaxis drug, taper speed, and treatment for refractory GVHD) to 300 new, non-trial patients. For example, a patient who received ATG prophylaxis and developed no or minimal GVHD should be prescribed a rapid taper of immunosuppression (since the 1-year OS among such patients is 76%) instead of a slow taper (since the 1-year OS for the latter trajectory is 60%). Thus, the blue line represents the proportion of 1000 simulations in which a rapid taper instead of a slow taper was correctly selected for the entire subset of trial patients who received ATG prophylaxis and developed no or only minimal GVHD. Similarly, the green line represents choosing standard second-line GVHD treatment over ATG GVHD treatment for patients who developed refractory GVHD after ATG prophylaxis. The red line represents choosing ATG second-line GVHD treatment over standard GVHD treatment for patients who developed refractory GVHD after standard prophylaxis. The yellow line represents recognition that under optimal subsequent treatment or taper speed, ATG prophylaxis is better than standard prophylaxis.

historical 'controls' (*i.e.* those patients treated at the institution prior to the adoption of the ATS). Two difficulties with this latter approach are the possibility that other technologies, procedures, or policies that are changed concurrent with the implementation of the ATS may affect outcomes and be incorrectly attributed to the ATS, and to maintain a true 'control' setting, all patients would be treated under the policy that was in place at their transplant date (ATS or usual/clinician-determined care). This nuance is necessary because if prophylaxis was chosen according to the ATS, acute GVHD second-line treatment ought to be chosen according to the ATS too. If GVHD or survival indices were to improve for patients transplanted under the ATS and if the improvement were to extinguish after withdrawing the ATS, we would have strong, albeit circumstantial, evidence of benefit. In summary, adoption of a particular ATS could only be endorsed if it is evaluated with the same rigor as other interventions.

16.4.2 Are SMARTs used in cancer?

Although SMARTs are most popular in the field of psychology [793, 794, 795, 796, 797], designs that successively randomize patients to initial treatments and then to subsequent treatments based on their individual responses are already widespread in oncology. Examples can be found in lung cancer [798], neuroblastoma [799], prostate cancer [788], myeloma [800], lymphoma [801, 802, 803, 804] and leukemia [805, 806] trials (although results of successive randomizations are often published separately, *e.g.*, LY-12 [805, 802]). Thus, the knowledge and infrastructure needed to conduct SMARTs in a cancer care context exist, and previously-published trials are yielding new insights when re-analyzed as SMARTs with ML approaches [787, 772, 807]. These SMARTs, ranging in size from 150 to 632 patients, varied considerably in their complexity, with most investigating two treatment options in each of two stages (four possible treatment pathways), and one considering four options at the first stage and three at the second (12 possible pathways). See Figure 16.5.

SMARTs in which patients are re-randomized to interventions as time progresses are relatively rare for cancer treatment studies; it is still more common for patients to be randomized to a planned sequence of treatments up-front. A current example of a SMART is a trial enrolling 240 patients with metastatic kidney cancer at MD Anderson Cancer Center. The goal is to compare 6 different two-drug sequences of everolimis, bevacizumab or pazopanib to learn which sequence yields the longest time to overall treatment failure (see ClinicalTrials.gov, NCT01217931). In this trial, patients are randomized to one of three initial treatments in the first stage, and non-responders are re-randomized to one of two treatments in the second stage.

Crucially, SMARTs require that both baseline and intermediate tailoring variables be accurately measured and captured. This could require additional attention and resources compared to traditional factorial trials, where data capture prior to successive randomizations might only have included depth of response. Observational data, such as registry data and prospective cohort studies, could also be analyzed with ML approaches and the results might be useful for developing ATSs that could then be tested in SMARTs (against alternative ATSs) or RCTs (against a single alternative ATS or usual care) [808]. This approach has successfully been taken using data from the Center for International Blood and Marrow Transplant Research to test the feasibility of analyzing the sequence of GVHD prophylaxis and treatment for refractory acute GVHD (study GV13-02) [809]. Observational – especially registry data – settings introduce sources of bias that are not encountered in

SMARTs because in the latter, the study sample is rebalanced thanks to randomization at each new decision stage. Analytic challenges encountered in GV13-02 include confounding by indication (*e.g.*, recipients of HLA-mismatched grafts and reduced intensity conditioning were more likely to receive ATG prophylaxis and more severe GVHD was associated both with ATG salvage treatment and poor survival), and estimating the impact of unmeasured confounders.

16.5   CHALLENGES AND OPPORTUNITIES

The personalized medicine paradigm recognizes heterogeneity in patient responses and seeks the systematic use of individual patient information to optimize that health care. While the idea of tailoring treatments to individual patients needs is not new, development of a formal framework for design and analysis of adaptive treatment strategies is still a young area of study and continues to evolve at a rapid pace. Even now, terminology continues to change, with the terms *personalized medicine* and *precision medicine* used interchangeably in some contexts, and carrying different meanings in other settings. Similarly, an ATS may also be referred to as a *dynamic treatment regime* (or regimen), an *individualized treatment rule*, or *targeted therapy*. Despite the as-yet unsettled nomenclature, the basic principle of selecting medical treatments that are tailored or targeted to individual patients at different points in their care is fundamental to the paradigm, however defined.

SMARTs allow systematic exploration of delayed effects and interactions between treatments at different stages of illness. However, while SMARTs are increasingly used in oncology and other branches of medicine and supportive care, they remain unfamiliar to many and thus require additional investments of time, learning, and money. A SMART design is uniquely suited to personalizing treatment options over time, and can lead to better patient retention. Adopting a SMART design might facilitate more rapid incorporation of biomarker-directed therapy and holds promise for advancing research in many areas of oncology. Bringing this approach into the mainstream is both a challenge and an opportunity. Currently, the adaptive treatment recommendations that are being made every day in clinical care are decided by each physician in isolation based on a combination of intuition, experience, and medical society guidelines about how to lace together results from different trials or cohort studies that examine only one decision point in the course of a long-lasting condition, and based on his or her experience from what is usually a far more limited number of patients than those included in the relevant medical literature or in patient registries. The danger of such an approach is that it could fail to detect delayed effects that might enhance or abrogate the benefit of a future treatment, fail to detect side effects that preclude the use of a needed future treatment, or fail to elicit valuable diagnostic information, such as the depth or rapidity of response, that allows personalized selection of the next treatment. The SMART approach offers an evidence-based way to personalize treatment while avoiding those pitfalls. In addition, prior to investing time and financial resources into a SMART or a traditional RCT, registry data analyzed with SMART-inspired methodology might be useful to identify which ATSs merit prospective evaluation.

Disease (Reference)	n	First Randomization	Eligibility criteria for second randomization	Second Randomization
Non-small cell lung cancer (Belani 2003)	401	3 different schedules/dose-levels of paclitaxel and carboplatin	Adequate response at week 16	Paclitaxel continuation therapy vs. observation
Neuroblastoma (Matthay 2009)	379	Before the last induction cycle, patients were randomized to one of 2 consolidation options: autologous purged bone marrow transplantation conditioned with myeloblative chemotherapy and total body irradiation vs. 3 cycles of intensive chemotherapy.	Completion of consolidation with absence of disease progression	13-cis-retinoic acid × 6 cycles vs. observation
Metastatic prostate cancer (Thall 2007)	150	One of 4 chemotherapy regimens (CVD vs. KA/VE vs. TEC vs. TEE)	Non-response (evaluated every 8 weeks)	One of the remaining 3 chemotherapy regimens to which the patient was not already exposed.
Multiple myeloma (Mateos 2010)	260	Induction with one of 2 regimens: VMP vs. VTP	Completion of 6 induction cycles	Maintenance therapy with bortezomib + prednisone vs. bortezomib + thalidomide
Relapsed or refractory aggressive non-hodgkin B-cell lymphoma (Crump 2014 and Kuruvilla 2013)	554	One of 2 chemotherapy regimens: GDP vs. DHAP	Recovery from autotransplant toxicity and progression-free status at 3-5 weeks post-transplant	Maintenance rituximab vs. observation
Untreated diffuse large B cell lymphoma (Habermann 2006)	632	Chemotherapy (CHOP) vs. chemo-immunotherapy (R-CHOP)	Adequate response	Maintenance rituximab vs. observation
Relapsed/resistant follicular lymphoma (van Oers 2006)	465	Chemotherapy (CHOP) vs. chemo-immunotherapy (R-CHOP)	Adequate response	Maintenance rituximab vs. observation
Acute myeloid leukemia (Stone 2001)	388	GM-CSF vs. placebo added to standard induction chemotherapy	Morphologic complete remission and fitness for further therapy	One of 2 consolidation regimens: Cytarabine vs. cytarabine + mitoxantrone

Abbreviations: CVD – cyclophosphamide, vincristine, dexamethasone; DHAP – dexamethasone, high-dose cytarabine, cisplatin; GDP –gemcitabine, dexamethasone, cisplatin; KA/VE – ketoconazole + doxorubicin/vinblastine + estramustine; GM-CSF – granulocyte colony-stimulating factor; R-CHOP: rituximab, cyclophosphamide, doxorubicin, vincristine, prednisone; TEC – paclitaxel, estramustine, carboplatin; TEE – paclitaxel, estramustine, etoposide ; VMP – Velcade (bortezomib), melphalan, prednisone; VTP – Velcade (bortezomib), thalidomide, prednisone.

Figure 16.5 Examples of sequentially randomized trials (SMART) in oncology from different cancer sites including lung, prostate, lymphoma, and leukemia.

GLOSSARY

Adaptive treatment strategy (ATS): A set of rules for determining effective treatments for individuals based on personal characteristics such as treatment and covariate history.

Confounding: The bias that occurs when the treatment and the outcome have a common cause that is not appropriately accounted for in an analysis.

Dynamic treatment regime: See *Adaptive treatment strategy.*

Interaction (statistical): Effect modification, where by the impact of one covariate (such as a treatment) on outcome depends on the value of another variable. In the ATS setting, we can informally say that interaction exists that is relevant to the development of an ATS when the effect of a treatment differs across levels of a clinical or demographic variable (*e.g.* a biomarker, cancer stage, age).

Prescriptive variable: See *Tailoring variable*

Q-learning: A recursive form of estimation based on a sequence of regressions that can be used to estimate an optimal ATS.

Sequential multiple assignment randomized trial (SMART): A randomized experimental design developed specifically to estimate time-varying adaptive interventions, whereby participants may require more than one stage of treatment, and are randomized to one of a set of possible treatments at each stage of intervention.

Tailoring variable: A personal characteristic, either fixed or time-varying, used to adapt treatment to an individual.

FURTHER READING

Bibhas Chakraborty (2011). Dynamic treatment regimes for managing chronic health conditions: a statistical perspective. *American Journal of Public Health*, 101: 40-45.

Bibhas Chakraborty and Erica E. M. Moodie (2013). *Statistical methods for dynamic treatment regimes: Reinforcement learning, causal inference, and personalized medicine.* Statistics for biology and health, Springer, New York, NY.

Michael R. Kosorok and Erica E. M. Moodie (2016). *Adaptive treatment strategies in practice: Planning trials and analyzing data for personalized medicine.* ASA-SIAM statistics and applied probability series, Society for Industrial and Applied Mathematics, Philadelphia, PA.

Bibliography

[1] George E. P. Box. Science and statistics. *Journal of the American Statistical Association*, 71(356):791–799, 1976.

[2] Kenneth P. Burnham, David Raymond Anderson, and Kenneth P. Burnham. *Model selection and multimodel inference : a practical information-theoretic approach.* Springer, New York, 2nd edition, 2002.

[3] Berger James O. Jefferys, William H. Ockham's razor and bayesian statistics. *American Scientist*, 80:64?72, 1991.

[4] Elliott Sober and Mike Steel. The contest between parsimony and likelihood. *Systematic Biology*, 53(4):644–653, 2004.

[5] Anca Bucur, Jasper van Leeuwen, Nikolaos Christodoulou, Kamana Sigdel, Katerina Argyri, Lefteris Koumakis, Norbert Graf, and Georgios Stamatakos. Workflow-driven clinical decision support for personalized oncology. *BMC Medical Informatics and Decision Making*, 16(Suppl 2):87, 2016.

[6] Elizabeth A. Kidd, Issam El Naqa, Barry A. Siegel, Farrokh Dehdashti, and Perry W. Grigsby. Fdg-pet-based prognostic nomograms for locally advanced cervical cancer. *Gynecologic Oncology*, 127(1):136–140, 2012.

[7] Lawrence B. Marks, Ellen D. Yorke, Andrew Jackson, Randall K. Ten Haken, Louis S. Constine, Avraham Eisbruch, Sren M. Bentzen, Jiho Nam, and Joseph O. Deasy. Use of normal tissue complication probability models in the clinic. *International Journal of Radiation Oncology ? Biology ? Physics*, 76(3):S10–S19, 2010.

[8] Philippe Lambin, Ruud G. P. M. van Stiphout, Maud H. W. Starmans, Emmanuel Rios-Velazquez, Georgi Nalbantov, Hugo J. W. L. Aerts, Erik Roelofs, Wouter van Elmpt, Paul C. Boutros, Pierluigi Granone, Vincenzo Valentini, Adrian C. Begg, Dirk De Ruysscher, and Andre Dekker. Predicting outcomes in radiation oncology ?multifactorial decision support systems. *Nature reviews. Clinical oncology*, 10(1):27–40, 2013.

[9] I. El Naqa. Perspectives on making big data analytics work for oncology. *Methods*, 111:32–44, 2016.

[10] Konstantina Kourou, Themis P. Exarchos, Konstantinos P. Exarchos, Michalis V. Karamouzis, and Dimitrios I. Fotiadis. Machine learning applications in cancer prognosis and prediction. *Computational and Structural Biotechnology Journal*, 13:8–17, 2015.

[11] Issam El Naqa, Joseph O. Deasy, Yi Mu, Ellen Huang, Andrew J. Hope, Patricia E. Lindsay, Aditya Apte, James Alaly, and Jeffrey D. Bradley. Datamining approaches for

modeling tumor control probability. *Acta oncologica (Stockholm, Sweden)*, 49(8):1363–1373, 2010.

[12] Dominique Barbolosi, Joseph Ciccolini, Bruno Lacarelle, Fabrice Barlesi, and Nicolas Andre. Computational oncology [mdash] mathematical modelling of drug regimens for precision medicine. *Nat Rev Clin Oncol*, 13(4):242–254, 2016.

[13] H. Nikjoo, D. Emfietzoglou, T. Liamsuwan, R. Taleei, D. Liljequist, and S. Uehara. Radiation track, dna damage and response-a review. *Rep Prog Phys*, 79(11):116601, 2016.

[14] J. F. Fowler. 21 years of biologically effective dose. *Br J Radiol*, 83(991):554–68, 2010.

[15] I. El Naqa, J.D. Bradley, P.E. Lindsay, A. I. Blanco, M. Vicic, A.J. Hope, and J.O. Deasy. Multi-variable modeling of radiotherapy outcomes including dose-volume and clinical factors. *Int J Radiat Oncol Biol Phys*, 64(4):1275–1286, 2006.

[16] Issam El Naqa, Ruijiang Li, and Martin J. Murphy. *Machine Learning in Radiation Oncology: Theory and Application*. Springer International Publishing, Switzerland, 1 edition, 2015.

[17] Naqa Issam El, L. Kerns Sarah, Coates James, Luo Yi, Speers Corey, M. L. West Catharine, S. Rosenstein Barry, and K. Ten Haken Randall. Radiogenomics and radiotherapy response modeling. *Physics in Medicine and Biology*, 2017.

[18] Issam El Naqa. The role of quantitative pet in predicting cancer treatment outcomes. *Clinical and Translational Imaging*, 2(4):305–320, 2014.

[19] John Wilder Tukey. *Exploratory data analysis*. Addison-Wesley series in behavioral science. Addison-Wesley Pub. Co., Reading, Mass., 1977.

[20] Bradley Efron and Robert Tibshirani. *An introduction to the bootstrap*. Monographs on statistics and applied probability. Chapman & Hall, New York, 1993.

[21] Lior Wolf and Stan Bileschi. Combining variable selection with dimensionality reduction. In *Proceedings of the 2005 IEEE Computer Society Conference on Computer Vision and Pattern Recognition (CVPR'05) - Volume 2 - Volume 02*, CVPR '05, pages 801–806, Washington, DC, USA, 2005. IEEE Computer Society.

[22] Trevor Hastie, Robert Tibshirani, and J. H. Friedman. *The elements of statistical learning : data mining, inference, and prediction*. Springer series in statistics,. Springer, New York, NY, 2nd edition, 2009.

[23] Randall C. Johnson, George W. Nelson, Jennifer L. Troyer, James A. Lautenberger, Bailey D. Kessing, Cheryl A. Winkler, and Stephen J. O'Brien. Accounting for multiple comparisons in a genome-wide association study (gwas). *BMC Genomics*, 11:724–724, 2010.

[24] James Coates, Luis Souhami, and Issam El Naqa. Big data analytics for prostate radiotherapy. *Frontiers in Oncology*, 6:149, 2016.

[25] R. Shouval, O. Bondi, H. Mishan, A. Shimoni, R. Unger, and A. Nagler. Application of machine learning algorithms for clinical predictive modeling: a data-mining approach in sct. *Bone Marrow Transplant*, 49(3):332–337, 2014.

[26] G. S. Collins, J. B. Reitsma, D. G. Altman, and K. G. Moons. Transparent reporting of a multivariable prediction model for individual prognosis or diagnosis (tripod): the tripod statement. *Ann Intern Med*, 162(1):55–63, 2015.

[27] B. Emami, J. Lyman, A. Brown, L. Coia, M. Goitein, J. E. Munzenrider, B. Shank, L. J. Solin, and M. Wesson. Tolerance of normal tissue to therapeutic irradiation. *Int J Radiat Oncol Biol Phys*, 21(1):109–22, 1991.

[28] M. T. Milano, L. S. Constine, and P. Okunieff. Normal tissue tolerance dose metrics for radiation therapy of major organs. *Semin Radiat Oncol*, 17(2):131–40, 2007.

[29] M. T. Milano, L. S. Constine, and P. Okunieff. Normal tissue toxicity after small field hypofractionated stereotactic body radiation. *Radiat Oncol*, 3:36, 2008.

[30] L. B. Marks, R. K. Ten Haken, and M. K. Martel. Guest editor's introduction to quantec: a users guide. *Int J Radiat Oncol Biol Phys*, 76(3 Suppl):S1–2, 2010.

[31] L. B. Marks, E. D. Yorke, A. Jackson, R. K. Ten Haken, L. S. Constine, A. Eisbruch, S. M. Bentzen, J. Nam, and J. O. Deasy. Use of normal tissue complication probability models in the clinic. *Int J Radiat Oncol Biol Phys*, 76(3 Suppl):S10–9, 2010.

[32] William F. Morgan. Human radiosensitivity. *International Journal of Radiation Biology*, 89(11):1002–1002, 2013.

[33] M. T. Milano, P. Rubin, and L. B. Marks. *Understanding and Predicting Radiation-Associated Normal Tissue Injury: A Global and Historical Perspective*, volume 1, pages 103–122. Springer, New York, NY, 2014.

[34] S. H. Benedict, K. M. Yenice, D. Followill, J. M. Galvin, W. Hinson, B. Kavanagh, P. Keall, M. Lovelock, S. Meeks, L. Papiez, T. Purdie, R. Sadagopan, M. C. Schell, B. Salter, D. J. Schlesinger, A. S. Shiu, T. Solberg, D. Y. Song, V. Stieber, R. Timmerman, W. A. Tome, D. Verellen, L. Wang, and F. F. Yin. Stereotactic body radiation therapy: the report of aapm task group 101. *Med Phys*, 37(8):4078–101, 2010.

[35] J. M. Robertson, D. H. Clarke, M. M. Pevzner, and R. C. Matter. Breast conservation therapy. severe breast fibrosis after radiation therapy in patients with collagen vascular disease. *Cancer*, 68(3):502–8, 1991.

[36] M. M. Morris and S. N. Powell. Irradiation in the setting of collagen vascular disease: acute and late complications. *J Clin Oncol*, 15(7):2728–35, 1997.

[37] V. Benk, A. Al-Herz, D. Gladman, M. Urowitz, and P. R. Fortin. Role of radiation therapy in patients with a diagnosis of both systemic lupus erythematosus and cancer. *Arthritis Rheum*, 53(1):67–72, 2005.

[38] M. E. Pinn, D. G. Gold, I. A. Petersen, T. G. Osborn, P. D. Brown, and R. C. Miller. Systemic lupus erythematosus, radiotherapy, and the risk of acute and chronic toxicity: the mayo clinic experience. *Int J Radiat Oncol Biol Phys*, 71(2):498–506, 2008.

[39] D. G. Gold, R. C. Miller, I. A. Petersen, and T. G. Osborn. Radiotherapy for malignancy in patients with scleroderma: The mayo clinic experience. *Int J Radiat Oncol Biol Phys*, 67(2):559–67, 2007.

[40] D. G. Gold, R. C. Miller, M. E. Pinn, T. G. Osborn, I. A. Petersen, and P. D. Brown. Chronic toxicity risk after radiotherapy for patients with systemic sclerosis (systemic scleroderma) or systemic lupus erythematosus: association with connective tissue disorder severity. *Radiother Oncol*, 87(1):127–31, 2008.

[41] A. Lin, E. Abu-Isa, K. A. Griffith, and E. Ben-Josef. Toxicity of radiotherapy in patients with collagen vascular disease. *Cancer*, 113(3):648–53, 2008.

[42] K. M. Creach, I. El Naqa, J. D. Bradley, J. R. Olsen, P. J. Parikh, R. E. Drzymala, C. Bloch, and C. G. Robinson. Dosimetric predictors of chest wall pain after lung stereotactic body radiotherapy. *Radiother Oncol*, 104(1):23–7, 2012.

[43] A. Safwat, S. M. Bentzen, I. Turesson, and J. H. Hendry. Deterministic rather than stochastic factors explain most of the variation in the expression of skin telangiectasia after radiotherapy. *Int J Radiat Oncol Biol Phys*, 52(1):198–204, 2002.

[44] J. M. Pollard and R. A. Gatti. Clinical radiation sensitivity with dna repair disorders: an overview. *Int J Radiat Oncol Biol Phys*, 74(5):1323–31, 2009.

[45] G. C. Barnett, C. E. Coles, R. M. Elliott, C. Baynes, C. Luccarini, D. Conroy, J. S. Wilkinson, J. Tyrer, V. Misra, R. Platte, S. L. Gulliford, M. R. Sydes, E. Hall, S. M. Bentzen, D. P. Dearnaley, N. G. Burnet, P. D. Pharoah, A. M. Dunning, and C. M. West. Independent validation of genes and polymorphisms reported to be associated with radiation toxicity: a prospective analysis study. *Lancet Oncol*, 13(1):65–77, 2012.

[46] C. J. Talbot, G. A. Tanteles, G. C. Barnett, N. G. Burnet, J. Chang-Claude, C. E. Coles, S. Davidson, A. M. Dunning, J. Mills, R. J. Murray, O. Popanda, P. Seibold, C. M. West, J. R. Yarnold, and R. P. Symonds. A replicated association between polymorphisms near tnfalpha and risk for adverse reactions to radiotherapy. *Br J Cancer*, 107(4):748–53, 2012.

[47] J. L. Lopez Guerra, Q. Wei, X. Yuan, D. Gomez, Z. Liu, Y. Zhuang, M. Yin, M. Li, L. E. Wang, J. D. Cox, and Z. Liao. Functional promoter rs2868371 variant of hspb1 associates with radiation-induced esophageal toxicity in patients with non-small-cell lung cancer treated with radio(chemo)therapy. *Radiother Oncol*, 101(2):271–7, 2011.

[48] Q. Pang, Q. Wei, T. Xu, X. Yuan, J. L. Lopez Guerra, L. B. Levy, Z. Liu, D. R. Gomez, Y. Zhuang, L. E. Wang, R. Mohan, R. Komaki, and Z. Liao. Functional promoter variant rs2868371 of hspb1 is associated with risk of radiation pneumonitis after chemoradiation for non-small cell lung cancer. *Int J Radiat Oncol Biol Phys*, 85(5):1332–9, 2013.

[49] H. Edvardsen, H. Landmark-Hoyvik, K. V. Reinertsen, X. Zhao, G. I. Grenaker-Alnaes, D. Nebdal, A. C. Syvanen, O. Rodningen, J. Alsner, J. Overgaard, A. L. Borresen-Dale, S. D. Fossa, and V. N. Kristensen. Snp in txnrd2 associated with radiation-induced fibrosis: a study of genetic variation in reactive oxygen species metabolism and signaling. *Int J Radiat Oncol Biol Phys*, 86(4):791–9, 2013.

[50] P. Seibold, S. Behrens, P. Schmezer, I. Helmbold, G. Barnett, C. Coles, J. Yarnold, C. J. Talbot, T. Imai, D. Azria, C. A. Koch, A. M. Dunning, N. Burnet, J. M. Bliss, R. P. Symonds, T. Rattay, T. Suga, S. L. Kerns, C. Bourgier, K. A. Vallis, M. L. Sautter-Bihl, J. Classen, J. Debus, T. Schnabel, B. S. Rosenstein, F. Wenz, C. M. West, O. Popanda, and J. Chang-Claude. Xrcc1 polymorphism associated with late

toxicity after radiation therapy in breast cancer patients. *Int J Radiat Oncol Biol Phys*, 92(5):1084–92, 2015.

[51] C. N. Andreassen, B. S. Rosenstein, S. L. Kerns, H. Ostrer, D. De Ruysscher, J. A. Cesaretti, G. C. Barnett, A. M. Dunning, L. Dorling, C. M. West, N. G. Burnet, R. Elliott, C. Coles, E. Hall, L. Fachal, A. Vega, A. Gomez-Caamano, C. J. Talbot, R. P. Symonds, K. De Ruyck, H. Thierens, P. Ost, J. Chang-Claude, P. Seibold, O. Popanda, M. Overgaard, D. Dearnaley, M. R. Sydes, D. Azria, C. A. Koch, M. Parliament, M. Blackshaw, M. Sia, M. J. Fuentes-Raspall, Y. Cajal T. Ramon, A. Barnadas, D. Vesprini, S. Gutierrez-Enriquez, M. Molla, O. Diez, J. R. Yarnold, J. Overgaard, S. M. Bentzen, J. Alsner, and Consortium International Radiogenomics. Individual patient data meta-analysis shows a significant association between the atm rs1801516 snp and toxicity after radiotherapy in 5456 breast and prostate cancer patients. *Radiother Oncol*, 2016.

[52] S. L. Kerns, H. Ostrer, R. Stock, W. Li, J. Moore, A. Pearlman, C. Campbell, Y. Shao, N. Stone, L. Kusnetz, and B. S. Rosenstein. Genome-wide association study to identify single nucleotide polymorphisms (snps) associated with the development of erectile dysfunction in african-american men after radiotherapy for prostate cancer. *Int J Radiat Oncol Biol Phys*, 78(5):1292–300, 2010.

[53] L. Fachal, A. Gomez-Caamano, G. C. Barnett, P. Peleteiro, A. M. Carballo, P. Calvo-Crespo, S. L. Kerns, M. Sanchez-Garcia, R. Lobato-Busto, L. Dorling, R. M. Elliott, D. P. Dearnaley, M. R. Sydes, E. Hall, N. G. Burnet, A. Carracedo, B. S. Rosenstein, C. M. West, A. M. Dunning, and A. Vega. A three-stage genome-wide association study identifies a susceptibility locus for late radiotherapy toxicity at 2q24.1. *Nat Genet*, 46(8):891–4, 2014.

[54] S. L. Kerns, L. Dorling, L. Fachal, S. Bentzen, P. D. Pharoah, D. R. Barnes, A. Gomez-Caamano, A. M. Carballo, D. P. Dearnaley, P. Peleteiro, S. L. Gulliford, E. Hall, K. Michailidou, A. Carracedo, M. Sia, R. Stock, N. N. Stone, M. R. Sydes, J. P. Tyrer, S. Ahmed, M. Parliament, H. Ostrer, B. S. Rosenstein, A. Vega, N. G. Burnet, A. M. Dunning, G. C. Barnett, C. M. West, and Consortium Radiogenomics. Meta-analysis of genome wide association studies identifies genetic markers of late toxicity following radiotherapy for prostate cancer. *EBioMedicine*, 10:150–63, 2016.

[55] G. C. Barnett, D. Thompson, L. Fachal, S. Kerns, C. Talbot, R. M. Elliott, L. Dorling, C. E. Coles, D. P. Dearnaley, B. S. Rosenstein, A. Vega, P. Symonds, J. Yarnold, C. Baynes, K. Michailidou, J. Dennis, J. P. Tyrer, J. S. Wilkinson, A. Gomez-Caamano, G. A. Tanteles, R. Platte, R. Mayes, D. Conroy, M. Maranian, C. Luccarini, S. L. Gulliford, M. R. Sydes, E. Hall, J. Haviland, V. Misra, J. Titley, S. M. Bentzen, P. D. Pharoah, N. G. Burnet, A. M. Dunning, and C. M. West. A genome wide association study (gwas) providing evidence of an association between common genetic variants and late radiotherapy toxicity. *Radiother Oncol*, 111(2):178–85, 2014.

[56] M. A. Mazurowski. Radiogenomics: what it is and why it is important. *J Am Coll Radiol*, 12(8):862–6, 2015.

[57] S. L. Kerns, C. M. West, C. N. Andreassen, G. C. Barnett, S. M. Bentzen, N. G. Burnet, A. Dekker, D. De Ruysscher, A. Dunning, M. Parliament, C. Talbot, A. Vega, and B. S. Rosenstein. Radiogenomics: the search for genetic predictors of radiotherapy response. *Future Oncol*, 10(15):2391–406, 2014.

[58] H. Pan, D. R. Simpson, L. K. Mell, A. J. Mundt, and J. D. Lawson. A survey of stereotactic body radiotherapy use in the united states. *Cancer*, 117(19):4566–72, 2011.

[59] D. L. Andolino, J. A. Forquer, M. A. Henderson, R. B. Barriger, R. H. Shapiro, J. G. Brabham, P. A. Johnstone, H. R. Cardenes, and A. J. Fakiris. Chest wall toxicity after stereotactic body radiotherapy for malignant lesions of the lung and liver. *Int J Radiat Oncol Biol Phys*, 80(3):692–7, 2011.

[60] N. E. Dunlap, J. Cai, G. B. Biedermann, W. Yang, S. H. Benedict, K. Sheng, T. E. Schefter, B. D. Kavanagh, and J. M. Larner. Chest wall volume receiving > 30 gy predicts risk of severe pain and/or rib fracture after lung stereotactic body radiotherapy. *Int J Radiat Oncol Biol Phys*, 76(3):796–801, 2010.

[61] R. W. Mutter, F. Liu, A. Abreu, E. Yorke, A. Jackson, and K. E. Rosenzweig. Dose-volume parameters predict for the development of chest wall pain after stereotactic body radiation for lung cancer. *Int J Radiat Oncol Biol Phys*, 82(5):1783–90, 2012.

[62] T. Shaikh and A. Turaka. Predictors and management of chest wall toxicity after lung stereotactic body radiotherapy. *Cancer Treat Rev*, 40(10):1215–20, 2014.

[63] F. Kimsey, J. McKay, J. Gefter, M. T. Milano, V. Moiseenko, J. Grimm, and R. Berg. Dose-response model for chest wall tolerance of stereotactic body radiation therapy. *Semin Radiat Oncol*, 26(2):129–34, 2016.

[64] J. Welsh, J. Thomas, D. Shah, P. K. Allen, X. Wei, K. Mitchell, S. Gao, P. Balter, R. Komaki, and J. Y. Chang. Obesity increases the risk of chest wall pain from thoracic stereotactic body radiation therapy. *Int J Radiat Oncol Biol Phys*, 81(1):91–6, 2011.

[65] E. M. Bongers, C. J. Haasbeek, F. J. Lagerwaard, B. J. Slotman, and S. Senan. Incidence and risk factors for chest wall toxicity after risk-adapted stereotactic radiotherapy for early-stage lung cancer. *J Thorac Oncol*, 6(12):2052–7, 2011.

[66] A. Nambu, H. Onishi, S. Aoki, T. Koshiishi, K. Kuriyama, T. Komiyama, K. Marino, M. Araya, R. Saito, L. Tominaga, Y. Maehata, E. Sawada, and T. Araki. Rib fracture after stereotactic radiotherapy on follow-up thin-section computed tomography in 177 primary lung cancer patients. *Radiat Oncol*, 6:137, 2011.

[67] M. Taremi, A. Hope, P. Lindsay, M. Dahele, S. Fung, T. G. Purdie, D. Jaffray, L. Dawson, and A. Bezjak. Predictors of radiotherapy induced bone injury (ribi) after stereotactic lung radiotherapy. *Radiat Oncol*, 7:159, 2012.

[68] A. Nambu, H. Onishi, S. Aoki, L. Tominaga, K. Kuriyama, M. Araya, R. Saito, Y. Maehata, T. Komiyama, K. Marino, T. Koshiishi, E. Sawada, and T. Araki. Rib fracture after stereotactic radiotherapy for primary lung cancer: prevalence, degree of clinical symptoms, and risk factors. *BMC Cancer*, 13:68, 2013.

[69] I. Thibault, A. Chiang, D. Erler, L. Yeung, I. Poon, A. Kim, B. Keller, F. Lochray, S. Jain, H. Soliman, and P. Cheung. Predictors of chest wall toxicity after lung stereotactic ablative radiotherapy. *Clin Oncol (R Coll Radiol)*, 28(1):28–35, 2016.

[70] R. Baker, G. Han, S. Sarangkasiri, M. DeMarco, C. Turke, C. W. Stevens, and T. J. Dilling. Clinical and dosimetric predictors of radiation pneumonitis in a large series of patients treated with stereotactic body radiation therapy to the lung. *Int J Radiat Oncol Biol Phys*, 85(1):190–5, 2013.

[71] R. B. Barriger, J. A. Forquer, J. G. Brabham, D. L. Andolino, R. H. Shapiro, M. A. Henderson, P. A. Johnstone, and A. J. Fakiris. A dose-volume analysis of radiation pneumonitis in non-small cell lung cancer patients treated with stereotactic body radiation therapy. *Int J Radiat Oncol Biol Phys*, 82(1):457–62, 2012.

[72] H. Liu, X. Zhang, Y. Y. Vinogradskiy, S. G. Swisher, R. Komaki, and J. Y. Chang. Predicting radiation pneumonitis after stereotactic ablative radiation therapy in patients previously treated with conventional thoracic radiation therapy. *Int J Radiat Oncol Biol Phys*, 84(4):1017–23, 2012.

[73] Y. Matsuo, K. Shibuya, M. Nakamura, M. Narabayashi, K. Sakanaka, N. Ueki, K. Miyagi, Y. Norihisa, T. Mizowaki, Y. Nagata, and M. Hiraoka. Dose–volume metrics associated with radiation pneumonitis after stereotactic body radiation therapy for lung cancer. *Int J Radiat Oncol Biol Phys*, 83(4):e545–9, 2012.

[74] J. Zhao, E. D. Yorke, L. Li, B. D. Kavanagh, X. A. Li, S. Das, M. Miften, A. Rimner, J. Campbell, J. Xue, A. Jackson, J. Grimm, M. T. Milano, and F. M. Spring Kong. Simple factors associated with radiation-induced lung toxicity after stereotactic body radiation therapy of the thorax: A pooled analysis of 88 studies. *Int J Radiat Oncol Biol Phys*, 95(5):1357–66, 2016.

[75] H. Bahig, E. Filion, T. Vu, J. Chalaoui, L. Lambert, D. Roberge, M. Gagnon, B. Fortin, D. Beliveau-Nadeau, D. Mathieu, and M. P. Campeau. Severe radiation pneumonitis after lung stereotactic ablative radiation therapy in patients with interstitial lung disease. *Pract Radiat Oncol*, 6(5):367–74, 2016.

[76] J. Wang, J. Cao, S. Yuan, W. Ji, D. Arenberg, J. Dai, P. Stanton, D. Tatro, R. K. Ten Haken, L. Wang, and F. M. Kong. Poor baseline pulmonary function may not increase the risk of radiation-induced lung toxicity. *Int J Radiat Oncol Biol Phys*, 85(3):798–804, 2013.

[77] T. Inoue, H. Shiomi, and R. J. Oh. Stereotactic body radiotherapy for stage i lung cancer with chronic obstructive pulmonary disease: special reference to survival and radiation-induced pneumonitis. *J Radiat Res*, 56(4):727–34, 2015.

[78] A. Takeda, E. Kunieda, T. Ohashi, Y. Aoki, Y. Oku, T. Enomoto, K. Nomura, and M. Sugiura. Severe copd is correlated with mild radiation pneumonitis following stereotactic body radiotherapy. *Chest*, 141(4):858–66, 2012.

[79] H. Yamashita, S. Kobayashi-Shibata, A. Terahara, K. Okuma, A. Haga, R. Wakui, K. Ohtomo, and K. Nakagawa. Prescreening based on the presence of ct-scan abnormalities and biomarkers (kl-6 and sp-d) may reduce severe radiation pneumonitis after stereotactic radiotherapy. *Radiat Oncol*, 5:32, 2010.

[80] A. Takeda, N. Sanuki, T. Enomoto, and E. Kunieda. Subclinical interstitial lung disease: is it a risk factor for fatal radiation pneumonitis following stereotactic body radiotherapy? *Lung Cancer*, 83(1):112, 2014.

[81] N. Ueki, Y. Matsuo, Y. Togashi, T. Kubo, K. Shibuya, Y. Iizuka, T. Mizowaki, K. Togashi, M. Mishima, and M. Hiraoka. Impact of pretreatment interstitial lung disease on radiation pneumonitis and survival after stereotactic body radiation therapy for lung cancer. *J Thorac Oncol*, 10(1):116–25, 2015.

[82] M. Ishijima, H. Nakayama, T. Itonaga, Y. Tajima, S. Shiraishi, M. Okubo, R. Mikami, and K. Tokuuye. Patients with severe emphysema have a low risk of radiation pneumonitis following stereotactic body radiotherapy. *Br J Radiol*, 88(1046):20140596, 2015.

[83] S. Ochiai, Y. Nomoto, Y. Yamashita, T. Inoue, S. Murashima, D. Hasegawa, Y. Kurita, Y. Watanabe, Y. Toyomasu, T. Kawamura, A. Takada, Noriko, S. Kobayashi, and H. Sakuma. The impact of emphysema on dosimetric parameters for stereotactic body radiotherapy of the lung. *J Radiat Res*, 57(5):555–566, 2016.

[84] L. A. Dawson, D. Normolle, J. M. Balter, C. J. McGinn, T. S. Lawrence, and R. K. Ten Haken. Analysis of radiation-induced liver disease using the lyman ntcp model. *Int J Radiat Oncol Biol Phys*, 53(4):810–21, 2002.

[85] C. C. Pan, B. D. Kavanagh, L. A. Dawson, X. A. Li, S. K. Das, M. Miften, and R. K. Ten Haken. Radiation-associated liver injury. *Int J Radiat Oncol Biol Phys*, 76(3 Suppl):S94–100, 2010.

[86] A. Mendez Romero, W. Wunderink, S. M. Hussain, J. A. De Pooter, B. J. Heijmen, P. C. Nowak, J. J. Nuyttens, R. P. Brandwijk, C. Verhoef, J. N. Ijzermans, and P. C. Levendag. Stereotactic body radiation therapy for primary and metastatic liver tumors: A single institution phase i-ii study. *Acta Oncol*, 45(7):831–7, 2006.

[87] M. T. Lee, J. J. Kim, R. Dinniwell, J. Brierley, G. Lockwood, R. Wong, B. Cummings, J. Ringash, R. V. Tse, J. J. Knox, and L. A. Dawson. Phase i study of individualized stereotactic body radiotherapy of liver metastases. *J Clin Oncol*, 27(10):1585–91, 2009.

[88] R. V. Tse, M. Hawkins, G. Lockwood, J. J. Kim, B. Cummings, J. Knox, M. Sherman, and L. A. Dawson. Phase i study of individualized stereotactic body radiotherapy for hepatocellular carcinoma and intrahepatic cholangiocarcinoma. *J Clin Oncol*, 26(4):657–64, 2008.

[89] J. C. Cheng, J. K. Wu, C. M. Huang, H. S. Liu, D. Y. Huang, S. H. Cheng, S. Y. Tsai, J. J. Jian, Y. M. Lin, T. I. Cheng, C. F. Horng, and A. T. Huang. Radiation-induced liver disease after three-dimensional conformal radiotherapy for patients with hepatocellular carcinoma: dosimetric analysis and implication. *Int J Radiat Oncol Biol Phys*, 54(1):156–62, 2002.

[90] J. C. Cheng, J. K. Wu, P. C. Lee, H. S. Liu, J. J. Jian, Y. M. Lin, J. L. Sung, and G. J. Jan. Biologic susceptibility of hepatocellular carcinoma patients treated with radiotherapy to radiation-induced liver disease. *Int J Radiat Oncol Biol Phys*, 60(5):1502–9, 2004.

[91] S. J. Shim, J. Seong, I. J. Lee, K. H. Han, C. Y. Chon, and S. H. Ahn. Radiation-induced hepatic toxicity after radiotherapy combined with chemotherapy for hepatocellular carcinoma. *Hepatol Res*, 37(11):906–13, 2007.

[92] J. Jung, S. M. Yoon, S. Y. Kim, B. Cho, J. H. Park, S. S. Kim, S. Y. Song, S. W. Lee, S. D. Ahn, E. K. Choi, and J. H. Kim. Radiation-induced liver disease after stereotactic body radiotherapy for small hepatocellular carcinoma: clinical and dose-volumetric parameters. *Radiat Oncol*, 8:249, 2013.

[93] S. X. Liang, X. D. Zhu, Z. Y. Xu, J. Zhu, J. D. Zhao, H. J. Lu, Y. L. Yang, L. Chen, A. Y. Wang, X. L. Fu, and G. L. Jiang. Radiation-induced liver disease in three-dimensional conformal radiation therapy for primary liver carcinoma: the risk factors and hepatic radiation tolerance. *Int J Radiat Oncol Biol Phys*, 65(2):426–34, 2006.

[94] S. H. Benedict, I. El Naqa, and E. E. Klein. Introduction to big data in radiation oncology: Exploring opportunities for research, quality assessment, and clinical care. *Int J Radiat Oncol Biol Phys*, 95(3):871–2, 2016.

[95] N. G. Burnet, R. M. Elliott, A. Dunning, and C. M. West. Radiosensitivity, radio-genomics and rapper. *Clin Oncol (R Coll Radiol)*, 18(7):525–8, 2006.

[96] A. Y. Ho, D. P. Atencio, S. Peters, R. G. Stock, S. C. Formenti, J. A. Cesaretti, S. Green, B. Haffty, K. Drumea, L. Leitzin, A. Kuten, D. Azria, M. Ozsahin, J. Overgaard, C. N. Andreassen, C. S. Trop, J. Park, and B. S. Rosenstein. Genetic predictors of adverse radiotherapy effects: the gene-pare project. *Int J Radiat Oncol Biol Phys*, 65(3):646–55, 2006.

[97] C. West, D. Azria, J. Chang-Claude, S. Davidson, P. Lambin, B. Rosenstein, D. De Ruysscher, C. Talbot, H. Thierens, R. Valdagni, A. Vega, and M. Yuille. The requite project: validating predictive models and biomarkers of radiotherapy toxicity to reduce side-effects and improve quality of life in cancer survivors. *Clin Oncol (R Coll Radiol)*, 26(12):739–42, 2014.

[98] S. L. Kerns, D. de Ruysscher, C. N. Andreassen, D. Azria, G. C. Barnett, J. Chang-Claude, S. Davidson, J. O. Deasy, A. M. Dunning, H. Ostrer, B. S. Rosenstein, C. M. West, and S. M. Bentzen. Strogar - strengthening the reporting of genetic association studies in radiogenomics. *Radiother Oncol*, 110(1):182–8, 2014.

[99] D. G. Altman, L. M. McShane, W. Sauerbrei, and S. E. Taube. Reporting recommendations for tumor marker prognostic studies (remark): explanation and elaboration. *PLoS Med*, 9(5):e1001216, 2012.

[100] L. M. McShane, D. G. Altman, W. Sauerbrei, S. E. Taube, M. Gion, G. M. Clark, and N. C. I. Eortc Working Group on Cancer Diagnostics Statistics Subcommittee of the. Reporting recommendations for tumor marker prognostic studies (remark). *J Natl Cancer Inst*, 97(16):1180–4, 2005.

[101] I. El Naqa, P. Grigsby, A. Apte, E. Kidd, E. Donnelly, D. Khullar, S. Chaudhari, D. Yang, M. Schmitt, R. Laforest, W. Thorstad, and J. O. Deasy. Exploring feature-based approaches in pet images for predicting cancer treatment outcomes. *Pattern Recognit*, 42(6):1162–1171, 2009.

[102] V. Kumar, Y. Gu, S. Basu, A. Berglund, S. A. Eschrich, M. B. Schabath, K. Forster, H. J. Aerts, A. Dekker, D. Fenstermacher, D. B. Goldgof, L. O. Hall, P. Lambin, Y. Balagurunathan, R. A. Gatenby, and R. J. Gillies. Radiomics: the process and the challenges. *Magn Reson Imaging*, 30(9):1234–48, 2012.

[103] P. Lambin, E. Rios-Velazquez, R. Leijenaar, S. Carvalho, R. G. van Stiphout, P. Granton, C. M. Zegers, R. Gillies, R. Boellard, A. Dekker, and H. J. Aerts. Radiomics: extracting more information from medical images using advanced feature analysis. *Eur J Cancer*, 48(4):441–6, 2012.

[104] Ludwig G. Strauss and Peter S. Conti. The applications of pet in clinical oncology. *J Nucl Med*, 32(4):623–648, 1991.

[105] S. M. Larson, Y. Erdi, T. Akhurst, M. Mazumdar, H. A. Macapinlac, R. D. Finn, C. Casilla, M. Fazzari, N. Srivastava, H. W. Yeung, J. L. Humm, J. Guillem, R. Downey, M. Karpeh, A. E. Cohen, and R. Ginsberg. Tumor treatment response based on visual and quantitative changes in global tumor glycolysis using pet-fdg imaging. the visual response score and the change in total lesion glycolysis. *Clin Positron Imaging*, 2(3):159–171, 1999.

[106] M. Vaidya, K. M. Creach, J. Frye, F. Dehdashti, J. D. Bradley, and I. El Naqa. Combined pet/ct image characteristics for radiotherapy tumor response in lung cancer. *Radiother Oncol*, 102(2):239–45, 2012.

[107] Anil K. Jain. *Fundamentals of digital image processing*. Prentice Hall, Englewood Cliffs, NJ, 1989.

[108] F. O'Sullivan, S. Roy, J. O'Sullivan, C. Vernon, and J. Eary. Incorporation of tumor shape into an assessment of spatial heterogeneity for human sarcomas imaged with fdg-pet. *Biostat*, 6(2):293–301, 2005.

[109] Finbarr O'Sullivan, Supratik Roy, and Janet Eary. A statistical measure of tissue heterogeneity with application to 3d pet sarcoma data. *Biostat*, 4(3):433–448, 2003.

[110] R. Haralick, K. Shanmugam, and I. Dinstein. Texture features for image classification. *IEEE Trans. on Sys. Man. and Cyb. SMC*, 3(6):610–621, 1973.

[111] Kenneth R. Castleman. *Digital image processing*. Prentice Hall, Englewood Cliffs, N.J., 1996.

[112] Jianguo Zhang and Tieniu Tan. Brief review of invariant texture analysis methods. *Pattern Recognition*, 35(3):735–747, 2002.

[113] G. Castellano, L. Bonilha, L. M. Li, and F. Cendes. Texture analysis of medical images. *Clinical Radiology*, 59(12):1061–1069, 2004.

[114] Sugama Chicklore, Vicky Goh, Musib Siddique, Arunabha Roy, PaulK Marsden, and GaryJ R. Cook. Quantifying tumour heterogeneity in 18f-fdg pet/ct imaging by texture analysis. *European Journal of Nuclear Medicine and Molecular Imaging*, 40(1):133–140, 2013.

[115] P. S. Tofts. Modeling tracer kinetics in dynamic gd-dtpa mr imaging. *J Magn Reson Imaging*, 7(1):91–101, 1997.

[116] Daniela Thorwarth, Susanne-Martina Eschmann, Felix Holzner, Frank Paulsen, and Markus Alber. Combined uptake of [18f]fdg and [18f]fmiso correlates with radiation therapy outcome in head-and-neck cancer patients. *Radiotherapy and Oncology*, 80(2):151–156, 2006.

[117] Daniela Thorwarth, Susanne-Martina Eschmann, Frank Paulsen, and Markus Alber. A model of reoxygenation dynamics of head-and-neck tumors based on serial 18f-fluoromisonidazole positron emission tomography investigations. *International Journal of Radiation Oncology*Biology*Physics*, 68(2):515–521, 2007.

[118] Steven P. Sourbron and David L. Buckley. On the scope and interpretation of the tofts models for dce-mri. *Magnetic Resonance in Medicine*, 66(3):735–745, 2011.

[119] A. J. Reader, J. C. Matthews, F. C. Sureau, C. Comtat, R. Trebossen, and I. Buvat. Iterative kinetic parameter estimation within fully 4d pet image reconstruction. In *Nuclear Science Symposium Conference Record, 2006. IEEE*, volume 3, pages 1752–1756, 2006.

[120] F. E. Turkheimer, J. A. Aston, M. C. Asselin, and R. Hinz. Multi-resolution bayesian regression in pet dynamic studies using wavelets. *Neuroimage*, 32(1):111–21, 2006.

[121] I. El Naqa, D. Low, J. Bradley, M. Vicic, and J Deasy. Deblurring of breathing motion artifacts in thoracic pet images by deconvolution methods. *Medical Physics*, 33(10):3587–3600, 2006.

[122] I. El Naqa, G. Suneja, P. E. Lindsay, A. J. Hope, J. R. Alaly, M. Vicic, J. D. Bradley, A. Apte, and J. O. Deasy. Dose response explorer: an integrated open-source tool for exploring and modelling radiotherapy dose-volume outcome relationships. *Phys Med Biol*, 51(22):5719–35, 2006.

[123] M. Vallieres, C. R. Freeman, S. R. Skamene, and I. El Naqa. A radiomics model from joint fdg-pet and mri texture features for the prediction of lung metastases in soft-tissue sarcomas of the extremities. *Phys Med Biol*, 60(14):5471–96, 2015.

[124] Berkman Sahiner, Heang-Ping Chan, and Lubomir Hadjiiski. Classifier performance prediction for computer-aided diagnosis using a limited dataset. *Medical Physics*, 35(4):1559–1570, 2008.

[125] Florent Tixier, Mathieu Hatt, Catherine Cheze Le Rest, Adrien Le Pogam, Laurent Corcos, and Dimitris Visvikis. Reproducibility of tumor uptake heterogeneity characterization through textural feature analysis in 18f-fdg pet. *Journal of Nuclear Medicine*, 53(5):693–700, 2012.

[126] N. M. Cheng, Y. H. Fang, and T. C. Yen. The promise and limits of pet texture analysis. *Ann Nucl Med*, 27(9):867–9, 2013.

[127] I. El Naqa, I. Kawrakow, M. Fippel, J. V. Siebers, P. E. Lindsay, M. V. Wickerhauser, M. Vicic, K. Zakarian, N. Kauffmann, and J. O. Deasy. A comparison of monte carlo dose calculation denoising techniques. *Phys Med Biol*, 50(5):909–22, 2005.

[128] H. Zaidi, M. Abdoli, C. L. Fuentes, and I. M. El Naqa. Comparative methods for pet image segmentation in pharyngolaryngeal squamous cell carcinoma. *Eur J Nucl Med Mol Imaging*, 39(5):881–91, 2012.

[129] Bahadir K. Gunturk and Xin Li. *Image Restoration: Fundamentals and Advances*. CRC Press, Taylor and Francis group, Boca Raton, FL, 2012.

[130] J. O. Deasy and I. El Naqa. Image-based modeling of normal tissue complication probability for radiation therapy. *Cancer Treat Res*, 139:215–56, 2008.

[131] Edward C. Halperin, Carlos A. Perez, and Luther W. Brady. *Perez and Brady's principles and practice of radiation oncology*. Wolters Kluwer Health/Lippincott Williams and Wilkins, Philadelphia, 5th edition, 2008.

[132] T. Bortfeld, R. Schmidt-Ullrich, W. De Neve, and D. Wazer. *Image-guided IMRT.* Springer-Verlag, Berlin, 2006.

[133] Steve Webb. *The physics of three-dimensional radiation therapy : conformal radiotherapy, radiosurgery, and treatment planning.* Series in medical physics. Institute of Physics Pub., Bristol, UK ; Philadelphia, 2001.

[134] Sren M. Bentzen, Louis S. Constine, Joseph O. Deasy, Avi Eisbruch, Andrew Jackson, Lawrence B. Marks, Randall K. Ten Haken, and Ellen D. Yorke. Quantitative analyses of normal tissue effects in the clinic (quantec): An introduction to the scientific issues. *International journal of radiation oncology, biology, physics,* 76(3 Suppl):S3–S9, 2010.

[135] A. Jackson, L. B. Marks, S. M. Bentzen, A. Eisbruch, E. D. Yorke, R. K. Ten Haken, L. S. Constine, and J. O. Deasy. The lessons of quantec: recommendations for reporting and gathering data on dose-volume dependencies of treatment outcome. *Int J Radiat Oncol Biol Phys,* 76(3 Suppl):S155–60, 2010.

[136] G. Gordon Steel. *Basic clinical radiobiology.* Arnold ; Oxford University Press, London New York, 3rd edition, 2002.

[137] Vitali Moissenko, J. O. Deasy, and Jacob Van Dyk. *Radiobiological Modeling for Treatment Planning,* volume 2, pages 185–220. Medical Physics Publishing, Madison, WI, 2005.

[138] I. El Naqa. *Outcomes Modeling,* book section 24, pages 257–275. CRC Press, Taylor and Francis, Boca Raton, FL, 2013.

[139] RK Ten Haken and ML Kessler. *3-D CRT Plan Evaluation,* pages 43–59. Advanced Medical Publishing, Madison, WI, 2001.

[140] A. I. Blanco, K. S. Chao, I. El Naqa, G. E. Franklin, K. Zakarian, M. Vicic, and J. O. Deasy. Dose-volume modeling of salivary function in patients with head-and-neck cancer receiving radiotherapy. *Int J Radiat Oncol Biol Phys,* 62(4):1055–69, 2005.

[141] J. Bradley, J. O. Deasy, S. Bentzen, and I. El-Naqa. Dosimetric correlates for acute esophagitis in patients treated with radiotherapy for lung carcinoma. *Int J Radiat Oncol Biol Phys,* 58(4):1106–13, 2004.

[142] A.J. Hope, P.E. Lindsay, I. El Naqa, J.D. Bradley, J. Alaly, M. Vicic, J.A. Purdy, and J.O. Deasy. Radiation pneumonitis risk based on clinical, dosimetric, and location related factors. *Int J Radiat Oncol Biol Phys,* 65(1):112–124, 2006.

[143] A. J. Hope, P. E. Lindsay, I. El Naqa, J. D. Bradley, M. Vicic, and J. O. Deasy. Clinical, dosimetric, and location-related factors to predict local control in non-small cell lung cancer. In *ASTRO 47TH ANNUAL MEETING,* volume 63, page S231, 2005.

[144] S. Levegrun, A. Jackson, M. J. Zelefsky, M. W. Skwarchuk, E. S. Venkatraman, W. Schlegel, Z. Fuks, S. A. Leibel, and C. C. Ling. Fitting tumor control probability models to biopsy outcome after three-dimensional conformal radiation therapy of prostate cancer: Pitfalls in deducing radiobiologic parameters for tumors from clinical data. *International Journal of Radiation Oncology Biology Physics,* 51(4):1064–1080, 2001.

[145] L. B. Marks. Dosimetric predictors of radiation-induced lung injury. *International Journal of Radiation Oncology Biology Physics,* 54(2):313–316, 2002.

[146] R. K. Ten Haken, M. K. Martel, M. L. Kessler, M. B. Hazuka, T. S. Lawrence, J. M. Robertson, A. T. Turrisi, and A. S. Lichter. Use of veff and iso-ntcp in the implementation of dose escalation protocols. *Int J Radiat Oncol Biol Phys*, 27(3):689–95, 1993.

[147] H. Edwin Romeijn, Ravindra K. Ahuja, James F. Dempsey, and Arvind Kumar. A new linear programming approach to radiation therapy treatment planning problems. *Operations Research*, 54(2):201–216, 2006.

[148] B. Jolles. Area factor in skin reaction. *Br. Emp. Cancer Campaign Rep*, 16(52), 1939.

[149] G.J. Kutcher and C. Burman. Calculation of complication probability factors for nonuniform normal tissue irradiation: the effective volume method. *Int J Radial Biol.*, 16:1623–30, 1989.

[150] R. Mohan, G. S. Mageras, B. Baldwin, L. J. Brewster, G. J. Kutcher, S. Leibel, C. M. Burman, C. C. Ling, and Z. Fuks. Clinically relevant optimization of 3-d conformal treatments. *Medical Physics*, 19(4):933–944, 1992.

[151] Niemierko. A generalized concept of equivalent uniform dose (eud). *Med. Phys.*, 26:1101, 1999.

[152] X. Allen Li, M. Alber, J. O. Deasy, A. Jackson, K. W. Ken Jee, L. B. Marks, M. K. Martel, C. Mayo, V. Moiseenko, A. E. Nahum, A. Niemierko, V. A. Semenenko, and E. D. Yorke. The use and qa of biologically related models for treatment planning: Short report of the tg-166 of the therapy physics committee of the aapm. *Med Phys*, 39(3):1386–409, 2012.

[153] I. Guyon and A. Elissee. An introduction to variable and feature selection. *Journal of Machine Learning Research*, 3:1157–1182, 2003.

[154] Laura A. Dawson, Matthew Biersack, Gina Lockwood, Avraham Eisbruch, Theodore S. Lawrence, and Randall K. Ten Haken. Use of principal component analysis to evaluate the partial organ tolerance of normal tissues to radiation. *International Journal of Radiation Oncology*Biology*Physics*, 62(3):829–837, 2005.

[155] R. Kennedy, Y. Lee, B. Van Roy, C.D. Reed, and R.P. Lippman. *Solving data mining problems through pattern recognition*. Prentice Hall, 1998.

[156] I. Guyon, J. Weston, S. Barnhill, and V. Vapnik. Gene selection for cancer classification using support vector machines. *Machine Learning*, 46(1-3):389–422, 2002.

[157] I. El Naqa, J.D. Bradley, P.E. Lindsay, A. I. Blanco, M. Vicic, A.J. Hope, and J.O. Deasy. Multi-variable modeling of radiotherapy outcomes including dose-volume and clinical factors. *Int J Radiat Oncol Biol Phys*, 64(4):1275–1286, 2006.

[158] P. E. Lindsay, I. El Naqa, A. J. Hope, M. Vicic, J. Cui, J. D. Bradley, and J. O. Deasy. Retrospective monte carlo dose calculations with limited beam weight information. *Med Phys*, 34(1):334–46, 2007.

[159] J. O. Deasy, A. I. Blanco, and V. H. Clark. Cerr: A computational environment for radiotherapy research. *Medical physics*, 30:979–985, 2003.

[160] Issam El Naqa, JO Deasy, Y. Mu, E. Huang, AJ. Hope, PE. Lindsay, A. Apte, J. Alaly, and JD Bradley. Datamining approaches for modeling tumor control probability. *Acta Oncol*, page accepted, 2009.

[161] J. Willner, K. Baier, E. Caragiani, A. Tschammler, and M. Flentje. Dose, volume, and tumor control prediction in primary radiotherapy of non-small-cell lung cancer. *Int J Radiat Oncol Biol Phys*, 52(2):382–9, 2002.

[162] Mary Kaye Martel, Randall K. Ten Haken, Mark B. Hazuka, Marc L. Kessler, Myla Strawderman, Andrew T. Turrisi, Theodore S. Lawrence, Benedick A. Fraass, and Allen S. Lichter. Estimation of tumor control probability model parameters from 3-d dose distributions of non-small cell lung cancer patients. *Lung Cancer*, 24(1):31–37, 1999.

[163] M. Mehta, R. Scrimger, R. Mackie, B. Paliwal, R. Chappell, and J. Fowler. A new approach to dose escalation in non-small-cell lung cancer. *Int J Radiat Oncol Biol Phys*, 49(1):23–33, 2001.

[164] O. Brodin, L. Lennartsson, and S. Nilsson. Single-dose and fractionated irradiation of four human lung cancer cell lines in vitro. *Acta Oncol*, 30(8):967–74, 1991.

[165] B. Sanchez-Nieto and A. E. Nahum. Bioplan: software for the biological evaluation of radiotherapy treatment plans. *Medical Dosimetry*, 25(2):71–76, 2000.

[166] R. M. Graham. Is cytology a reliable prognosticator of radiosensitivity? *JAMA*, 193:825, 1965.

[167] C. West and J. Hendry. Prediction of radiotherapy response using sf2: is it methodology or mythology? *Radiother. Oncol.*, 31.

[168] I. Eke and N. Cordes. Radiobiology goes 3d: how ecm and cell morphology impact on cell survival after irradiation.

[169] T Sato, R Watanabe, Y Kase, C Tsuruoka, M Suzuki, Y Furusawa, and K Niita. Analysis of cell-survival fractions for heavy-ion irradiations based on microdosimetric kinetic model implemented in the particle and heavy ion transport code system. *Radiat Prot Dosim*, 143(2-4):491–496, 2011.

[170] Meritxell Huch, Paola Bonfanti, Sylvia F Boj, Toshiro Sato, Cindy J M Loomans, Marc van de Wetering, Mozhdeh Sojoodi, Vivian S W Li, Jurian Schuijers, Ana Gracanin, Femke Ringnalda, Harry Begthel, Karien Hamer, Joyce Mulder, Johan H van Es, Eelco de Koning, Robert G J Vries, Harry Heimberg, and Hans Clevers. Unlimited in vitro expansion of adult bi-potent pancreas progenitors through the Lgr5/R-spondin axis. *The EMBO journal*, 32(20):2708–21, oct 2013.

[171] Dong Gao, Ian Vela, Andrea Sboner, Phillip J Iaquinta, Wouter R Karthaus, Anuradha Gopalan, Catherine Dowling, Jackline N Wanjala, Eva A Undvall, Vivek K Arora, John Wongvipat, Myriam Kossai, Sinan Ramazanoglu, Luendreo P Barboza, Wei Di, Zhen Cao, Qi Fan Zhang, Inna Sirota, Leili Ran, Theresa Y MacDonald, Himisha Beltran, Juan-Miguel Mosquera, Karim A Touijer, Peter T Scardino, Vincent P Laudone, Kristen R Curtis, Dana E Rathkopf, Michael J Morris, Daniel C Danila, Susan F Slovin, Stephen B Solomon, James A Eastham, Ping Chi, Brett Carver, Mark A Rubin, Howard I Scher, Hans Clevers, Charles L Sawyers, and Yu Chen. Organoid cultures derived from patients with advanced prostate cancer. *Cell*, 159(1):176–87, sep 2014.

[172] H Chuck Zhang and Calvin J Kuo. Personalizing pancreatic cancer organoids with hPSCs. *Nature medicine*, 21(11):1249–51, nov 2015.

[173] Norman Sachs and Hans Clevers. Organoid cultures for the analysis of cancer phenotypes. *Current opinion in genetics & development*, 24:68–73, feb 2014.

[174] Hans Clevers and Eric Bender. Q&A: Hans Clevers. Banking on organoids. *Nature*, 521(7551):S15, may 2015.

[175] James T Neal and Calvin J Kuo. Organoids as Models for Neoplastic Transformation. *Annual review of pathology*, 11:199–220, may 2016.

[176] Peter W Nagle, Nynke A Hosper, Emily M Ploeg, Marc-Jan van Goethem, Sytze Brandenburg, Johannes A Langendijk, Roland K Chiu, and Robert P Coppes. The InVitro Response of Tissue Stem Cells to Irradiation With Different Linear Energy Transfers. *International journal of radiation oncology, biology, physics*, 95(1):103–11, may 2016.

[177] Mami Matano, Shoichi Date, Mariko Shimokawa, Ai Takano, Masayuki Fujii, Yuki Ohta, Toshiaki Watanabe, Takanori Kanai, and Toshiro Sato. Modeling colorectal cancer using CRISPR-Cas9-mediated engineering of human intestinal organoids. *Nature medicine*, 21(3):256–62, mar 2015.

[178] E W Bradley, P C Chan, and S J Adelstein. The radiotoxicity of iodine-125 in mammalian cells. I. Effects on the survival curve of radioiodine incorporated into DNA. *Radiation research*, 64(3):555–63, dec 1975.

[179] R F Martin and W A Haseltine. Range of radiochemical damage to DNA with decay of iodine-125. *Science (New York, N.Y.)*, 213(4510):896–8, aug 1981.

[180] Sren M Bentzen. Preventing or reducing late side effects of radiation therapy: radiobiology meets molecular pathology. *Nature reviews. Cancer*, 6(9):702–13, sep 2006.

[181] H R Withers, J M Taylor, and B Maciejewski. Treatment volume and tissue tolerance. *International journal of radiation oncology, biology, physics*, 14(4):751–9, apr 1988.

[182] V V Moiseenko, J J Battista, R P Hill, E L Travis, and J Van Dyk. In-field and out-of-field effects in partial volume lung irradiation in rodents: possible correlation between early dna damage and functional endpoints. *International journal of radiation oncology, biology, physics*, 48(5):1539–48, dec 2000.

[183] Hendrik P. Bijl, Peter Van Luijk, Rob P. Coppes, Jacobus M. Schippers, A. W T Konings, and Albert J. Van Der Kogel. Regional differences in radiosensitivity across the rat cervical spinal cord. *International Journal of Radiation Oncology Biology Physics*, 61(2):543–551, feb 2005.

[184] Peter van Luijk, Sarah Pringle, Joseph O Deasy, Vitali V Moiseenko, Hette Faber, Allan Hovan, Mirjam Baanstra, Hans P van der Laan, Roel G J Kierkels, Arjen van der Schaaf, Max J Witjes, Jacobus M Schippers, Sytze Brandenburg, Johannes A Langendijk, Jonn Wu, and Robert P Coppes. Sparing the region of the salivary gland containing stem cells preserves saliva production after radiotherapy for head and neck cancer. *Science translational medicine*, 7(305):305ra147, sep 2015.

[185] Alena Novakova-Jiresova, Peter van Luijk, Harry van Goor, Harm H Kampinga, and Robert P Coppes. Pulmonary radiation injury: identification of risk factors associated with regional hypersensitivity. *Cancer research*, 65(9):3568–76, may 2005.

[186] Robert P. Coppes, Christina T. Muijs, Hette Faber, Sascha Gross, Jacobus M. Schippers, Sytze Brandenburg, Johannes A. Langendijk, and Peter Van Luijk. Volume-Dependent Expression of In-Field and Out-of-Field Effects in the Proton-Irradiated Rat Lung. *International Journal of Radiation Oncology Biology Physics*, 81(1):262–269, sep 2011.

[187] G. Ghobadi, B. Bartelds, S. J. van der Veen, M. G. Dickinson, S. Brandenburg, R. M. F. Berger, J. a. Langendijk, R. P. Coppes, and P. van Luijk. Lung irradiation induces pulmonary vascular remodelling resembling pulmonary arterial hypertension. *Thorax*, 67(4):334–341, apr 2012.

[188] Isabelle M A Lombaert, Jeanette F Brunsting, Pieter K Wierenga, Hette Faber, Monique A Stokman, Tineke Kok, Willy H Visser, Harm H Kampinga, Gerald de Haan, and Robert P Coppes. Rescue of salivary gland function after stem cell transplantation in irradiated glands. *PloS one*, 3(4):e2063, jan 2008.

[189] Antonius W T Konings, Femmy Cotteleer, Hette Faber, Peter van Luijk, Harm Meertens, and Rob P Coppes. Volume effects and region-dependent radiosensitivity of the parotid gland. *International journal of radiation oncology, biology, physics*, 62(4):1090–5, jul 2005.

[190] Peter van Luijk, Alena Novakova-Jiresova, Hette Faber, Jacobus M Schippers, Harm H Kampinga, Harm Meertens, and Rob P Coppes. Radiation damage to the heart enhances early radiation-induced lung function loss. *Cancer research*, 65(15):6509–11, aug 2005.

[191] Georgi Nalbantov, Bas Kietselaer, Katrien Vandecasteele, Cary Oberije, Maaike Berbee, Esther Troost, Anne-Marie Dingemans, Angela van Baardwijk, Kim Smits, Andre Dekker, Johan Bussink, Dirk De Ruysscher, Yolande Lievens, and Philippe Lambin. Cardiac comorbidity is an independent risk factor for radiation-induced lung toxicity in lung cancer patients. *Radiotherapy and oncology : journal of the European Society for Therapeutic Radiology and Oncology*, 109(1):100–6, oct 2013.

[192] Yuichi Ozawa, Takefumi Abe, Minako Omae, Takashi Matsui, Masato Kato, Hirotsugu Hasegawa, Yasunori Enomoto, Takeaki Ishihara, Naoki Inui, Kazunari Yamada, Koshi Yokomura, and Takafumi Suda. Impact of Preexisting Interstitial Lung Disease on Acute, Extensive Radiation Pneumonitis: Retrospective Analysis of Patients with Lung Cancer. *PloS one*, 10(10):e0140437, jan 2015.

[193] Sadayuki Murayama, Sadayuki Muryama, Tamaki Akamine, Shuji Sakai, Yasuji Oshiro, Yasumasa Kakinohana, Hiroyasu Soeda, Takafumi Toita, and Genki Adachi. Risk factor of radiation pneumonitis: assessment with velocity-encoded cine magnetic resonance imaging of pulmonary artery. *Journal of computer assisted tomography*, 28(2):204–8, jan.

[194] Ivo Beetz, Cornelis Schilstra, Arjen Van Der Schaaf, Edwin R. Van Den Heuvel, Patricia Doornaert, Peter Van Luijk, Arjan Vissink, Bernard F A M Van Der Laan, Charles R. Leemans, Henk P. Bijl, Miranda E M C Christianen, Roel J H M Steenbakkers, and Johannes A. Langendijk. NTCP models for patient-rated xerostomia

and sticky saliva after treatment with intensity modulated radiotherapy for head and neck cancer: The role of dosimetric and clinical factors. *Radiotherapy and Oncology*, 105(1):101–106, oct 2012.

[195] Stephanie T H Peeters, Mischa S Hoogeman, Wilma D Heemsbergen, Augustinus A M Hart, Peter C M Koper, and Joos V Lebesque. Rectal bleeding, fecal incontinence, and high stool frequency after conformal radiotherapy for prostate cancer: normal tissue complication probability modeling. *International journal of radiation oncology, biology, physics*, 66(1):11–9, sep 2006.

[196] Bob Kuska. Beer, bethesda, and biology: How ?genomics? came into being. *Journal of the National Cancer Institute*, 90(2):93, 1998.

[197] Richard P. Horgan and Louise C. Kenny. ?omic? technologies: genomics, transcriptomics, proteomics and metabolomics. *The Obstetrician & Gynaecologist*, 13(3):189–195, 2011.

[198] Catharine West and Barry S. Rosenstein. Establishment of a radiogenomics consortium. *International Journal of Radiation Oncology*Biology*Physics*, 76(5):1295–1296, 2010.

[199] F. Crick. Central dogma of molecular biology. *Nature*, 227(5258):561–3, 1970.

[200] C. B. Yoo and P. A. Jones. Epigenetic therapy of cancer: past, present and future. *Nat Rev Drug Discov.*, 5, 2006.

[201] Biomarkers Definitions Working Group. Biomarkers and surrogate endpoints: preferred definitions and conceptual framework. *Clin Pharmacol Ther*, 69(3):89–95, 2001.

[202] Karla V. Ballman. Biomarker: Predictive or prognostic? *Journal of Clinical Oncology*, 33(33):3968–3971, 2015.

[203] C. N. Oldenhuis, S. F. Oosting, J. A. Gietema, and E. G. de Vries. Prognostic versus predictive value of biomarkers in oncology. *Eur J Cancer*, 44(7):946–53, 2008.

[204] S. L. Kerns, S. Kundu, J. H. Oh, S. K. Singhal, M. Janelsins, L. B. Travis, J. O. Deasy, A. C. Janssens, H. Ostrer, M. Parliament, N. Usmani, and B. S. Rosenstein. The prediction of radiotherapy toxicity using single nucleotide polymorphism-based models: A step toward prevention. *Semin Radiat Oncol*, 25(4):281–91, 2015.

[205] H. Mei, S. Sun, Y. Bai, Y. Chen, R. Chai, and H. Li. Reduced mtdna copy number increases the sensitivity of tumor cells to chemotherapeutic drugs. *Cell Death Dis*, 6:e1710, 2015.

[206] A. Shlien and D. Malkin. Copy number variations and cancer. *Genome Med*, 1(6):62, 2009.

[207] G. Yavas, M. Koyuturk, M. Ozsoyoglu, M. P. Gould, and T. LaFramboise. An optimization framework for unsupervised identification of rare copy number variation from snp array data. *Genome Biol*, 10(10):R119, 2009.

[208] Steven A. McCarroll, Alan Huett, Petric Kuballa, Shannon D. Chilewski, Aimee Landry, Philippe Goyette, Michael C. Zody, Jennifer L. Hall, Steven R. Brant, Judy H. Cho, Richard H. Duerr, Mark S. Silverberg, Kent D. Taylor, John D. Rioux, David

Altshuler, Mark J. Daly, and Ramnik J. Xavier. Deletion polymorphism upstream of irgm associated with altered irgm expression and crohn's disease. *Nat Genet*, 40(9):1107–1112, 2008.

[209] Rafael de Cid, Eva Riveira-Munoz, Patrick L. J. M. Zeeuwen, Jason Robarge, Wilson Liao, Emma N. Dannhauser, Emiliano Giardina, Philip E. Stuart, Rajan Nair, Cynthia Helms, Georgia Escaramis, Ester Ballana, Gemma Martin-Ezquerra, Martin den Heijer, Marijke Kamsteeg, Irma Joosten, Evan E. Eichler, Conxi Lazaro, Ramon M. Pujol, Lluis Armengol, Goncalo Abecasis, James T. Elder, Giuseppe Novelli, John A. L. Armour, Pui-Yan Kwok, Anne Bowcock, Joost Schalkwijk, and Xavier Estivill. Deletion of the late cornified envelope lce3b and lce3c genes as a susceptibility factor for psoriasis. *Nat Genet*, 41(2):211–215, 2009.

[210] Y. Qiao, C. Tyson, M. Hrynchak, E. Lopez-Rangel, J. Hildebrand, S. Martell, C. Fawcett, L. Kasmara, K. Calli, C. Harvard, X. Liu, J. J. Holden, S. M. Lewis, and E. Rajcan-Separovic. Clinical application of 2.7m cytogenetics array for cnv detection in subjects with idiopathic autism and/or intellectual disability. *Clin Genet*, 83(2):145–54, 2013.

[211] D. Pinto, A. T. Pagnamenta, L. Klei, R. Anney, D. Merico, R. Regan, J. Conroy, T. R. Magalhaes, C. Correia, B. S. Abrahams, J. Almeida, E. Bacchelli, G. D. Bader, A. J. Bailey, G. Baird, A. Battaglia, T. Berney, N. Bolshakova, S. Bolte, P. F. Bolton, T. Bourgeron, S. Brennan, J. Brian, S. E. Bryson, A. R. Carson, G. Casallo, J. Casey, B. H. Chung, L. Cochrane, C. Corsello, E. L. Crawford, A. Crossett, C. Cytrynbaum, G. Dawson, M. de Jonge, R. Delorme, I. Drmic, E. Duketis, F. Duque, A. Estes, P. Farrar, B. A. Fernandez, S. E. Folstein, E. Fombonne, C. M. Freitag, J. Gilbert, C. Gillberg, J. T. Glessner, J. Goldberg, A. Green, J. Green, S. J. Guter, H. Hakonarson, E. A. Heron, M. Hill, R. Holt, J. L. Howe, G. Hughes, V. Hus, R. Igliozzi, C. Kim, S. M. Klauck, A. Kolevzon, O. Korvatska, V. Kustanovich, C. M. Lajonchere, J. A. Lamb, M. Laskawiec, M. Leboyer, A. Le Couteur, B. L. Leventhal, A. C. Lionel, X. Q. Liu, C. Lord, L. Lotspeich, S. C. Lund, E. Maestrini, W. Mahoney, C. Mantoulan, C. R. Marshall, H. McConachie, C. J. McDougle, J. McGrath, W. M. McMahon, A. Merikangas, O. Migita, N. J. Minshew, G. K. Mirza, J. Munson, S. F. Nelson, C. Noakes, A. Noor, G. Nygren, G. Oliveira, K. Papanikolaou, J. R. Parr, B. Parrini, T. Paton, A. Pickles, M. Pilorge, et al. Functional impact of global rare copy number variation in autism spectrum disorders. *Nature*, 466(7304):368–72, 2010.

[212] Roland P. Kuiper, Marjolijn J. L. Ligtenberg, Nicoline Hoogerbrugge, and Ad Geurts van Kessel. Germline copy number variation and cancer risk. *Current Opinion in Genetics and Development*, 20(3):282–289, 2010.

[213] W. Liu, J. Sun, G. Li, Y. Zhu, S. Zhang, S. T. Kim, F. Wiklund, K. Wiley, S. D. Isaacs, P. Stattin, J. Xu, D. Duggan, J. D. Carpten, W. B. Isaacs, H. Gronberg, S. L. Zheng, and B. L. Chang. Association of a germ-line copy number variation at 2p24.3 and risk for aggressive prostate cancer. *Cancer Res*, 69(6):2176–9, 2009.

[214] Feng Zhang, Wenli Gu, Matthew E. Hurles, and James R. Lupski. Copy number variation in human health, disease, and evolution. *Annual Review of Genomics and Human Genetics*, 10(1):451–481, 2009.

[215] Barbara E. Stranger, Matthew S. Forrest, Mark Dunning, Catherine E. Ingle, Claude Beazley, Natalie Thorne, Richard Redon, Christine P. Bird, Anna de Grassi, Charles

Lee, Chris Tyler-Smith, Nigel Carter, Stephen W. Scherer, Simon Tavar?, Panagiotis Deloukas, Matthew E. Hurles, and Emmanouil T. Dermitzakis. Relative impact of nucleotide and copy number variation on gene expression phenotypes. *Science*, 315(5813):848–853, 2007.

[216] J. Coates, A. K. Jeyaseelan, N. Ybarra, M. David, S. Faria, L. Souhami, F. Cury, M. Duclos, and I. El Naqa. Contrasting analytical and data-driven frameworks for radiogenomic modeling of normal tissue toxicities in prostate cancer. *Radiother Oncol*, 2015.

[217] A. Eustace, N. Mani, P. N. Span, J. J. Irlam, J. Taylor, G. N. Betts, H. Denley, C. J. Miller, J. J. Homer, A. M. Rojas, P. J. Hoskin, F. M. Buffa, A. L. Harris, J. H. Kaanders, and C. M. West. A 26-gene hypoxia signature predicts benefit from hypoxia-modifying therapy in laryngeal cancer but not bladder cancer. *Clin Cancer Res*, 19(17):4879–88, 2013.

[218] L. Yang, J. Taylor, A. Eustace, J. Irlam, H. Denley, P. J. Hoskin, J. Alsner, F. M. Buffa, A. L. Harris, A. Choudhury, and C. M. West. A gene signature for selecting benefit from hypoxia modification of radiotherapy for high risk bladder cancer patients. *Clin Cancer Res*, 2017.

[219] Wei Wang and Yun-ping Luo. Micrornas in breast cancer: oncogene and tumor suppressors with clinical potential. *Journal of Zhejiang University. Science. B*, 16(1):18–31, 2015.

[220] F. M. Buffa, C. Camps, L. Winchester, C. E. Snell, H. E. Gee, H. Sheldon, M. Taylor, A. L. Harris, and J. Ragoussis. microrna-associated progression pathways and potential therapeutic targets identified by integrated mrna and microrna expression profiling in breast cancer. *Cancer Res*, 71(17):5635–45, 2011.

[221] X. Huang, Q. T. Le, and A. J. Giaccia. Mir-210–micromanager of the hypoxia pathway. *Trends Mol Med*, 16(5):230–7, 2010.

[222] J. Cai, H. Guan, L. Fang, Y. Yang, X. Zhu, J. Yuan, J. Wu, and M. Li. Microrna-374a activates wnt/beta-catenin signaling to promote breast cancer metastasis. *J Clin Invest*, 123(2):566–79, 2013.

[223] Paul Bertone, Viktor Stolc, Thomas E. Royce, Joel S. Rozowsky, Alexander E. Urban, Xiaowei Zhu, John L. Rinn, Waraporn Tongprasit, Manoj Samanta, Sherman Weissman, Mark Gerstein, and Michael Snyder. Global identification of human transcribed sequences with genome tiling arrays. *Science*, 306(5705):2242–2246, 2004.

[224] ValerieBrd O?Leary, SaakVictor Ovsepian, LauraGarcia Carrascosa, FabianAndreas Buske, Vanja Radulovic, Maximilian Niyazi, Simone Moertl, Matt Trau, MichaelJohn Atkinson, and Nataa Anastasov. ¡em¿particle,¡/em¿ a triplex-forming long ncrna, regulates locus-specific methylation in response to low-dose irradiation. *Cell Reports*, 11(3):474–485, 2015.

[225] Youyou Zhang, Qun He, Zhongyi Hu, Yi Feng, Lingling Fan, Zhaoqing Tang, Jiao Yuan, Weiwei Shan, Chunsheng Li, Xiaowen Hu, Janos L. Tanyi, Yi Fan, Qihong Huang, Kathleen Montone, Chi V. Dang, and Lin Zhang. Long noncoding rna linp1 regulates repair of dna double-strand breaks in triple-negative breast cancer. *Nat Struct Mol Biol*, 23(6):522–530, 2016.

[226] Abul K. Abbas, Andrew H. Lichtman, and Shiv Pillai. *Cellular and molecular immunology.* Saunders Elsevier, Philadelphia, 6th edition, 2007.

[227] Feng-Ming Kong, Mary Kay Washington, Randy L. Jirtle, and Mitchell S. Anscher. Plasma transforming growth factor-[beta]1 reflects disease status in patients with lung cancer after radiotherapy: a possible tumor marker. *Lung Cancer*, 16(1):47–59, 1996.

[228] M. S. Anscher, F. M. Kong, and R. L. Jirtle. The relevance of transforming growth factor beta 1 in pulmonary injury after radiation therapy. *Lung Cancer*, 19(2):109–20, 1998.

[229] F. M. Kong, M. S. Anscher, T. A. Sporn, M. K. Washington, R. Clough, M. H. Barcellos-Hoff, and R. L. Jirtle. Loss of heterozygosity at the mannose 6-phosphate insulin-like growth factor 2 receptor (m6p/igf2r) locus predisposes patients to radiation-induced lung injury. *Int J Radiat Oncol Biol Phys*, 49(1):35–41, 2001.

[230] M. S. Anscher, L. B. Marks, T. D. Shafman, R. Clough, H. Huang, A. Tisch, M. Munley, J. E. Herndon, J. Garst, J. Crawford, and R. L. Jirtle. Risk of long-term complications after tfg-beta1-guided very-high-dose thoracic radiotherapy. *Int J Radiat Oncol Biol Phys*, 56(4):988–95, 2003.

[231] R. P. Hill. Radiation effects on the respiratory system. *BJR Suppl*, 27:75–81, 2005.

[232] D. E. Hallahan, L. Geng, and Y. Shyr. Effects of intercellular adhesion molecule 1 (icam-1) null mutation on radiation-induced pulmonary fibrosis and respiratory insufficiency in mice. *J Natl Cancer Inst*, 94(10):733–41, 2002.

[233] C. Linard, C. Marquette, J. Mathieu, A. Pennequin, D. Clarencon, and D. Mathe. Acute induction of inflammatory cytokine expression after gamma-irradiation in the rat: effect of an nf-kappab inhibitor. *Int J Radiat Oncol Biol Phys*, 58(2):427–34, 2004.

[234] M. F. Chen, P. C. Keng, P. Y. Lin, C. T. Yang, S. K. Liao, and W. C. Chen. Caffeic acid phenethyl ester decreases acute pneumonitis after irradiation in vitro and in vivo. *BMC Cancer*, 5:158, 2005.

[235] Matthew R. Jones, Benjamin T. Simms, Michal M. Lupa, Mariya S. Kogan, and Joseph P. Mizgerd. Lung nf-kappab activation and neutrophil recruitment require il-1 and tnf receptor signaling during pneumococcal pneumonia. *J Immunol*, 175(11):7530–7535, 2005.

[236] C. J. Johnston, T. W. Wright, P. Rubin, and J. N. Finkelstein. Alterations in the expression of chemokine mrna levels in fibrosis-resistant and -sensitive mice after thoracic irradiation. *Exp Lung Res*, 24(3):321–37, 1998.

[237] Y. Chen, P. Okunieff, and S. A. Ahrendt. Translational research in lung cancer. *Semin Surg Oncol*, 21(3):205–19, 2003.

[238] C. J. Johnston, B. Piedboeuf, P. Rubin, J. P. Williams, R. Baggs, and J. N. Finkelstein. Early and persistent alterations in the expression of interleukin-1 alpha, interleukin-1 beta and tumor necrosis factor alpha mrna levels in fibrosis-resistant and sensitive mice after thoracic irradiation. *Radiat Res*, 145(6):762–7, 1996.

[239] Y. Chen, O. Hyrien, J. Williams, P. Okunieff, T. Smudzin, and P. Rubin. Interleukin (il)-1a and il-6: applications to the predictive diagnostic testing of radiation pneumonitis. *Int J Radiat Oncol Biol Phys*, 62(1):260–6, 2005.

[240] J. P. Hart, G. Broadwater, Z. Rabbani, B. J. Moeller, R. Clough, D. Huang, G. A. Sempowski, M. Dewhirst, S. V. Pizzo, Z. Vujaskovic, and M. S. Anscher. Cytokine profiling for prediction of symptomatic radiation-induced lung injury. *Int J Radiat Oncol Biol Phys*, 63(5):1448–54, 2005.

[241] D. Arpin, D. Perol, J. Y. Blay, L. Falchero, L. Claude, S. Vuillermoz-Blas, I. Martel-Lafay, C. Ginestet, L. Alberti, D. Nosov, B. Etienne-Mastroianni, V. Cottin, M. Perol, J. C. Guerin, J. F. Cordier, and C. Carrie. Early variations of circulating interleukin-6 and interleukin-10 levels during thoracic radiotherapy are predictive for radiation pneumonitis. *J Clin Oncol*, 23(34):8748–56, 2005.

[242] J. H. Oh, J. Craft, R. Al Lozi, M. Vaidya, Y. Meng, J. O. Deasy, J. D. Bradley, and I. El Naqa. A bayesian network approach for modeling local failure in lung cancer. *Phys Med Biol*, 56(6):1635–51, 2011.

[243] Caroline H. Johnson, Julijana Ivanisevic, and Gary Siuzdak. Metabolomics: beyond biomarkers and towards mechanisms. *Nat Rev Mol Cell Biol*, 17(7):451–459, 2016.

[244] Emily G. Armitage and Coral Barbas. Metabolomics in cancer biomarker discovery: Current trends and future perspectives. *Journal of Pharmaceutical and Biomedical Analysis*, 87:1–11, 2014.

[245] D. Kim, B. P. Fiske, K. Birsoy, E. Freinkman, K. Kami, R. L. Possemato, Y. Chudnovsky, M. E. Pacold, W. W. Chen, J. R. Cantor, L. M. Shelton, D. Y. Gui, M. Kwon, S. H. Ramkissoon, K. L. Ligon, S. W. Kang, M. Snuderl, M. G. Vander Heiden, and D. M. Sabatini. Shmt2 drives glioma cell survival in ischaemia but imposes a dependence on glycine clearance. *Nature*, 520(7547):363–7, 2015.

[246] Habtom W. Ressom, Jun Feng Xiao, Leepika Tuli, Rency S. Varghese, Bin Zhou, Tsung-Heng Tsai, Mohammad R. Nezami Ranjbar, Yi Zhao, Jinlian Wang, Cristina Di Poto, Amrita K. Cheema, Mahlet G. Tadesse, Radoslav Goldman, and Kirti Shetty. Utilization of metabolomics to identify serum biomarkers for hepatocellular carcinoma in patients with liver cirrhosis. *Analytica Chimica Acta*, 743:90–100, 2012.

[247] Simon R. Lord, Neel Patel, Dan Liu, John Fenwick, Fergus Gleeson, Francesca Buffa, and Adrian L. Harris. Neoadjuvant window studies of metformin and biomarker development for drugs targeting cancer metabolism. *JNCI Monographs*, 2015(51):81–86, 2015.

[248] Laila M. Poisson, Adnan Munkarah, Hala Madi, Indrani Datta, Sharon Hensley-Alford, Calvin Tebbe, Thomas Buekers, Shailendra Giri, and Ramandeep Rattan. A metabolomic approach to identifying platinum resistance in ovarian cancer. *Journal of Ovarian Research*, 8(1):13, 2015.

[249] Carl Wibom, Izabella Surowiec, Lina Mrn, Per Bergstrm, Mikael Johansson, Henrik Antti, and A. Tommy Bergenheim. Metabolomic patterns in glioblastoma and changes during radiotherapy: A clinical microdialysis study. *Journal of Proteome Research*, 9(6):2909–2919, 2010.

[250] Q. Yan. Biomedical informatics methods in pharmacogenomics. *Methods Mol Med*, 108:459–86, 2005.

[251] Ganesh A. Viswanathan, Jeremy Seto, Sonali Patil, German Nudelman, and Stuart C. Sealfon. Getting started in biological pathway construction and analysis. *PLoS Computational Biology*, 4(2):e16, 2008.

[252] Lily Wang, Bing Zhang, Russell D. Wolfinger, and Xi Chen. An integrated approach for the analysis of biological pathways using mixed models. *PLoS Genetics*, 4(7):e1000115, 2008.

[253] A. Subramanian, P. Tamayo, V. K. Mootha, S. Mukherjee, B. L. Ebert, M. A. Gillette, A. Paulovich, S. L. Pomeroy, T. R. Golub, E. S. Lander, and J. P. Mesirov. Gene set enrichment analysis: a knowledge-based approach for interpreting genome-wide expression profiles. *Proc Natl Acad Sci U S A*, 102(43):15545–50, 2005.

[254] Corey Speers, Shuang Zhao, Meilan Liu, Harry Bartelink, Lori J. Pierce, and Felix Y. Feng. Development and validation of a novel radiosensitivity signature in human breast cancer. *Clinical Cancer Research*, 21(16):3667–3677, 2015.

[255] S. Lee, N. Ybarra, K. Jeyaseelan, S. Faria, N. Kopek, P. Brisebois, J. D. Bradley, C. Robinson, J. Seuntjens, and I. El Naqa. Bayesian network ensemble as a multivariate strategy to predict radiation pneumonitis risk. *Med Phys*, 42(5):2421–30, 2015.

[256] Y. Luo, I. El Naqa, D. L. McShan, D. Ray, I. Lohse, M. M. Matuszak, D. Owen, S. Jolly, T. S. Lawrence, F. S. Kong, and R. K. Ten Haken. Unraveling biophysical interactions of radiation pneumonitis in non-small-cell lung cancer via bayesian network analysis. *Radiother Oncol*, 2017.

[257] J. Niederer and J. R. Cunningham. The response of cells in culture to fractionated radiation: a theoretical approach. *Phys Med Biol*, 21(5):823–39, 1976.

[258] G. W. Barendsen, H. M. Walter, J. F. Fowler, and D. K. Bewley. Effects of different ionizing radiations on human cells in tissue culture. iii. experiments with cyclotron-accelerated alpha-particles and deuterons. *Radiat Res*, 18:106–19, 1963.

[259] A. Cole, W. G. Cooper, F. Shonka, P. M. Corry, R. M. Humphrey, and A. T. Ansevin. Dna scission in hamster cells and isolated nuclei studied by low-voltage electron beam irradiation. *Radiat Res*, 60(1):1–33, 1974.

[260] A. Cole, F. Shonka, P. Corry, and W. G. Cooper. Cho cell repair of single-strand and double-strand dna breaks induced by gamma- and alpha-radiations. *Basic Life Sci*, 5B:665–76, 1975.

[261] D. T. Goodhead. Initial events in the cellular effects of ionizing radiations: clustered damage in dna. *Int J Radiat Biol*, 65(1):7–17, 1994.

[262] V. V. Moiseenko, R. N. Hamm, A. J. Waker, and W. V. Prestwich. The cellular environment in computer simulations of radiation-induced damage to dna. *Radiat Environ Biophys*, 37(3):167–72, 1998.

[263] I. El Naqa, P. Pater, and J. Seuntjens. Monte carlo role in radiobiological modelling of radiotherapy outcomes. *Phys Med Biol*, 57(11):R75–97, 2012.

[264] H. Nikjoo, S. Uehara, W. E. Wilson, M. Hoshi, and D. T. Goodhead. Track structure in radiation biology: theory and applications. *Int J Radiat Biol*, 73(4):355–64, 1998.

[265] W. Friedland, M. Dingfelder, P. Kundrat, and P. Jacob. Track structures, dna targets and radiation effects in the biophysical monte carlo simulation code partrac. *Mutat Res*, 711(1-2):28–40, 2011.

[266] V. A. Semenenko, J. E. Turner, and T. B. Borak. Norec, a monte carlo code for simulating electron tracks in liquid water. *Radiat Environ Biophys*, 42(3):213–7, 2003.

[267] M. A. Bernal, M. C. Bordage, J. M. Brown, M. Davidkova, E. Delage, Z. El Bitar, S. A. Enger, Z. Francis, S. Guatelli, V. N. Ivanchenko, M. Karamitros, I. Kyriakou, L. Maigne, S. Meylan, K. Murakami, S. Okada, H. Payno, Y. Perrot, I. Petrovic, Q. T. Pham, A. Ristic-Fira, T. Sasaki, V. Stepan, H. N. Tran, C. Villagrasa, and S. Incerti. Track structure modeling in liquid water: A review of the geant4-dna very low energy extension of the geant4 monte carlo simulation toolkit. *Phys Med*, 31(8):861–74, 2015.

[268] H. Nikjoo, D. T. Goodhead, D. E. Charlton, and H. G. Paretzke. Energy deposition in small cylindrical targets by monoenergetic electrons. *Int J Radiat Biol*, 60(5):739–56, 1991.

[269] S. J. Karnas, V. V. Moiseenko, E. Yu, P. Truong, and J. J. Battista. Monte carlo simulations and measurement of dna damage from x-ray-triggered auger cascades in iododeoxyuridine (iudr). *Radiat Environ Biophys*, 40(3):199–206, 2001.

[270] V. V. Moiseenko, R. N. Hamm, A. J. Waker, and W. V. Prestwich. Modelling dna damage induced by different energy photons and tritium beta-particles. *Int J Radiat Biol*, 74(5):533–50, 1998.

[271] W. Friedland and P. Kundrat. Track structure based modelling of chromosome aberrations after photon and alpha-particle irradiation. *Mutat Res*, 756(1-2):213–23, 2013.

[272] J. F. Fowler. The linear-quadratic formula and progress in fractionated radiotherapy. *Br J Radiol*, 62(740):679–94, 1989.

[273] D. G. Catcheside, D. E. Lea, and J. M. Thoday. The production of chromosome structural changes in tradescantia microspores in relation to dosage, intensity and temperature. *J Genet*, 47:137–49, 1946.

[274] M. N. Cornforth and J. S. Bedford. A quantitative comparison of potentially lethal damage repair and the rejoining of interphase chromosome breaks in low passage normal human fibroblasts. *Radiat Res*, 111(3):385–405, 1987.

[275] D. J. Brenner and D. E. Herbert. The use of the linear-quadratic model in clinical radiation oncology can be defended on the basis of empirical evidence and theoretical argument. *Med Phys*, 24(8):1245–8, 1997.

[276] V. V. Moiseenko, A. A. Edwards, and N. Nikjoo. Modelling the kinetics of chromosome exchange formation in human cells exposed to ionising radiation. *Radiat Environ Biophys*, 35(1):31–5, 1996.

[277] J. P. Kirkpatrick, D. J. Brenner, and C. G. Orton. Point/counterpoint. the linear-quadratic model is inappropriate to model high dose per fraction effects in radiosurgery. *Med Phys*, 36(8):3381–4, 2009.

[278] M. Zaider. There is no mechanistic basis for the use of the linear-quadratic expression in cellular survival analysis. *Med Phys*, 25(5):791–2, 1998.

[279] D. J. Brenner, L. R. Hlatky, P. J. Hahnfeldt, Y. Huang, and R. K. Sachs. The linear-quadratic model and most other common radiobiological models result in similar predictions of time-dose relationships. *Radiat Res*, 150(1):83–91, 1998.

[280] S. A. Roberts and J. H. Hendry. A realistic closed-form radiobiological model of clinical tumor-control data incorporating intertumor heterogeneity. *Int J Radiat Oncol Biol Phys*, 41(3):689–99, 1998.

[281] S. Webb and A. E. Nahum. A model for calculating tumour control probability in radiotherapy including the effects of inhomogeneous distributions of dose and clonogenic cell density. *Phys Med Biol*, 38(6):653–66, 1993.

[282] A. M. Kellerer and H. H. Rossi. Generalized formulation of dual radiation action. *Radiation Research*, 75(3):471–488, 1978.

[283] S. B. Curtis. Lethal and potentially lethal lesions induced by radiation–a unified repair model. *Radiat Res*, 106(2):252–70, 1986.

[284] C. A. Tobias. The repair-misrepair model in radiobiology: comparison to other models. *Radiat Res Suppl*, 8:S77–95, 1985.

[285] R. K. Sachs, P. Hahnfeld, and D. J. Brenner. The link between low-let dose-response relations and the underlying kinetics of damage production/repair/misrepair. *Int J Radiat Biol*, 72(4):351–74, 1997.

[286] L. Herr, T. Friedrich, M. Durante, and M. Scholz. A comparison of kinetic photon cell survival models. *Radiat Res*, 184(5):494–508, 2015.

[287] G. Iliakis. The role of dna double strand breaks in ionizing radiation-induced killing of eukaryotic cells. *Bioessays*, 13(12):641–8, 1991.

[288] M. Durante, J. S. Bedford, D. J. Chen, S. Conrad, M. N. Cornforth, A. T. Natarajan, D. C. van Gent, and G. Obe. From dna damage to chromosome aberrations: joining the break. *Mutat Res*, 756(1-2):5–13, 2013.

[289] D. Alloni, A. Campa, W. Friedland, L. Mariotti, and A. Ottolenghi. Track structure, radiation quality and initial radiobiological events: considerations based on the partrac code experience. *Int J Radiat Biol*, 88(1-2):77–86, 2012.

[290] W. Friedland, H. G. Paretzke, F. Ballarini, A. Ottolenghi, G. Kreth, and C. Cremer. First steps towards systems radiation biology studies concerned with dna and chromosome structure within living cells. *Radiat Environ Biophys*, 47(1):49–61, 2008.

[291] T. Friedrich, M. Durante, and M. Scholz. Modeling cell survival after photon irradiation based on double-strand break clustering in megabase pair chromatin loops. *Radiat Res*, 178(5):385–94, 2012.

[292] J. F. Fowler. Is repair of dna strand break damage from ionizing radiation second-order rather than first-order? a simpler explanation of apparently multiexponential repair. *Radiat Res*, 152(2):124–36, 1999.

[293] S. V. Costes, I. Chiolo, J. M. Pluth, M. H. Barcellos-Hoff, and B. Jakob. Spatiotemporal characterization of ionizing radiation induced dna damage foci and their relation to chromatin organization. *Mutat Res*, 704(1-3):78–87, 2010.

[294] W. Sontag. Comparison of six different models describing survival of mammalian cells after irradiation. *Radiat Environ Biophys*, 29(3):185–201, 1990.

[295] D. E. Charlton, H. Nikjoo, and J. L. Humm. Calculation of initial yields of single- and double-strand breaks in cell nuclei from electrons, protons and alpha particles. *Int J Radiat Biol*, 56(1):1–19, 1989.

[296] W. Friedland, P. Jacob, and P. Kundrat. Stochastic simulation of dna double-strand break repair by non-homologous end joining based on track structure calculations. *Radiat Res*, 173(5):677–88, 2010.

[297] R. K. Sachs and D. J. Brenner. Effect of let on chromosomal aberration yields. i. do long-lived, exchange-prone double strand breaks play a role? *Int J Radiat Biol*, 64(6):677–88, 1993.

[298] D. T. Goodhead. The initial physical damage produced by ionizing radiations. *Int J Radiat Biol*, 56(5):623–34, 1989.

[299] J. F. Ward. The yield of dna double-strand breaks produced intracellularly by ionizing radiation: a review. *Int J Radiat Biol*, 57(6):1141–50, 1990.

[300] D. Alloni, G. Baiocco, G. Babini, W. Friedland, P. Kundrat, L. Mariotti, and A. Ottolenghi. Energy dependence of the complexity of dna damage induced by carbon ions. *Radiat Prot Dosimetry*, 166(1-4):86–90, 2015.

[301] A. A. Edwards, V. V. Moiseenko, and H. Nikjoo. Modelling of dna breaks and the formation of chromosome aberrations. *Int J Radiat Biol*, 66(5):633–7, 1994.

[302] T. Friedrich, M. Durante, and M. Scholz. Modeling cell survival after irradiation with ultrasoft x rays using the giant loop binary lesion model. *Radiat Res*, 181(5):485–94, 2014.

[303] L. Herr, I. Shuryak, T. Friedrich, M. Scholz, M. Durante, and D. J. Brenner. New insight into quantitative modeling of dna double-strand break rejoining. *Radiat Res*, 184(3):280–95, 2015.

[304] M. Scholz, A. M. Kellerer, W. Kraft-Weyrather, and G. Kraft. Computation of cell survival in heavy ion beams for therapy. the model and its approximation. *Radiat Environ Biophys*, 36(1):59–66, 1997.

[305] R. B. Hawkins. A microdosimetric-kinetic model of cell death from exposure to ionizing radiation of any let, with experimental and clinical applications. *Int J Radiat Biol*, 69(6):739–55, 1996.

[306] R. B. Hawkins. A microdosimetric-kinetic theory of the dependence of the rbe for cell death on let. *Med Phys*, 25(7 Pt 1):1157–70, 1998.

[307] R. B. Hawkins. A microdosimetric-kinetic model for the effect of non-poisson distribution of lethal lesions on the variation of rbe with let. *Radiat Res*, 160(1):61–9, 2003.

[308] R. B. Hawkins and T. Inaniwa. A microdosimetric-kinetic model for cell killing by protracted continuous irradiation including dependence on let i: repair in cultured mammalian cells. *Radiat Res*, 180(6):584–94, 2013.

[309] D. J. Carlson, R. D. Stewart, V. A. Semenenko, and G. A. Sandison. Combined use of monte carlo dna damage simulations and deterministic repair models to examine putative mechanisms of cell killing. *Radiat Res*, 169(4):447–59, 2008.

[310] M. C. Frese, V. K. Yu, R. D. Stewart, and D. J. Carlson. A mechanism-based approach to predict the relative biological effectiveness of protons and carbon ions in radiation therapy. *Int J Radiat Oncol Biol Phys*, 83(1):442–50, 2012.

[311] F. Kamp, G. Cabal, A. Mairani, K. Parodi, J. J. Wilkens, and D. J. Carlson. Fast biological modeling for voxel-based heavy ion treatment planning using the mechanistic repair-misrepair-fixation model and nuclear fragment spectra. *Int J Radiat Oncol Biol Phys*, 93(3):557–68, 2015.

[312] V. A. Semenenko and R. D. Stewart. Fast monte carlo simulation of dna damage formed by electrons and light ions. *Phys Med Biol*, 51(7):1693–706, 2006.

[313] R. D. Stewart, V. K. Yu, A. G. Georgakilas, C. Koumenis, J. H. Park, and D. J. Carlson. Effects of radiation quality and oxygen on clustered dna lesions and cell death. *Radiat Res*, 176(5):587–602, 2011.

[314] A. Mairani, I. Dokic, G. Magro, T. Tessonnier, F. Kamp, D. J. Carlson, M. Ciocca, F. Cerutti, P. R. Sala, A. Ferrari, T. T. Bohlen, O. Jakel, K. Parodi, J. Debus, A. Abdollahi, and T. Haberer. Biologically optimized helium ion plans: calculation approach and its in vitro validation. *Phys Med Biol*, 61(11):4283–99, 2016.

[315] C. C. Ling, L. E. Gerweck, M. Zaider, and E. Yorke. Dose-rate effects in external beam radiotherapy redux. *Radiother Oncol*, 95(3):261–8, 2010.

[316] J. M. Brown, D. J. Carlson, and D. J. Brenner. The tumor radiobiology of srs and sbrt: are more than the 5 rs involved? *Int J Radiat Oncol Biol Phys*, 88(2):254–62, 2014.

[317] J. P. Kirkpatrick, J. J. Meyer, and L. B. Marks. The linear-quadratic model is inappropriate to model high dose per fraction effects in radiosurgery. *Semin Radiat Oncol*, 18(4):240–3, 2008.

[318] T. Sheu, J. Molkentine, M. K. Transtrum, T. A. Buchholz, H. R. Withers, H. D. Thames, and K. A. Mason. Use of the lq model with large fraction sizes results in underestimation of isoeffect doses. *Radiother Oncol*, 109(1):21–5, 2013.

[319] C. W. Song, L. C. Cho, J. Yuan, K. E. Dusenbery, R. J. Griffin, and S. H. Levitt. Radiobiology of stereotactic body radiation therapy/stereotactic radiosurgery and the linear-quadratic model. *Int J Radiat Oncol Biol Phys*, 87(1):18–9, 2013.

[320] J. M. Brown, D. J. Brenner, and D. J. Carlson. Dose escalation, not "new biology," can account for the efficacy of stereotactic body radiation therapy with non-small cell lung cancer. *Int J Radiat Oncol Biol Phys*, 85(5):1159–60, 2013.

[321] I. Shuryak, D. J. Carlson, J. M. Brown, and D. J. Brenner. High-dose and fractionation effects in stereotactic radiation therapy: Analysis of tumor control data from 2965 patients. *Radiother Oncol*, 115(3):327–34, 2015.

[322] D. J. Brenner. The linear-quadratic model is an appropriate methodology for determining isoeffective doses at large doses per fraction. *Semin Radiat Oncol*, 18(4):234–9, 2008.

[323] R. Timmerman, R. Paulus, J. Galvin, J. Michalski, W. Straube, J. Bradley, A. Fakiris, A. Bezjak, G. Videtic, D. Johnstone, J. Fowler, E. Gore, and H. Choy. Stereotactic body radiation therapy for inoperable early stage lung cancer. *JAMA*, 303(11):1070–6, 2010.

[324] J. F. Fowler, J. S. Welsh, and S. P. Howard. Loss of biological effect in prolonged fraction delivery. *Int J Radiat Oncol Biol Phys*, 59(1):242–9, 2004.

[325] H. J. Park, R. J. Griffin, S. Hui, S. H. Levitt, and C. W. Song. Radiation-induced vascular damage in tumors: implications of vascular damage in ablative hypofractionated radiotherapy (sbrt and srs). *Radiat Res*, 177(3):311–27, 2012.

[326] J. F. Fowler. Development of radiobiology for oncology-a personal view. *Phys Med Biol*, 51(13):R263–86, 2006.

[327] M. Guerrero and M. Carlone. Mechanistic formulation of a lineal-quadratic-linear (lql) model: split-dose experiments and exponentially decaying sources. *Med Phys*, 37(8):4173–81, 2010.

[328] M. Guerrero and X. A. Li. Extending the linear-quadratic model for large fraction doses pertinent to stereotactic radiotherapy. *Phys Med Biol*, 49(20):4825–35, 2004.

[329] Michael Joiner and Albert van der Kogel. *Basic clinical radiobiology*. Hodder Arnold ;, London, 4th edition, 2009.

[330] C. Park, L. Papiez, S. Zhang, M. Story, and R. D. Timmerman. Universal survival curve and single fraction equivalent dose: useful tools in understanding potency of ablative radiotherapy. *Int J Radiat Oncol Biol Phys*, 70(3):847–52, 2008.

[331] J. Z. Wang, Z. Huang, S. S. Lo, W. T. Yuh, and N. A. Mayr. A generalized linear-quadratic model for radiosurgery, stereotactic body radiation therapy, and high-dose rate brachytherapy. *Sci Transl Med*, 2(39):39ra48, 2010.

[332] M. Astrahan. Some implications of linear-quadratic-linear radiation dose-response with regard to hypofractionation. *Med Phys*, 35(9):4161–72, 2008.

[333] M. Guckenberger. Dose and fractionation in stereotactic body radiation therapy for stage i non-small cell lung cancer: Lessons learned and where do we go next? *Int J Radiat Oncol Biol Phys*, 93(4):765–8, 2015.

[334] M. Guckenberger, R. J. Klement, M. Allgauer, S. Appold, K. Dieckmann, I. Ernst, U. Ganswindt, R. Holy, U. Nestle, M. Nevinny-Stickel, S. Semrau, F. Sterzing, A. Wittig, N. Andratschke, and M. Flentje. Applicability of the linear-quadratic formalism for modeling local tumor control probability in high dose per fraction stereotactic body radiotherapy for early stage non-small cell lung cancer. *Radiother Oncol*, 109(1):13–20, 2013.

[335] G. R. Borst, M. Ishikawa, J. Nijkamp, M. Hauptmann, H. Shirato, G. Bengua, R. Onimaru, A. de Josien Bois, J. V. Lebesque, and J. J. Sonke. Radiation pneumonitis after hypofractionated radiotherapy: evaluation of the lq(l) model and different dose parameters. *Int J Radiat Oncol Biol Phys*, 77(5):1596–603, 2010.

[336] A. E. Scheenstra, M. M. Rossi, J. S. Belderbos, E. M. Damen, J. V. Lebesque, and J. J. Sonke. Alpha/beta ratio for normal lung tissue as estimated from lung cancer patients treated with stereotactic body and conventionally fractionated radiation therapy. *Int J Radiat Oncol Biol Phys*, 88(1):224–8, 2014.

[337] N. Mehta, C. R. King, N. Agazaryan, M. Steinberg, A. Hua, and P. Lee. Stereotactic body radiation therapy and 3-dimensional conformal radiotherapy for stage i non-small cell lung cancer: A pooled analysis of biological equivalent dose and local control. *Pract Radiat Oncol*, 2(4):288–95, 2012.

[338] D. J. Carlson, P. J. Keall, Jr. Loo, B. W., Z. J. Chen, and J. M. Brown. Hypofractionation results in reduced tumor cell kill compared to conventional fractionation for tumors with regions of hypoxia. *Int J Radiat Oncol Biol Phys*, 79(4):1188–95, 2011.

[339] N. Ohri, B. Piperdi, M. K. Garg, W. R. Bodner, R. Gucalp, R. Perez-Soler, S. M. Keller, and C. Guha. Pre-treatment fdg-pet predicts the site of in-field progression following concurrent chemoradiotherapy for stage iii non-small cell lung cancer. *Lung Cancer*, 87(1):23–7, 2015.

[340] I. Joye, C. M. Deroose, V. Vandecaveye, and K. Haustermans. The role of diffusion-weighted mri and (18)f-fdg pet/ct in the prediction of pathologic complete response after radiochemotherapy for rectal cancer: a systematic review. *Radiother Oncol*, 113(2):158–65, 2014.

[341] N. A. Mayr, Z. Huang, J. Z. Wang, S. S. Lo, J. M. Fan, J. C. Grecula, S. Sammet, C. L. Sammet, G. Jia, J. Zhang, M. V. Knopp, and W. T. Yuh. Characterizing tumor heterogeneity with functional imaging and quantifying high-risk tumor volume for early prediction of treatment outcome: cervical cancer as a model. *Int J Radiat Oncol Biol Phys*, 83(3):972–9, 2012.

[342] S. S. Korreman, S. Ulrich, S. Bowen, M. Deveau, S. M. Bentzen, and R. Jeraj. Feasibility of dose painting using volumetric modulated arc optimization and delivery. *Acta Oncol*, 49(7):964–71, 2010.

[343] S. M. Bentzen and V. Gregoire. Molecular imaging-based dose painting: a novel paradigm for radiation therapy prescription. *Semin Radiat Oncol*, 21(2):101–10, 2011.

[344] J. van Loon, M. H. Janssen, M. Ollers, H. J. Aerts, L. Dubois, M. Hochstenbag, A. M. Dingemans, R. Lalisang, B. Brans, B. Windhorst, G. A. van Dongen, H. Kolb, J. Zhang, D. De Ruysscher, and P. Lambin. Pet imaging of hypoxia using [18f]hx4: a phase i trial. *Eur J Nucl Med Mol Imaging*, 37(9):1663–8, 2010.

[345] O. J. Kelada and D. J. Carlson. Molecular imaging of tumor hypoxia with positron emission tomography. *Radiat Res*, 181(4):335–49, 2014.

[346] D. J. Carlson, R. D. Stewart, and V. A. Semenenko. Effects of oxygen on intrinsic radiation sensitivity: A test of the relationship between aerobic and hypoxic linear-quadratic (lq) model parameters. *Med Phys*, 33(9):3105–15, 2006.

[347] B. G. Wouters and J. M. Brown. Cells at intermediate oxygen levels can be more important than the "hypoxic fraction" in determining tumor response to fractionated radiotherapy. *Radiat Res*, 147(5):541–50, 1997.

[348] R. Ruggieri, N. Stavreva, S. Naccarato, and P. Stavrev. Applying a hypoxia-incorporating tcp model to experimental data on rat sarcoma. *Int J Radiat Oncol Biol Phys*, 83(5):1603–8, 2012.

[349] Y. Shibamoto, A. Miyakawa, S. Otsuka, and H. Iwata. Radiobiology of hypofractionated stereotactic radiotherapy: what are the optimal fractionation schedules? *J Radiat Res*, 57 Suppl 1:i76–i82, 2016.

[350] R. Ruggieri, N. Stavreva, S. Naccarato, and P. Stavrev. Computed 88modelling. *Phys Med Biol*, 58(13):4611–20, 2013.

[351] E. Lindblom, A. Dasu, and I. Toma-Dasu. Optimal fractionation in radiotherapy for non-small cell lung cancer–a modelling approach. *Acta Oncol*, 54(9):1592–8, 2015.

[352] J. Jeong, K. I. Shoghi, and J. O. Deasy. Modelling the interplay between hypoxia and proliferation in radiotherapy tumour response. *Phys Med Biol*, 58(14):4897–919, 2013.

[353] J. Jeong and J. O. Deasy. Modeling the relationship between fluorodeoxyglucose uptake and tumor radioresistance as a function of the tumor microenvironment. *Comput Math Methods Med*, 2014:847162, 2014.

[354] Arjien van der Schaaf, Johannes Albertus Langendijk, Claudio Fiorino, and Tiziana Rancati. Embracing phenomenological approaches to normal tissue complication probability modeling: A question of method. *International Journal of Radiation Oncology*Biology*Physics*, 91(3):468–471, 2015.

[355] Ewout W. Steyerberg. *Clinical prediction models : a practical approach to development, validation, and updating*. Statistics for biology and health. Springer, New York, NY, 2009.

[356] Frank Harrell. *Regression Modeling Strategies: With Applications to Linear Models, Logistic and Ordinal Regression, and Survival Analysis*. Springer, New York, NY, 2015.

[357] Gareth James, Daniela Witten, Trevor Hastie, and Robert Tibshirani. *An introduction to statistical learning : with applications in R*. Springer texts in statistics,. Springer, New York, 2013.

[358] Fred C. Pampel. *Logistic Regression: A Primer*. SAGE, 2000.

[359] Development Core Team. *R: A language and environment for statistical computing*. 2008.

[360] R. Valdagni, P. Allavena, B. Avuzzi, N. Bedini, T. Magnani, S. Morlino, S. Pesce, T. Rancati, S. Villa, and N. Zaffaroni. 202: Mediators associated to the inflammatory response in prostate cancer patients undergoing rt: preliminary results. *Radiotherapy and Oncology*, 110:S99, 2014.

[361] R Core Team et al. R: A language and environment for statistical computing. 2013.

[362] DJ Chapman and AE Nahum. *Radiotherapy Treatment Planning: Linear-Quadratic Radiobiology*. CRC Press, 2016.

[363] D. W. Hosmer, T. Hosmer, S. Le Cessie, and S. Lemeshow. A comparison of goodness-of-fit tests for the logistic regression model. *Stat Med*, 16(9):965–80, 1997.

[364] P. Peduzzi, J. Concato, A. R. Feinstein, and T. R. Holford. Importance of events per independent variable in proportional hazards regression analysis. ii. accuracy and precision of regression estimates. *J Clin Epidemiol*, 48(12):1503–10, 1995.

[365] P. Peduzzi, J. Concato, E. Kemper, T. R. Holford, and A. R. Feinstein. A simulation study of the number of events per variable in logistic regression analysis. *J Clin Epidemiol*, 49(12):1373–9, 1996.

[366] E. W. Steyerberg, A. J. Vickers, N. R. Cook, T. Gerds, M. Gonen, N. Obuchowski, M. J. Pencina, and M. W. Kattan. Assessing the performance of prediction models: a framework for traditional and novel measures. *Epidemiology*, 21(1):128–38, 2010.

[367] Hirotugu Akaike. A new look at the statistical model identification. *Automatic Control, IEEE Transactions on*, 19(6):716–723, 1974.

[368] Robert Tibshirani. Regression shrinkage and selection via the lasso: a retrospective. *Journal of the Royal Statistical Society Series B*, 73(3):273–282, 2011.

[369] F. Palorini, T. Rancati, C. Cozzarini, I. Improta, V. Carillo, B. Avuzzi, V. Casanova Borca, A. Botti, C. Degli Esposti, P. Franco, E. Garibaldi, G. Girelli, C. Iotti, A. Maggio, M. Palombarini, A. Pierelli, E. Pignoli, V. Vavassori, R. Valdagni, and C. Fiorino. Multi-variable models of large international prostate symptom score worsening at the end of therapy in prostate cancer radiotherapy. *Radiother Oncol*, 118(1):92–8, 2016.

[370] J. A. Sloan, M. Halyard, I. El Naqa, and C. Mayo. Lessons from large-scale collection of patient-reported outcomes: Implications for big data aggregation and analytics. *Int J Radiat Oncol Biol Phys*, 95(3):922–9, 2016.

[371] W. S. Mcculloch and W. Pitts. A logical calculus of the ideas immanent in nervous activity (reprinted from bulletin of mathematical biophysics, vol 5, pg 115-133, 1943). *Bulletin of Mathematical Biology*, 52(1-2):99–115, 1990.

[372] W. H. Wolberg and O. L. Mangasarian. Multisurface method of pattern separation for medical diagnosis applied to breast cytology. *Proc Natl.Acad.Sci.U.S.A*, 87(23):9193–9196, 1990.

[373] J. Kang, R. Schwartz, J. Flickinger, and S. Beriwal. Machine learning approaches for predicting radiation therapy outcomes: A clinician's perspective. *Int J Radiat Oncol Biol Phys*, 93(5):1127–35, 2015.

[374] D. Posada and T. R. Buckley. Model selection and model averaging in phylogenetics: Advantages of akaike information criterion and bayesian approaches over likelihood ratio tests. *Systematic Biology*, 53(5):793–808, 2004.

[375] J. D. Bauer, A. Jackson, M. Skwarchuk, and M. Zelefsky. Principal component, varimax rotation and cost analysis of volume effects in rectal bleeding in patients treated with 3d-crt for prostate cancer. *Phys Med Biol*, 51(20):5105–23, 2006.

[376] M. A. Benadjaoud, P. Blanchard, B. Schwartz, J. Champoudry, R. Bouaita, D. Lefkopoulos, E. Deutsch, I. Diallo, H. Cardot, and F. de Vathaire. Functional data analysis in ntcp modeling: a new method to explore the radiation dose-volume effects. *Int J Radiat Oncol Biol Phys*, 90(3):654–63, 2014.

[377] A. Kneip and K. J. Utikal. Inference for density families using functional principal component analysis. *Journal of the American Statistical Association*, 96(454):519–532, 2001.

[378] S. L. Gulliford, M. Partridge, M. R. Sydes, S. Webb, P. M. Evans, and D. P. Dearnaley. Parameters for the lyman kutcher burman (lkb) model of normal tissue complication probability (ntcp) for specific rectal complications observed in clinical practise. *Radiother Oncol*, 102(3):347–51, 2012.

[379] J. A. Dean, K. H. Wong, H. Gay, L. C. Welsh, A. B. Jones, U. Schick, J. H. Oh, A. Apte, K. L. Newbold, S. A. Bhide, K. J. Harrington, J. O. Deasy, C. M. Nutting, and S. L. Gulliford. Functional data analysis applied to modeling of severe acute mucositis and dysphagia resulting from head and neck radiation therapy. *Int J Radiat Oncol Biol Phys*, 96(4):820–831, 2016.

[380] D. E. Rumelhart. Learning representations by back-propagating errors. *Nature*, 323:533–536, 1986.

[381] J. Chen, J. Chen, H. Y. Ding, Q. S. Pan, W. D. Hong, G. Xu, F. Y. Yu, and Y. M. Wang. Use of an artificial neural network to construct a model of predicting deep fungal infection in lung cancer patients. *Asian Pac J Cancer Prev*, 16(12):5095–9, 2015.

[382] S Tomatis, T Rancati, C Fiorino, V Vavassori, G Fellin, E Cagna, F A Mauro, G Girelli, A Monti, M Baccolini, G Naldi, C Bianchi, L Menegotti, M Pasquino, M Stasi, and R Valdagni. Late rectal bleeding after 3d-crt for prostate cancer: development of a neural-network-based predictive model. *Physics in Medicine and Biology*, 57(5):1399, 2012.

[383] A. Pella, R. Cambria, M. Riboldi, B. A. Jereczek-Fossa, C. Fodor, D. Zerini, A. E. Torshabi, F. Cattani, C. Garibaldi, G. Pedroli, G. Baroni, and R. Orecchia. Use of machine learning methods for prediction of acute toxicity in organs at risk following prostate radiotherapy. *Med Phys*, 38(6):2859–67, 2011.

[384] F. Buettner, S. L. Gulliford, S. Webb, and M. Partridge. Using dose-surface maps to predict radiation-induced rectal bleeding: a neural network approach. *Phys Med Biol*, 54(17):5139–53, 2009.

[385] S. Chen, S. Zhou, F. F. Yin, L. B. Marks, and S. K. Das. Investigation of the support vector machine algorithm to predict lung radiation-induced pneumonitis. *Medical Physics*, 34(10):3808–3814, 2007.

[386] K. Nie, L. Shi, Q. Chen, X. Hu, S. K. Jabbour, N. Yue, T. Niu, and X. Sun. Rectal cancer: Assessment of neoadjuvant chemoradiation outcome based on radiomics of multiparametric mri. *Clin Cancer Res*, 2016.

[387] W. Sun, T. B. Tseng, J. Zhang, and W. Qian. Enhancing deep convolutional neural network scheme for breast cancer diagnosis with unlabeled data. *Comput Med Imaging Graph*, 2016.

[388] S. Chen, S. Zhou, J. Zhang, F. F. Yin, L. B. Marks, and S. K. Das. A neural network model to predict lung radiation-induced pneumonitis. *Med Phys*, 34(9):3420–7, 2007.

[389] R. J. Klement, M. Allgauer, S. Appold, K. Dieckmann, I. Ernst, U. Ganswindt, R. Holy, U. Nestle, M. Nevinny-Stickel, S. Semrau, F. Sterzing, A. Wittig, N. Andratschke, and M. Guckenberger. Support vector machine-based prediction of local tumor control after stereotactic body radiation therapy for early-stage non-small cell lung cancer. *Int J Radiat Oncol Biol Phys*, 88(3):732–8, 2014.

[390] N. V. Chawla, K. W. Bowyer, L. O. Hall, and W. P. Kegelmeyer. Smote: Synthetic minority over-sampling technique. *Journal of Artificial Intelligence Research*, 16:321–357, 2002.

[391] Z. Zhou, M. Folkert, N. Cannon, P. Iyengar, K. Westover, Y. Zhang, H. Choy, R. Timmerman, J. Yan, X. J. Xie, S. Jiang, and J. Wang. Predicting distant failure in early stage nsclc treated with sbrt using clinical parameters. *Radiother Oncol*, 119(3):501–4, 2016.

[392] L. Zhang, Y. Zhong, B. Huang, J. Gong, and P. Li. Dimensionality reduction based on clonal selection for hyperspectral imagery. *Ieee Transactions on Geoscience and Remote Sensing*, 45(12):4172–4186, 2007.

[393] W. Y. Loh. Classification and regression trees. *Wiley Interdisciplinary Reviews-Data Mining and Knowledge Discovery*, 1(1):14–23, 2011.

[394] T Hastie, R Tibshirani, and J Friedman. *The Elements of Statistical Learning: Data Mining, Inference, and Prediction*. Springer Series in Statistics. Springer, 2nd edition, 2011.

[395] J. D. Ospina, J. Zhu, C. Chira, A. Bossi, J. B. Delobel, V. Beckendorf, B. Dubray, J. L. Lagrange, J. C. Correa, A. Simon, O. Acosta, and R. de Crevoisier. Random forests to predict rectal toxicity following prostate cancer radiation therapy. *Int J Radiat Oncol Biol Phys*, 89(5):1024–31, 2014.

[396] E. Charniak. Bayesian networks without tears. *Ai Magazine*, 12(4):50–63, 1991.

[397] M. K. Cowles and B. P. Carlin. Markov chain monte carlo convergence diagnostics: A comparative review. *Journal of the American Statistical Association*, 91(434):883–904, 1996.

[398] K. Jayasurya, G. Fung, S. Yu, C. Dehing-Oberije, D. De Ruysscher, A. Hope, W. De Neve, Y. Lievens, P. Lambin, and A. L. Dekker. Comparison of bayesian network and support vector machine models for two-year survival prediction in lung cancer patients treated with radiotherapy. *Med Phys*, 37(4):1401–7, 2010.

[399] N. Yahya, M. A. Ebert, M. Bulsara, M. J. House, A. Kennedy, D. J. Joseph, and J. W. Denham. Statistical-learning strategies generate only modestly performing predictive models for urinary symptoms following external beam radiotherapy of the prostate: A comparison of conventional and machine-learning methods. *Med Phys*, 43(5):2040, 2016.

[400] Ron Kohavi. A study of cross-validation and bootstrap for accuracy estimation and model selection. In *Proceedings of the 14th International Joint Conference on Artificial Intelligence - Volume 2*, IJCAI'95, pages 1137–1143, San Francisco, CA, USA, 1995. Morgan Kaufmann Publishers Inc.

[401] Eric J. Hall and Amato J. Giaccia. *Radiobiology for the radiologist*. Wolters Kluwer Health/Lippincott Williams & Wilkins, Philadelphia, 7th edition, 2012.

[402] S. C. Darby, D. J. Cutter, M. Boerma, L. S. Constine, L. F. Fajardo, K. Kodama, K. Mabuchi, L. B. Marks, F. A. Mettler, L. J. Pierce, K. R. Trott, E. T. Yeh, and R. E. Shore. Radiation-related heart disease: current knowledge and future prospects. *Int J Radiat Oncol Biol Phys*, 76(3):656–65, 2010.

[403] Y. Zhang, Y. Feng, W. Wang, C. Yang, and P. Wang. An expanded multi-scale monte carlo simulation method for personalized radiobiological effect estimation in radiotherapy: a feasibility study. *Sci Rep*, 7:45019, 2017.

[404] P. Lazarakis, M. U. Bug, E. Gargioni, S. Guatelli, H. Rabus, and A. B. Rosenfeld. Comparison of nanodosimetric parameters of track structure calculated by the monte carlo codes geant4-dna and ptra. *Phys Med Biol*, 57(5):1231–50, 2012.

[405] H. Nikjoo, S. Uehara, D. Emfietzoglou, and F. A. Cucinotta. Track-structure codes in radiation research. *Radiation Measurements*, 41(9-10):1052–1074, 2006.

[406] A. Verkhovtsev, E. Surdutovich, and A. V. Solov'yov. Multiscale approach predictions for biological outcomes in ion-beam cancer therapy. *Sci Rep*, 6:27654, 2016.

[407] F. A. Cucinotta, J. M. Pluth, J. A. Anderson, J. V. Harper, and P. O'Neill. Biochemical kinetics model of dsb repair and induction of gamma-h2ax foci by non-homologous end joining. *Radiat Res*, 169(2):214–22, 2008.

[408] R. Taleei and H. Nikjoo. Repair of the double-strand breaks induced by low energy electrons: a modelling approach. *Int J Radiat Biol*, 88(12):948–53, 2012.

[409] W. Ruhm, M. Eidemuller, and J. C. Kaiser. Biologically-based mechanistic models of radiation-related carcinogenesis applied to epidemiological data. *Int J Radiat Biol*, pages 1–25, 2017.

[410] I. Plante and F. A. Cucinotta. Energy deposition and relative frequency of hits of cylindrical nanovolume in medium irradiated by ions: Monte carlo simulation of tracks structure. *Radiat Environ Biophys*, 49(1):5–13, 2010.

[411] S. Incerti, M. Douglass, S. Penfold, S. Guatelli, and E. Bezak. Review of geant4-dna applications for micro and nanoscale simulations. *Phys Med*, 32(10):1187–1200, 2016.

[412] W. Friedland, E. Schmitt, P. Kundrat, M. Dingfelder, G. Baiocco, S. Barbieri, and A. Ottolenghi. Comprehensive track-structure based evaluation of dna damage by light ions from radiotherapy-relevant energies down to stopping. *Sci Rep*, 7:45161, 2017.

[413] P. Bernhardt, W. Friedland, P. Jacob, and H. G. Paretzke. Modeling of ultrasoft x-ray induced dna damage using structured higher order dna targets. *International Journal of Mass Spectrometry*, 223(1-3):579–597, 2003.

[414] M. Matsumoto and T. Nishimura. Mersenne twister: a 623-dimensionally equidistributed uniform pseudo-random number generator. *ACM Transactions on Modeling and Computer Simulation (TOMACS)*, 8(1):3–30, 1998.

[415] M. Dingfelder, D. Hantke, M. Inokuti, and H. G. Paretzke. Electron inelastic-scattering cross sections in liquid water. *Radiation Physics and Chemistry*, 53(1):1–18, 1998.

[416] M. Dingfelder, M. Inokuti, and H. G. Paretzke. Inelastic-collision cross sections of liquid water for interactions of energetic protons. *Radiation Physics and Chemistry*, 59(3):255–275, 2000.

[417] D.E. Cullen, J.H. Hubbell, and L. Kissel. Epdl97: the evaluated photo data library '97 version. Report, Lawrence Livermore National Laboratory, 1997.

[418] W. Friedland, M. Dingfelder, P. Jacob, and H. G. Paretzke. Calculated dna double-strand break and fragmentation yields after irradiation with he ions. *Radiation Physics and Chemistry*, 72(2-3):279–286, 2005.

[419] S.T. Perkins, D.E. Cullen, M.H. Chen, J. Rathkopf, J Scofield, and J.H. Hubbell. Tables and graphs of atomic subshell and relaxation data derived from the llnl evaluated atomic data library (eadl), z = 1–100. Report, 1991.

[420] J. Meesungnoen, M. Benrahmoune, A. Filali-Mouhim, S. Mankhetkorn, and J. P. Jay-Gerin. Monte carlo calculation of the primary radical and molecular yields of liquid water radiolysis in the linear energy transfer range 0.3-6.5 kev/micrometer: application to 137cs gamma rays. *Radiat Res*, 155(2):269–78, 2001.

[421] E. Schmitt, W. Friedland, P. Kundrat, M. Dingfelder, and A. Ottolenghi. Cross-section scaling for track structure simulations of low-energy ions in liquid water. *Radiat Prot Dosimetry*, 166(1-4):15–8, 2015.

[422] H.W. Barkas. *Nuclear Research Emulsions*. Academic Press, New York, 1963.

[423] Peter Siegmund, Andreas Schinner, and Helmut Paul. Errata and addenda for icru report 73, stopping of ions heavier than helium. *Journal of the ICRU*, 5(1), 2009.

[424] J. F. Ziegler, M. D. Ziegler, and J. P. Biersack. Srim - the stopping and range of ions in matter (2010). *Nuclear Instruments & Methods in Physics Research Section B-Beam Interactions with Materials and Atoms*, 268(11-12):1818–1823, 2010.

[425] Maximilian S. Kreipl, Werner Friedland, and Herwig G. Paretzke. Time- and space-resolved monte carlo study of water radiolysis for photon, electron and ion irradiation. *Radiation and environmental biophysics*, 48(1):11–20, 2009.

[426] F. Ballarini, M. Biaggi, M. Merzagora, A. Ottolenghi, M. Dingfelder, W. Friedland, P. Jacob, and H. G. Paretzke. Stochastic aspects and uncertainties in the prechemical and chemical stages of electron tracks in liquid water: a quantitative analysis based on monte carlo simulations. *Radiat Environ Biophys*, 39(3):179–88, 2000.

[427] R.M. Noyes. *Effects of diffusion rate on chemical kinetics*, volume 1, pages 129–160. Pergamon, New York, 1961.

[428] R. N. Hamm, J. E. Turner, and M. G. Stabin. Monte carlo simulation of diffusion and reaction in water radiolysis–a study of reactant 'jump through' and jump distances. *Radiat Environ Biophys*, 36(4):229–34, 1998.

[429] Y. Frongillo, T. Goulet, M. J. Fraser, V. Cobut, J. P. Patau, and J. P. Jay-Gerin. Monte carlo simulation of fast electron and proton tracks in liquid water - ii. nonhomogeneous chemistry. *Radiation Physics and Chemistry*, 51(3):245–254, 1998.

[430] R.H. Bisby, R.B. Cundall, H.E. Sims, and W.G. Burns. Linear energy transfer (let) effects in the radiation-induced inactivation of papain. *Faraday Discusssions of the Chemical Society*, 63:237–247, 1977.

[431] W.G. Burns and H.E. Sims. Effect of radiation type in water radiolysis. *Journal of the Chemical Society, Faraday Transactions 1: Physical Chemistry in Condensed Phases*, 77:2803–2813, 1981.

[432] A. J. Elliot, M. P. Chenier, and D. C. Ouellette. Temperature dependence of g values for h2o and d2o irradiated with low linear energy transf radiation. *Journal of the Chemical Society, Faraday Transactions*, 89(8):1193–1197, 1993.

[433] K. Luger, A. W. Mader, R. K. Richmond, D. F. Sargent, and T. J. Richmond. Crystal structure of the nucleosome core particle at 2.8 a resolution. *Nature*, 389(6648):251–60, 1997.

[434] Karolin Luger, Mekonnen L. Dechassa, and David J. Tremethick. New insights into nucleosome and chromatin structure: an ordered state or a disordered affair? *Nature reviews Molecular cell biology*, 13(7):436–47, 2012.

[435] Sergei A. Grigoryev, Gaurav Arya, Sarah Correll, Christopher L. Woodcock, and Tamar Schlick. Evidence for heteromorphic chromatin fibers from analysis of nucleosome interactions. *Proceedings of the National Academy of Sciences of the United States of America*, 106(32):13317–22, 2009.

[436] W. Friedland, P. Jacob, H. G. Paretzke, and T. Stork. Monte carlo simulation of the production of short dna fragments by low-linear energy transfer radiation using higher-order dna models. *Radiat Res*, 150(2):170–82, 1998.

[437] A. Valota, F. Ballarini, W. Friedland, P. Jacob, A. Ottolenghi, and H. G. Paretzke. Modelling study on the protective role of oh radical scavengers and dna higher-order structures in induction of single- and double-strand break by gamma-radiation. *Int J Radiat Biol*, 79(8):643–53, 2003.

[438] D. Alloni, A. Campa, W. Friedland, L. Mariotti, and A. Ottolenghi. Integration of monte carlo simulations with pfge experimental data yields constant rbe of 2.3 for dna double-strand break induction by nitrogen ions between 125 and 225 kev/mum let. *Radiat Res*, 179(6):690–7, 2013.

[439] E. Hoglund and B. Stenerlow. Induction and rejoining of dna double-strand breaks in normal human skin fibroblasts after exposure to radiation of different linear energy transfer: Possible roles of track structure and chromatin organization. *Radiation Research*, 155(6):818–825, 2001.

[440] B. Rydberg, W. R. Holley, I. S. Mian, and A. Chatterjee. Chromatin conformation in living cells: support for a zig-zag model of the 30 nm chromatin fiber. *J Mol Biol*, 284(1):71–84, 1998.

[441] E. Hoglund. *DNA Fragmentation in Cultured Cells Exposed to High Linear Energy Transfer Radiation*. Thesis, 2000.

[442] E. Hoglund, E. Blomquist, J. Carlsson, and B. Stenerlow. Dna damage induced by radiation of different linear energy transfer: initial fragmentation. *Int J Radiat Biol*, 76(4):539–47, 2000.

[443] T. Friedrich, U. Scholz, T. Elsasser, M. Durante, and M. Scholz. Systematic analysis of rbe and related quantities using a database of cell survival experiments with ion beam irradiation. *J Radiat Res*, 54(3):494–514, 2013.

[444] E. Weterings and D. J. Chen. The endless tale of non-homologous end-joining. *Cell Res*, 18(1):114–24, 2008.

[445] P. O. Mari, B. I. Florea, S. P. Persengiev, N. S. Verkaik, H. T. Bruggenwirth, M. Modesti, G. Giglia-Mari, K. Bezstarosti, J. A. Demmers, T. M. Luider, A. B. Houtsmuller, and D. C. van Gent. Dynamic assembly of end-joining complexes requires interaction between ku70/80 and xrcc4. *Proc Natl Acad Sci U S A*, 103(49):18597–602, 2006.

[446] N. Uematsu, E. Weterings, K. Yano, K. Morotomi-Yano, B. Jakob, G. Taucher-Scholz, P. O. Mari, D. C. van Gent, B. P. Chen, and D. J. Chen. Autophosphorylation of dna-pkcs regulates its dynamics at dna double-strand breaks. *J Cell Biol*, 177(2):219–29, 2007.

[447] S. Girst, V. Hable, G. A. Drexler, C. Greubel, C. Siebenwirth, M. Haum, A. A. Friedl, and G. Dollinger. Subdiffusion supports joining of correct ends during repair of dna double-strand breaks. *Scientific reports*, 3:2511, 2013.

[448] G. Du, G. A. Drexler, W. Friedland, C. Greubel, V. Hable, R. Krucken, A. Kugler, L. Tonelli, A. A. Friedl, and G. Dollinger. Spatial dynamics of dna damage response protein foci along the ion trajectory of high-let particles. *Radiat Res*, 176(6):706–15, 2011.

[449] B. Stenerlow, E. Hoglund, J. Carlsson, and E. Blomquist. Rejoining of dna fragments produced by radiations of different linear energy transfer. *Int J Radiat Biol*, 76(4):549–57, 2000.

[450] W. Friedland, P. Kundrat, and P. Jacob. Stochastic modelling of dsb repair after photon and ion irradiation. *Int J Radiat Biol*, 88(1-2):129–36, 2012.

[451] T. E. Schmid, C. Greubel, V. Hable, O. Zlobinskaya, D. Michalski, S. Girst, C. Siebenwirth, E. Schmid, M. Molls, G. Multhoff, and G. Dollinger. Low let protons focused to submicrometer shows enhanced radiobiological effectiveness. *Phys Med Biol*, 57(19):5889–907, 2012.

[452] T. E. Schmid, W. Friedland, C. Greubel, S. Girst, J. Reindl, C. Siebenwirth, K. Ilicic, E. Schmid, G. Multhoff, E. Schmitt, P. Kundrat, and G. Dollinger. Sub-micrometer 20mev protons or 45mev lithium spot irradiation enhances yields of dicentric chromosomes due to clustering of dna double-strand breaks. *Mutat Res Genet Toxicol Environ Mutagen*, 793:30–40, 2015.

[453] F. M. Lyng, C. B. Seymour, and C. Mothersill. Initiation of apoptosis in cells exposed to medium from the progeny of irradiated cells: a possible mechanism for bystander-induced genomic instability? *Radiation research*, 157(4):365–70, 2002.

[454] Zhengfeng Liu, Carmel E. Mothersill, Fiona E. McNeill, Fiona M. Lyng, Soo Hyun Byun, Colin B. Seymour, and William V. Prestwich. A dose threshold for a medium transfer bystander effect for a human skin cell line. *Radiation research*, 166(1 Pt 1):19–23, 2006.

[455] Pavel Kundrat and Werner Friedland. Track structure calculations on intracellular targets responsible for signal release in bystander experiments with transfer of irradiated cell-conditioned medium. *International journal of radiation biology*, 88(1-2):98–102, 2012.

[456] P. Kundrat and W. Friedland. Mechanistic modelling of radiation-induced bystander effects. *Radiat Prot Dosimetry*, 166(1-4):148–51, 2015.

[457] G. Bauer. Low dose radiation and intercellular induction of apoptosis: potential implications for the control of oncogenesis. *Int J Radiat Biol*, 83(11-12):873–88, 2007.

[458] Pavel Kundrat and Werner Friedland. Non-linear response of cells to signals leads to revised characteristics of bystander effects inferred from their modelling. *International journal of radiation biology*, 88(10):743–50, 2012.

[459] D. I. Portess, G. Bauer, M. A. Hill, and P. O'Neill. Low-dose irradiation of nontransformed cells stimulates the selective removal of precancerous cells via intercellular induction of apoptosis. *Cancer Res*, 67(3):1246–53, 2007.

[460] P. Kundrat and W. Friedland. Enhanced release of primary signals may render intercellular signalling ineffective due to spatial aspects. *Sci Rep*, 6:33214, 2016.

[461] Alberto D'Onofrio and Alberto Gandolfi. *Mathematical Oncology 2013*. Modeling and simulation in science, engineering and technology,. Birkhauser, New York, 2014.

[462] Dominik Wodarz and Natalia L. Komarova. *Dynamics of cancer : mathematical foundations of oncology*. World Scientific, Hackensack New Jersey, 2014.

[463] P. M. Altrock, L. L. Liu, and F. Michor. The mathematics of cancer: integrating quantitative models. *Nat Rev Cancer*, 15(12):730–45, 2015.

[464] T. S. Deisboeck, Z. Wang, P. Macklin, and V. Cristini. Multiscale cancer modeling. *Annu Rev Biomed Eng*, 13:127–55, 2011.

[465] Douglas Hanahan and RobertA Weinberg. Hallmarks of cancer: The next generation. *Cell*, 144(5):646–674.

[466] Anna Bill-Axelson, Lars Holmberg, Mirja Ruutu, Michael Hggman, Swen-Olof Andersson, Stefan Bratell, Anders Spngberg, Christer Busch, Stig Nordling, Hans Garmo, Juni Palmgren, Hans-Olov Adami, Bo Johan Norln, and Jan-Erik Johansson. Radical prostatectomy versus watchful waiting in early prostate cancer. *New England Journal of Medicine*, 352(19):1977–1984, 2005.

[467] J. Evans, S. Ziebland, and A. R. Pettitt. Incurable, invisible and inconclusive: watchful waiting for chronic lymphocytic leukaemia and implications for doctor-patient communication. *Eur J Cancer Care (Engl)*, 21(1):67–77, 2012.

[468] Bruce A. Chabner and Thomas G. Roberts. Chemotherapy and the war on cancer. *Nat Rev Cancer*, 5(1):65–72, 2005.

[469] Taline Khoukaz. Administration of anti-egfr therapy: A practical review. *Seminars in Oncology Nursing*, 22:20–27, 2006.

[470] Tanguy Y. Seiwert, Joseph K. Salama, and Everett E. Vokes. The concurrent chemoradiation paradigm[mdash]general principles. *Nat Clin Prac Oncol*, 4(2):86–100, 2007.

[471] S. Novello, F. Barlesi, R. Califano, T. Cufer, S. Ekman, M. Giaj Levra, K. Kerr, S. Popat, M. Reck, S. Senan, G. V. Simo, J. Vansteenkiste, and S. Peters. Metastatic non-small-cell lung cancer: Esmo clinical practice guidelines for diagnosis, treatment and follow-up? *Annals of Oncology*, 27(suppl 5):v1–v27, 2016.

[472] Elizabeth I. Buchbinder and Anupam Desai. Ctla-4 and pd-1 pathways: Similarities, differences, and implications of their inhibition. *American Journal of Clinical Oncology*, 39(1):98–106, 2016.

[473] Weiping Zou, Jedd D. Wolchok, and Lieping Chen. Pd-l1 (b7-h1) and pd-1 pathway blockade for cancer therapy: Mechanisms, response biomarkers, and combinations. *Science Translational Medicine*, 8(328):328rv4–328rv4, 2016.

[474] Paul C. Tumeh, Christina L. Harview, Jennifer H. Yearley, I. Peter Shintaku, Emma J. M. Taylor, Lidia Robert, Bartosz Chmielowski, Marko Spasic, Gina Henry, Voicu Ciobanu, Alisha N. West, Manuel Carmona, Christine Kivork, Elizabeth Seja, Grace Cherry, Antonio J. Gutierrez, Tristan R. Grogan, Christine Mateus, Gorana Tomasic, John A. Glaspy, Ryan O. Emerson, Harlan Robins, Robert H. Pierce, David A. Elashoff, Caroline Robert, and Antoni Ribas. Pd-1 blockade induces responses by inhibiting adaptive immune resistance. *Nature*, 515(7528):568–571, 2014.

[475] Barbara Melosky, Quincy Chu, Rosalyn Juergens, Natasha Leighl, Deanna McLeod, and Vera Hirsh. Pointed progress in second-line advanced non?small-cell lung cancer: The rapidly evolving field of checkpoint inhibition. *Journal of Clinical Oncology*, 2016.

[476] Edward B. Garon. Cancer immunotherapy trials not immune from imprecise selection of patients. *New England Journal of Medicine*, 376(25):2483–2485, 2017.

[477] Weijie Ma, Barbara M. Gilligan, Jianda Yuan, and Tianhong Li. Current status and perspectives in translational biomarker research for pd-1/pd-l1 immune checkpoint blockade therapy. *Journal of Hematology & Oncology*, 9:47, 2016.

[478] Vittorio Cristini and John Lowengrub. *Multiscale modeling of cancer : an integrated experimental and mathematical modeling approach*. Cambridge University Press, Cambridge ; New York, 2010.

[479] S. M. Wise, J. S. Lowengrub, and V. Cristini. An adaptive multigrid algorithm for simulating solid tumor growth using mixture models. *Mathematical and computer modelling*, 53(1-2):1–20, 2011.

[480] Paul Macklin, Mary E. Edgerton, John S. Lowengrub, and Vittorio Cristini. *Discrete cell modeling*, book section 6, pages 88–122. Cambridge University Press, Cambridge, UK, 2010.

[481] Franois Graner and James A. Glazier. Simulation of biological cell sorting using a two-dimensional extended potts model. *Physical Review Letters*, 69(13):2013–2016, 1992.

[482] Andrs Szab and Rocland M. Merks. Cellular potts modeling of tumor growth, tumor invasion, and tumor evolution. *Frontiers in Oncology*, 3(87), 2013.

[483] J. Chmielecki, J. Foo, G. R. Oxnard, K. Hutchinson, K. Ohashi, R. Somwar, L. Wang, K. R. Amato, M. Arcila, M. L. Sos, N. D. Socci, A. Viale, E. de Stanchina, M. S. Ginsberg, R. K. Thomas, M. G. Kris, A. Inoue, M. Ladanyi, V. A. Miller, F. Michor, and W. Pao. Optimization of dosing for egfr-mutant non-small cell lung cancer with evolutionary cancer modeling. *Sci Transl Med*, 3(90):90ra59, 2011.

[484] Lei Tang, Anne L. van de Ven, Dongmin Guo, Vivi Andasari, Vittorio Cristini, King C. Li, and Xiaobo Zhou. Computational modeling of 3d tumor growth and angiogenesis for chemotherapy evaluation. *PLOS ONE*, 9(1):e83962, 2014.

[485] A. Konstorum, A. T. Vella, A. J. Adler, and R. C. Laubenbacher. Addressing current challenges in cancer immunotherapy with mathematical and computational modelling. *J R Soc Interface*, 14(131), 2017.

[486] F. Pappalardo, I. Martinez Forero, M. Pennisi, A. Palazon, I. Melero, and S. Motta. Simb16: modeling induced immune system response against b16-melanoma. *PLoS One*, 6(10):e26523, 2011.

[487] R. Serre, S. Benzekry, L. Padovani, C. Meille, N. Andre, J. Ciccolini, F. Barlesi, X. Muracciole, and D. Barbolosi. Mathematical modeling of cancer immunotherapy and its synergy with radiotherapy. *Cancer Res*, 76(17):4931–40, 2016.

[488] Lisette de Pillis, K. Renee Fister, Weiqing Gu, Craig Collins, Michael Daub, David Gross, James Moore, and Benjamin Preskill. Mathematical model creation for cancer chemo-immunotherapy. *Computational and Mathematical Methods in Medicine*, 10(3), 2009.

[489] M. H. Swat, G. L. Thomas, J. M. Belmonte, A. Shirinifard, D. Hmeljak, and J. A. Glazier. Multi-scale modeling of tissues using compucell3d. *Methods Cell Biol*, 110:325–66, 2012.

[490] Thomas Sterlin, Christoph Kolb, Hartmut Dickhaus, Dirk Jger, and Niels Grabe. Bridging the scales: semantic integration of quantitative sbml in graphical multicellular models and simulations with episim and copasi. *Bioinformatics*, 29(2):223–229, 2013.

[491] Gary R. Mirams, Christopher J. Arthurs, Miguel O. Bernabeu, Rafel Bordas, Jonathan Cooper, Alberto Corrias, Yohan Davit, Sara-Jane Dunn, Alexander G. Fletcher, Daniel G. Harvey, Megan E. Marsh, James M. Osborne, Pras Pathmanathan, Joe Pitt-Francis, James Southern, Nejib Zemzemi, and David J. Gavaghan. Chaste: An open source c++ library for computational physiology and biology. *PLOS Computational Biology*, 9(3):e1002970, 2013.

[492] Ahmadreza Ghaffarizadeh, Samuel H. Friedman, Shannon M. Mumenthaler, and Paul Macklin. Physicell: an open source physics-based cell simulator for 3-d multicellular systems. *bioRxiv*, 2016.

[493] C. C. Ling and X. A. Li. Over the next decade the success of radiation treatment planning will be judged by the immediate biological response of tumor cells rather than by surrogate measures such as dose maximization and uniformity. *Medical physics*, 32(7):2189–92, 2005.

[494] M. V. Graham, J. A. Purdy, B. Emami, W. Harms, W. Bosch, M. A. Lockett, and C. A. Perez. Clinical dose-volume histogram analysis for pneumonitis after 3d treatment for non-small cell lung cancer (nsclc). *Int J Radiat Oncol Biol Phys*, 45(2):323–9, 1999.

[495] J. O. Deasy. Multiple local minima in radiotherapy optimization problems with dose-volume constraints. *Med Phys*, 24(7):1157–61, 1997.

[496] Q. Wu and R. Mohan. Multiple local minima in imrt optimization based on dose-volume criteria. *Med Phys*, 29(7):1514–27, 2002.

[497] X. S. Qi, V. A. Semenenko, and X. A. Li. Improved critical structure sparing with biologically based imrt optimization. *Medical physics*, 36(5):1790–9, 2009.

[498] V. A. Semenenko, B. Reitz, E. Day, X. S. Qi, M. Miften, and X. A. Li. Evaluation of a commercial biologically based imrt treatment planning system. *Med Phys*, 35(12):5851–60, 2008.

[499] A. Niemierko. Reporting and analyzing dose distributions: a concept of equivalent uniform dose. *Med Phys*, 24(1):103–10, 1997.

[500] R. K. Sachs and D. J. Brenner. The mechanistic basis of the linear-quadratic formalism. *Med Phys*, 25(10):2071–3, 1998.

[501] Niemierko A Okunieff P, Morgan D and Suit HD. Radiation dose-response of human tumors. *Int J Radiat Oncol Biol Phys*, 1995.

[502] R. Nath, W. S. Bice, W. M. Butler, Z. Chen, A. S. Meigooni, V. Narayana, M. J. Rivard, and Y. Yu. Aapm recommendations on dose prescription and reporting methods for permanent interstitial brachytherapy for prostate cancer: Report of task group 137. *Med Phys*, 36(11):5310–5322, 2009.

[503] J T Lyman. Complication probability as assessed from dose-volume histograms. *Radiat Res Suppl*, 8(2):S13–9, 1985.

[504] C. S. Hamilton, L. Y. Chan, D. L. McElwain, and J. W. Denham. A practical evaluation of five dose-volume histogram reduction algorithms. *Radiother Oncol*, 24(4):251–60, 1992.

[505] L. Cozzi, F. M. Buffa, and A. Fogliata. Comparative analysis of dose volume histogram reduction algorithms for normal tissue complication probability calculations. *Acta Oncol*, 39(2):165–71, 2000.

[506] Kllman P. Brahme A, Lind BK. Physical and biological dose optimization using inverse radiation therapy planning. Report, Department of Radiation Physics, Karolinska Institute and University of Stockholm, 1991.

[507] L. B. Marks. Extrapolating hypofractionated radiation schemes from radiosurgery data: regarding hall et al., ijrobp 21:819-824; 1991 and hall and brenner, ijrobp 25:381-385; 1993. *Int J Radiat Oncol Biol Phys*, 32(1):274–6, 1995.

[508] E. J. Hall and D. J. Brenner. The radiobiology of radiosurgery: rationale for different treatment regimes for avms and malignancies. *Int J Radiat Oncol Biol Phys*, 25(2):381–5, 1993.

[509] R. G. Dale. The application of the linear-quadratic dose-effect equation to fractionated and protracted radiotherapy. *Br J Radiol.*

[510] T. T. Puck and P. I. Marcus. Action of x-rays on mammalian cells. *The Journal of experimental medicine*, 103(5):653–66, 1956.

[511] B. Fertil, I. Reydellet, and P. J. Deschavanne. A benchmark of cell survival models using survival curves for human cells after completion of repair of potentially lethal damage. *Radiation research*, 138(1):61–9, 1994.

[512] JF. Fowler. Linear quadratics is alive and well: In regard to park et al. (int j radiat oncol biol phys 2008;70:847?852). *Int J Radiat Oncol Biol Phys*, 72:957, 2008.

[513] G. W. Barendsen. Dose fractionation, dose rate and iso-effect relationships for normal tissue responses. *Int J Radiat Oncol Biol Phys*, 8(11):1981–97, 1982.

[514] M. Carlone, D. Wilkins, B. Nyiri, and P. Raaphorst. Tcp isoeffect analysis using a heterogeneous distribution of radiosensitivity. *Med Phys*, 31(5):1176–82, 2004.

[515] R. G. Dale. The application of the linear-quadratic model to fractionated radiotherapy when there is incomplete normal tissue recovery between fractions, and possible implications for treatments involving multiple fractions per day. *Br J Radiol*, 59(705):919–27, 1986.

[516] A. Tai, B. Erickson, K. A. Khater, and X. A. Li. Estimate of radiobiologic parameters from clinical data for biologically based treatment planning for liver irradiation. *Int J Radiat Oncol Biol Phys*, 70(3):900–7, 2008.

[517] A. Tai, F. Liu, E. Gore, and X. A. Li. An analysis of tumor control probability of stereotactic body radiation therapy for lung cancer with a regrowth model. *Phys Med Biol*, 61(10):3903–13, 2016.

[518] I. C. Gibbs. Spinal and paraspinal lesions: the role of stereotactic body radiotherapy. *Front Radiat Ther Oncol*, 40:407–14, 2007.

[519] J. J. Nuyttens, V. Moiseenko, M. McLaughlin, S. Jain, S. Herbert, and J. Grimm. Esophageal dose tolerance in patients treated with stereotactic body radiation therapy. *Semin Radiat Oncol*, 26(2):120–8, 2016.

[520] J. J. Nuyttens, N. C. van der Voort van Zyp, J. Praag, S. Aluwini, R. J. van Klaveren, C. Verhoef, P. M. Pattynama, and M. S. Hoogeman. Outcome of four-dimensional stereotactic radiotherapy for centrally located lung tumors. *Radiother Oncol*, 102(3):383–7, 2012.

[521] K. L. Stephans, T. Djemil, C. Diaconu, C. A. Reddy, P. Xia, N. M. Woody, J. Greskovich, V. Makkar, and G. M. Videtic. Esophageal dose tolerance to hypofractionated stereotactic body radiation therapy: risk factors for late toxicity. *Int J Radiat Oncol Biol Phys*, 90(1):197–202, 2014.

[522] T. A. LaCouture, J. Xue, G. Subedi, Q. Xu, J. T. Lee, G. Kubicek, and S. O. Asbell. Small bowel dose tolerance for stereotactic body radiation therapy. *Semin Radiat Oncol*, 26(2):157–64, 2016.

[523] B. M. Barney, S. N. Markovic, N. N. Laack, R. C. Miller, J. N. Sarkaria, O. K. Macdonald, H. J. Bauer, and K. R. Olivier. Increased bowel toxicity in patients treated with a vascular endothelial growth factor inhibitor (vegfi) after stereotactic body radiation therapy (sbrt). *Int J Radiat Oncol Biol Phys*, 87(1):73–80, 2013.

[524] J. J. Nuyttens and M. van de Pol. The cyberknife radiosurgery system for lung cancer. *Expert Rev Med Devices*, 9(5):465–75, 2012.

[525] A. Rashid, S. D. Karam, B. Rashid, J. H. Kim, D. Pang, W. Jean, J. Grimm, and S. P. Collins. Multisession radiosurgery for hearing preservation. *Semin Radiat Oncol*, 26(2):105–11, 2016.

[526] F. C. Timmer, P. E. Hanssens, A. E. van Haren, J. J. Mulder, C. W. Cremers, A. J. Beynon, J. J. van Overbeeke, and K. Graamans. Gamma knife radiosurgery for vestibular schwannomas: results of hearing preservation in relation to the cochlear radiation dose. *Laryngoscope*, 119(6):1076–81, 2009.

[527] R. D. Timmerman. An overview of hypofractionation and introduction to this issue of seminars in radiation oncology. *Semin Radiat Oncol*, 18(4):215–22, 2008.

[528] J. Wang, Y. Y. Chen, A. Tai, X. L. Chen, S. M. Huang, C. Yang, Y. Bao, N. W. Li, X. W. Deng, C. Zhao, M. Chen, and X. A. Li. Sensorineural hearing loss after combined intensity modulated radiation therapy and cisplatin-based chemotherapy for nasopharyngeal carcinoma. *Transl Oncol*, 8(6):456–62, 2015.

[529] C. Goldsmith, P. Price, T. Cross, S. Loughlin, I. Cowley, and N. Plowman. Dose-volume histogram analysis of stereotactic body radiotherapy treatment of pancreatic cancer: A focus on duodenal dose constraints. *Semin Radiat Oncol*, 26(2):149–56, 2016.

[530] S. Nishimura, A. Takeda, N. Sanuki, S. Ishikura, Y. Oku, Y. Aoki, E. Kunieda, and N. Shigematsu. Toxicities of organs at risk in the mediastinal and hilar regions following stereotactic body radiotherapy for centrally located lung tumors. *J Thorac Oncol*, 9(9):1370–6, 2014.

[531] M. Alber and F. Nusslin. An objective function for radiation treatment optimization based on local biological measures. *Phys Med Biol*, 44(2):479–93, 1999.

[532] Marcus Alber. A concept for the optimization of radiotherapy.

[533] M. Alber, M. Birkner, and F. Nusslin. Tools for the analysis of dose optimization: Ii. sensitivity analysis. *Phys Med Biol*, 47(19):N265–70, 2002.

[534] Rehbinder H Lf J Hrdemark B, Liander A and Robinson D. P3imrt biological optimization and eud.

[535] B. K. Lind, P. Mavroidis, S. Hyodynmaa, and C. Kappas. Optimization of the dose level for a given treatment plan to maximize the complication-free tumor cure. *Acta Oncol*, 38(6):787–98, 1999.

[536] B. Sanchez-Nieto and A. E. Nahum. The delta-tcp concept: a clinically useful measure of tumor control probability. *Int J Radiat Oncol Biol Phys*, 44(2):369–80, 1999.

[537] B. K. Lind, L. M. Persson, M. R. Edgren, I. Hedlof, and A. Brahme. Repairable-conditionally repairable damage model based on dual poisson processes. *Radiat Res*, 160(3):366–75, 2003.

[538] MD Jacobs P, Nelson A and Liu I. Biological effective dose and tumor control probability modeling using the mim software suite.

[539] Carlone Marco, Wilkins David, and Raaphorst Peter. The modified linear-quadratic model of guerrero and li can be derived from a mechanistic basis and exhibits linear-quadratic-linear behaviour. *Physics in Medicine & Biology*, 50(10):L9, 2005.

[540] R. Singh, H. Al-Hallaq, C. A. Pelizzari, G. P. Zagaja, A. Chen, and A. B. Jani. Dosimetric quality endpoints for low-dose-rate prostate brachytherapy using biological effective dose (bed) vs. conventional dose. *Med Dosim*, 28(4):255–9, 2003.

[541] Jack F. Fowler. Brief summary of radiobiological principles in fractionated radiotherapy. *Seminars in Radiation Oncology*, 2(1):16–21.

[542] Turesson T. ?gren A, Brahme A. Optimization of uncomplicated control for head and neck tumors. *International Journal of Radiation Oncology Biology Physics*, 19(4):1077–1085, 1990.

[543] Ding K Winton T Bezjak A Seymour L Shepherd FA Leighl NB. Jang RW, Le Matre A. Quality-adjusted time without symptoms or toxicity analysis of adjuvant chemotherapy in non-small-cell lung cancer: an analysis of the national cancer institute of canada clinical trials group jbr.10 trial. *Journal of Clinical Oncology*, 27(26):4268–4273, 2009.

[544] Biomarkers Definitions Working Group. Biomarkers and surrogate endpoints: preferred definitions and conceptual framework. *Clin Pharmacol Ther*, 69(3):89–95, 2001.

[545] Shedden K Hayman JA Yuan S Ritter T Ten Haken RK Lawrence TS Kong FM. Stenmark MH, Cai XW. Combining physical and biologic parameters to predict radiation-induced lung toxicity in patients with non-small-cell lung cancer treated with definitive radiation therapy. *International Journal of Radiation Oncology Biology Physics*, 84(2).217–222, 2012.

[546] Schipper M Stringer KA Cai X Hayman JA Yu J Lawrence TS Kong FM. Yuan ST, Ellingrod VL. Genetic variations in tgfbeta1, tpa, and ace and radiation-induced thoracic toxicities in patients with non-small-cell lung cancer. *Journal of Thoracic Oncology*, 8(2):208–213, 2013.

[547] Liu Z Wang LE Tucker SL Mao L Wang XS Martel M Komaki R Cox JD Milas L Wei Q. Yuan X, Liao Z. Single nucleotide polymorphism at rs1982073:t869c of the tgfbeta1 gene is associated with the risk of radiation pneumonitis in patients with non-small-cell lung cancer treated with definitive radiotherapy. *Journal of Clinical Oncology*, 27(20):3370–3378, 2009.

[548] TenHaken R Matuzak M Kong FM Schipper MJ, Taylor JMG and Lawrence T. Personalized dose selection in radiation therapy using statistical models for toxicity and efficacy with dose and biomarkers as covariates. *Stat Med*, 33(30):5330–90, 2014.

[549] P.F. Thall and J.D. Cook. Dose-finding based on efficacy-toxicity tradeoffs. *Biometrics*, 60:684–693, 2004.

[550] Y. Yuan and G. Yin. Bayesian dose finding by jointly modelling toxicity and efficacy as time-to-event outcomes. *J Roy Stat Soc Ser C*, 58:719–736, 2009.

[551] Nguyen HQ Yuan Y and Thall PF. *Bayesian Designs for Phase I-II Clinical Trials*. CRC Press, New York, 2016.

[552] Thall FP. Bekele NB. Dose-finding based on multiple toxicities in a soft tissue sarcoma trial. *Journal of the American Statistical Association*, 99:26–35, 2004.

[553] C A Tobias, H O Anger, and J H Lawrence. Radiological use of high energy deuterons and alpha particles. *Am J Roentgenol Radium Ther Nucl Med*, 67(1):1–27, 1952.

[554] H Bethe. Zur theorie des durchgangs schneller korpuskularstrahlen durch materie. *Annalen der Physik*, 397(3):325 – 400, 1930.

[555] J F Ziegler. Comments on ICRU report no. 49: stopping powers and ranges for protons and alpha particles. *Radiat Res*, 152(2):219–222, 1999.

[556] Aimee McNamara, Changran Geng, Robert Turner, Jose Ramos Mendez, Joseph Perl, Kathryn Held, Bruce Faddegon, Harald Paganetti, and Jan Schuemann. Validation of the radiobiology toolkit TOPAS-nBio in simple DNA geometries. *Physica medica : PM : an international journal devoted to the applications of physics to medicine and biology : official journal of the Italian Association of Biomedical Physics (AIFB)*, 33:207–215, January 2017.

[557] S Incerti, M Douglass, S Penfold, S Guatelli, and E Bezak. Review of Geant4-DNA applications for micro and nanoscale simulations. *Phy Med*, 32(10):1187 – 1200, 2016.

[558] I Plante and F A Cucinotta. Multiple CPU computing: the example of the code RITRACKS. In *Computational Intelligence Methods for Bioinformatics and Biostatistics*, pages 12–25. Springer Berlin Heidelberg, Berlin, Heidelberg, 2012.

[559] W Friedland, M Dingfelder, P Kundrat, and P Jacob. Track structures, DNA targets and radiation effects in the biophysical Monte Carlo simulation code PARTRAC. *Mutat Res*, 711(1-2):28–40, 2011.

[560] D Schulz-Ertner and H Tsujii. Particle radiation therapy using proton and heavier ion beams. *J Clin Oncol*, 25(8):953–964, 2007.

[561] D R Olsen, O S Bruland, G Frykholm, and I N Norderhaug. Proton therapy - a systematic review of clinical effectiveness. *Radiother Oncol*, 83(2):123–132, 2007.

[562] A M Allen, T Pawlicki, L Dong, E Fourkal, M Buyyounouski, K Cengel, J Plastaras, M K Bucci, T I Yock, L Bonilla, R Price, E E Harris, and A A Konski. An evidence based review of proton beam therapy: the report of ASTRO's emerging technology committee. *Radiother Oncol*, 103(1):8–11, 2012.

[563] M Engelsman, M Schwarz, and L Dong. Physics controversies in proton therapy. *Semin Radiat Oncol*, 23(2):88–96, 2013.

[564] L E Smith, S Nagar, G J Kim, and W F Morgan. Radiation-induced genomic instability: radiation quality and dose response. *Health Phys*, 85(1):23–29, July 2003.

[565] S Girdhani, R Sachs, and L Hlatky. Biological effects of proton radiation: what we know and don't know. *Radiat Res*, 179(3):257–272, 2013.

[566] M Durante. New challenges in high-energy particle radiobiology. *Brit J Radiol*, 87(1035):20130626, 2014.

[567] D J Thomas. ICRU report 85: fundamental quantities and units for ionizing radiation. *Radiat Prot Dosim*, 150(4):550–552, 2012.

[568] J J Wilkens and U Oelfke. Analytical linear energy transfer calculations for proton therapy. *Med Phys*, 30(5):806–815, 2003.

[569] J J Wilkens and U Oelfke. A phenomenological model for the relative biological effectiveness in therapeutic proton beams. *Phys Med Biol*, 49(13):2811–2825, 2004.

[570] J Kempe, I Gudowska, and A Brahme. Depth absorbed dose and LET distributions of therapeutic ^{1}H, ^{4}He, ^{7}Li, and ^{12}C beams. *Med Phys*, 34(1):183–10, 2007.

[571] J Perl, J Shin, J Schuemann, B Faddegon, and H Paganetti. TOPAS: An innovative proton Monte Carlo platform for research and clinical applications. *Med Phys*, 39(11):6818, 2012.

[572] C Grassberger, A Trofimov, A Lomax, and H Paganetti. Variations in linear energy transfer within clinical proton therapy fields and the potential for biological treatment planning. *Int J Radiat Oncol*, 80(5):1559–1566, 2011.

[573] C Grassberger and H Paganetti. Elevated LET components in clinical proton beams. *Phys Med Biol*, 56(20):6677–6691, 2011.

[574] B G Wouters, G K Y Lam, U Oelfke, K Gardey, R E Durand, and L D Skarsgard. Measurements of relative biological effectiveness of the 70 MeV proton beam at TRIUMF using Chinese hamster V79 cells and the high-precision cell sorter assay. *Radiat Res*, 146(2):159, 1996.

[575] H Paganetti and M Goitein. Biophysical modelling of proton radiation effects based on amorphous track models. *Internat J Radiat Biol*, 77(9):911–928, 2001.

[576] L E Gerweck and S V Kozin. Relative biological effectiveness of proton beams in clinical therapy. *Radiother Oncol*, 50(2):135–142, 1999.

[577] N Tilly, J Johansson, U Isacsson, J Medin, E Blomquist, E Grusell, and B Glimelius. The influence of RBE variations in a clinical proton treatment plan for a hypopharynx cancer. *Phys Med Biol*, 50(12):2765–2777, 2005.

[578] A Carabe, M Moteabbed, N Depauw, J Schuemann, and H Paganetti. Range uncertainty in proton therapy due to variable biological effectiveness. *Phys Med Biol*, 57(5):1159–1172, 2012.

[579] M Wedenberg, B K Lind, and B Hrdemark. A model for the relative biological effectiveness of protons: The tissue specific parameter α/β of photons is a predictor for the sensitivity to LET changes. *Acta Oncol*, 52(3):580–588, 2013.

[580] Y Chen and S Ahmad. Empirical model estimation of relative biological effectiveness for proton beam therapy. *Radiat Prot Dosim*, 149(2):116–123, 2012.

[581] A L McNamara, J Schuemann, and H Paganetti. A phenomenological relative biological effectiveness (RBE) model for proton therapy based on all published in vitro cell survival data. *Phys Med Biol*, 60(21):8399–8416, 2015.

[582] J J Butts and R Katz. Theory of RBE for heavy ion bombardment of dry enzymes and viruses. *Radiat Res*, 30(4):855–871, 1967.

[583] R Katz, B Ackerson, M Homayoonfar, and S C Sharma. Inactivation of cells by heavy ion bombardment. *Radiat Res*, 47(2):402–425, 1971.

[584] H Paganetti, A Niemierko, M Ancukiewicz, L E Gerweck, M Goitein, J S Loeffler, and H D Suit. Relative biological effectiveness (RBE) values for proton beam therapy. *Internat J Radiat Oncol*, 53(2):407–421, 2002.

[585] H Paganetti. Relative biological effectiveness (RBE) values for proton beam therapy. Variations as a function of biological endpoint, dose, and linear energy transfer. *Phys Med Biol*, 59(22):R419–R472, 2014.

[586] H Paganetti. Relating proton treatments to photon treatments via the relative biological effectiveness - should we revise current clinical practice? *Int J Radiat Oncol*, 91(5):892–895, 2015.

[587] T. Underwood, D. Giantsoudi, M. Moteabbed, A. Zietman, J. Efstathiou, H. Paganetti, and H. M. Lu. Can we advance proton therapy for prostate? considering alternative beam angles and relative biological effectiveness variations when comparing against intensity modulated radiation therapy. *Int J Radiat Oncol Biol Phys*, 95(1):454–64, 2016.

[588] C. R. Peeler, D. Mirkovic, U. Titt, P. Blanchard, J. R. Gunther, A. Mahajan, R. Mohan, and D. R. Grosshans. Clinical evidence of variable proton biological effectiveness in pediatric patients treated for ependymoma. *Radiother Oncol*, 121(3):395–401, 2016.

[589] D. Giantsoudi, J. Adams, S. M. MacDonald, and H. Paganetti. Proton treatment techniques for posterior fossa tumors: Consequences for linear energy transfer and dose-volume parameters for the brainstem and organs at risk. *Int J Radiat Oncol Biol Phys*, 97(2):401–410, 2017.

[590] H Paganetti. Interpretation of proton relative biological effectiveness using lesion induction, lesion repair, and cellular dose distribution. 32:2548–2556, 2005.

[591] R G Dale and B Jones. The assessment of RBE effects using the concept of biologically effective dose. *Int J Radiat Oncol*, 43(3):639–645, 1999.

[592] A Carabe-Fernandez, R G Dale, and B Jones. The incorporation of the concept of minimum RBE (RBEmin) into the linear-quadratic model and the potential for improved radiobiological analysis of high-LET treatments. *Int J Radiat Biol*, 83(1):27–39, 2007.

[593] A Carabe-Fernandez, R G Dale, J W Hopewell, B Jones, and H Paganetti. Fractionation effects in particle radiotherapy: implications for hypo-fractionation regimes. *Phys Med Biol*, 55(19):5685–5700, 2010.

[594] C A Tobias, F Ngo, E A Blakely, and T C Yang. The repair-misrepair model of cell survival. In *Radiation Biology and Cancer Research*, pages 195–230. Raven press, Baltimore, 1980.

[595] R B Hawkins. The relationship between the sensitivity of cells to high-energy photons and the RBE of particle radiation used in radiotherapy. *Radiat Res*, 172(6):761–776, 2009.

[596] R B Hawkins. Mammalian cell killing by ultrasoft X-rays and high-energy radiation: an extension of the MK model. *Radiat Res*, 166(2):431–442, 2006.

[597] Y Kase, T Kanai, N Matsufuji, Y Furusawa, T Elsasser, and M Scholz. Biophysical calculation of cell survival probabilities using amorphous track structure models for heavy-ion irradiation. *Phys Med Biol*, 53(1):37–59, 2008.

[598] R. Katz, Jr. Fullerton, B. G., R. A. Roth, and S. C. Sharma. Simplified rbe-dose calculations for mixed radiation fields. *Health Phys*, 30(1):148–50, 1976.

[599] T Elsasser and M Scholz. Cluster effects within the local effect model. *Radiat Res*, 167(3):319–329, 2007.

[600] T Elsasser, M Kramer, and M Scholz. Accuracy of the local effect model for the prediction of biologic effects of carbon ion beams in vitro and in vivo. *Int J Radiat Oncol*, 71(3):866–872, 2008.

[601] T Elsasser, W K Weyrather, T Friedrich, M Durante, G Iancu, M Kramer, G Kragl, S Brons, M Winter, K-J Weber, and M Scholz. Quantification of the relative biological effectiveness for ion beam radiotherapy: direct experimental comparison of proton and carbon ion beams and a novel approach for treatment planning. *Int J Radiat Oncol*, 78(4):1177–1183, 2010.

[602] M Goitein. Calculation of the uncertainty in the dose delivered during radiation therapy. *Med Phys*, 12(5):608, 1985.

[603] H Paganetti. Range uncertainties in proton therapy and the role of Monte Carlo simulations. *Phys Med Biol*, 57(11):R99–R117, 2012.

[604] J Schuemann, S Dowdell, C Grassberger, C H Min, and H Paganetti. Site-specific range uncertainties caused by dose calculation algorithms for proton therapy. *Phys Med Biol*, 59(15):4007–4031, 2014.

[605] X Jia, J Schuemann, H Paganetti, and S B Jiang. GPU-based fast Monte Carlo dose calculation for proton therapy. *Phys Med Biol*, 57(23):7783–7797, 2012.

[606] W C H S Tseung, J Ma, and C Beltran. A fast GPU-based Monte Carlo simulation of proton transport with detailed modeling of nonelastic interactions. *Med Phys*, 42(6):2967–2978, 2015.

[607] D Giantsoudi, J Schuemann, X Jia, S Dowdell, S Jiang, and H Paganetti. Validation of a GPU-based Monte Carlo code (gPMC) for proton radiation therapy: clinical cases study. *Phys Med Biol*, 60(6):2257–2269, 2015.

[608] J Schuemann, D Giantsoudi, C Grassberger, M Moteabbed, C H Min, and H Paganetti. Assessing the clinical impact of approximations in analytical dose calculations for proton therapy. *Int J Radiat Oncol*, 92(5):1157–1164, 2015.

[609] C Geng, M Moteabbed, Y Xie, J Schuemann, T Yock, and H Paganetti. Assessing the radiation-induced second cancer risk in proton therapy for pediatric brain tumors: the impact of employing a patient-specific aperture in pencil beam scanning. *Phys Med Biol*, 61(1):12–22, 2016.

[610] T Inaniwa, N Kanematsu, Y Hara, T Furukawa, M Fukahori, M Nakao, and T Shirai. Implementation of a triple Gaussian beam model with subdivision and redefinition against density heterogeneities in treatment planning for scanned carbon-ion radiotherapy. *Phys Med Biol*, 59(18):5361, 2014.

[611] Changran Geng, Juliane Daartz, Kimberley Lam-Tin-Cheung, Marc Bussiere, Helen A Shih, Harald Paganetti, and Jan Schuemann. Limitations of analytical dose calculations for small field proton radiosurgery. *Physics in Medicine and Biology*, 62(1):246–257, January 2017.

[612] A Carabe, S Espana, C Grassberger, and H Paganetti. Clinical consequences of relative biological effectiveness variations in proton radiotherapy of the prostate, brain and liver. *Phys Med Biol*, 58(7):2103–2117, 2013.

[613] D A Granville and G O Sawakuchi. Comparison of linear energy transfer scoring techniques in Monte Carlo simulations of proton beams. *Phys Med Biol*, 60(14):N283–91, 2015.

[614] L Polster, J Schuemann, I Rinaldi, L Burigo, A L McNamara, R D Stewart, A Attili, D J Carlson, T Sato, J R Mendez, B Faddegon, J Perl, and H Paganetti. Extension of TOPAS for the simulation of proton radiation effects considering molecular and cellular endpoints. *Phys Med Biol*, 60(13):5053–5070, 2015.

[615] S Girdhani, R Sachs, and L Hlatky. Biological effects of proton radiation: an update. *Radiat Prot Dosim*, 166(1-4):334–338, 2015.

[616] I S Sofia Vala, L R Martins, N Imaizumi, R J Nunes, J Rino, F Kuonen, L M Carvalho, C Ruegg, I M Grillo, J T Barata, M Mareel, and S C R Santos. Low doses of ionizing radiation promote tumor growth and metastasis by enhancing angiogenesis. *PLOS ONE*, 5(6):e11222, 2010.

[617] B J Moeller, Y Cao, C Y Li, and M W Dewhirst. Radiation activates HIF-1 to regulate vascular radiosensitivity in tumors: role of reoxygenation, free radicals, and stress granules. *Cancer cell*, 5(5):429–441, 2004.

[618] G H Jang, J-H Ha, T-L Huh, and Y M Lee. Effect of proton beam on blood vessel formation in early developing zebrafish (Danio rerio) embryos. *Arch Pharm Res*, 31(6):779–785, 2008.

[619] S R Boyd, A Gittos, M Richter, J L Hungerford, R D Errington, and I A Cree. Proton beam therapy and iris neovascularisation in uveal melanoma. *Eye*, 20(7):832–836, 2006.

[620] S Girdhani, C Lamont, P Hahnfeldt, A Abdollahi, and L Hlatky. Proton irradiation suppresses angiogenic genes and impairs cell invasion and tumor growth. *Radiat Res*, 178(1):33–45, 2012.

[621] F Kamlah, J Hanze, A Arenz, U Seay, D Hasan, J Juricko, B Bischoff, O R Gottschald, C Fournier, G Taucher-Scholz, M Scholz, W Seeger, R Engenhart-Cabillic, and F Rose. Comparison of the effects of carbon ion and photon irradiation on the angiogenic response in human lung adenocarcinoma cells. *Int J Radiat Oncol*, 80(5):1541–1549, 2011.

[622] Y Liu, Y Liu, C Sun, L Gan, L Zhang, A Mao, Y Du, R Zhou, and H Zhang. Carbon ion radiation inhibits glioma and endothelial cell migration induced by secreted VEGF. *PLOS ONE*, 9(6):e98448–8, 2014.

[623] T Ogata, T Teshima, K Kagawa, Y Hishikawa, Y Takahashi, A Kawaguchi, Y Suzumoto, K Nojima, Y Furusawa, and N Matsuura. Particle irradiation suppresses metastatic potential of cancer cells. *Cancer Res*, 65(1):113–120, 2005.

[624] B Romanowska-Dixon, M Elas, J Swakon, U Sowa, M Ptaszkiewicz, M Szczygieϒ, M Krzykawska, P Olko, and K Urbanska. Metastasis inhibition after proton beam, β- and γ-irradiation of melanoma growing in the hamster eye. *Acta Biochim Pol*, 60(3):307–311, 2013.

[625] D S Gridley, R B Bonnet, D A Bush, C Franke, G A Cheek, J D Slater, and J M Slater. Time course of serum cytokines in patients receiving proton or combined photon/proton beam radiation for resectable but medically inoperable non-small-cell lung cancer. *Int J Radiat Oncol*, 60(3):759–766, 2004.

[626] A Matsunaga, Y Ueda, S Yamada, Y Harada, H Shimada, M Hasegawa, H Tsujii, T Ochiai, and Y Yonemitsu. Carbon-ion beam treatment induces systemic antitumor immunity against murine squamous cell carcinoma. *Cancer*, 116(15):3740–3748, 2010.

[627] M Durante, N Reppingen, and K D Held. Immunologically augmented cancer treatment using modern radiotherapy. *Trends Mol Med*, 19(9):565–582, 2013.

[628] T Shimokawa, L Ma, K Ando, K Sato, and T Imai. The future of combining carbon-ion radiotherapy with immunotherapy: evidence and progress in mouse models. *International Journal of Particle Therapy*, 3(1):61–70, 2016.

[629] C Glowa, C P Karger, S Brons, D Zhao, R P Mason, P E Huber, J Debus, and P Peschke. Carbon ion radiotherapy decreases the impact of tumor heterogeneity on radiation response in experimental prostate tumors. *Cancer lett*, 378(2):97–103, 2016.

[630] J F Fowler. 21 years of biologically effective dose. *Brit J Radiol*, 83(991):554–568, 2010.

[631] A Niemierko. A generalized concept of equivalent uniform dose (EUD). 26(1100), 1999.

[632] J Unkelbach, D Craft, E Salari, J Ramakrishnan, and T Bortfeld. The dependence of optimal fractionation schemes on the spatial dose distribution. *Phys Med Biol*, 58(1):159–167, 2013.

[633] L B Marks, E D Yorke, A Jackson, R K Ten Haken, L S Constine, A Eisbruch, S M Bentzen, J Nam, and J O Deasy. Use of normal tissue complication probability models in the clinic. *Int J Radiat Oncol*, 76(3 Suppl):S10–9, 2010.

[634] C C Pan, B D Kavanagh, L A Dawson, X A Li, S K Das, M Miften, and R K Ten Haken. Radiation-associated liver injury. *Int J Radiat Oncol*, 76(3):S94–S100, 2010.

[635] L B Marks, S M Bentzen, J O Deasy, F-M S Kong, J D Bradley, I S Vogelius, I El Naqa, J L Hubbs, J V Lebesque, R D Timmerman, M K Martel, and A Jackson. Radiation dose - volume effects in the lung. *Int J Radiat Oncol*, 76(3):S70–S76, 2010.

[636] Y R Lawrence, X A Li, I El Naqa, C A Hahn, L B Marks, T E Merchant, and A P Dicker. Radiation dose - volume effects in the brain. *Int J Radiat Oncol*, 76(3):S20–S27, 2010.

[637] A Jackson, R K Ten Haken, J M Robertson, M L Kessler, G J Kutcher, and T S Lawrence. Analysis of clinical complication data for radiation hepatitis using a parallel architecture model. *Int J Radiat Oncol*, 31(4):883–891, 1995.

[638] R Kohno, K Hotta, S Nishioka, K Matsubara, R Tansho, and T Suzuki. Clinical implementation of a GPU-based simplified Monte Carlo method for a treatment planning system of proton beam therapy. *Phys Med Biol*, 56(22):N287–94, 2011.

[639] W C H S Tseung, J Ma, C R Kreofsky, D J Ma, and Chris B. Clinically applicable Monte Carlo-based biological dose optimization for the treatment of head and neck cancers with spot-scanning proton therapy. *Int J Radiat Oncol*, 95(5):1535–1543, 2016.

[640] M Testa, J Schuemann, H M Lu, J Shin, B Faddegon, J Perl, and H Paganetti. Experimental validation of the TOPAS Monte Carlo system for passive scattering proton therapy. *Medical Physics*, 40(12):121719, 2013.

[641] Benjamin Clasie, Nicolas Depauw, Maurice Fransen, Carles Gomà, Hamid Reza Panahandeh, Joao Seco, Jacob B Flanz, and Hanne M Kooy. Golden beam data for proton pencil-beam scanning. *Physics in Medicine and Biology*, 57(5):1147–1158, March 2012.

[642] David C Hall, Anastasia Makarova, Harald Paganetti, and Bernard Gottschalk. Validation of nuclear models in Geant4 using the dose distribution of a 177 MeV proton pencil beam. *Physics in Medicine and Biology*, 61(1):N1–N10, January 2016.

[643] Clemens Grassberger, Juliane Daartz, Stephen Dowdell, Thomas Ruggieri, Greg Sharp, and Harald Paganetti. Quantification of proton dose calculation accuracy in the lung. *International journal of radiation oncology, biology, physics*, 89(2):424–430, June 2014.

[644] Hellen Gelband, Prabhat Jha, Rengaswamy Sankaranarayanan, and Susan Horton. *Disease Control Priorities, (Volume 3): Cancer*. World Bank Publications, 2015.

[645] Joanie Mayer Hope and Bhavana Pothuri. The role of palliative surgery in gynecologic cancer cases. *The oncologist*, 18(1):73–79, 2013.

[646] Ryan M Thomas, Mark J Truty, Graciela M Nogueras-Gonzalez, Jason B Fleming, Jean-Nicolas Vauthey, Peter WT Pisters, Jeffrey E Lee, David C Rice, Wayne L Hofstetter, Robert A Wolff, et al. Selective reoperation for locally recurrent or metastatic pancreatic ductal adenocarcinoma following primary pancreatic resection. *Journal of Gastrointestinal Surgery*, 16(9):1696–1704, 2012.

[647] Andreas Karachristos and Nestor F Esnaola. Surgical management of pancreatic neoplasms: What's new? *Current gastroenterology reports*, 16(8):1–8, 2014.

[648] Oliver Strobel, Werner Hartwig, Thilo Hackert, Ulf Hinz, Viktoria Berens, Lars Grenacher, Frank Bergmann, Jurgen Debus, Dirk Jager, Markus Buchler, et al. Re-resection for isolated local recurrence of pancreatic cancer is feasible, safe, and associated with encouraging survival. *Annals of surgical oncology*, 20(3):964–972, 2013.

[649] Rafif E Mattar, Faisal Al-alem, Eve Simoneau, and Mazen Hassanain. Preoperative selection of patients with colorectal cancer liver metastasis for hepatic resection. *World journal of gastroenterology*, 22(2):567, 2016.

[650] Alessando Cucchetti, Alessandro Ferrero, Matteo Cescon, Matteo Donadon, Nadia Russolillo, Giorgio Ercolani, Giacomo Stacchini, Federico Mazzotti, Guido Torzilli, and Antonio Daniele Pinna. Cure model survival analysis after hepatic resection for colorectal liver metastases. *Annals of surgical oncology*, 22(6):1908–1914, 2015.

[651] Skye C Mayo, Mechteld C De Jong, Carlo Pulitano, Brian M Clary, Srinevas K Reddy, T Clark Gamblin, Scott A Celinksi, David A Kooby, Charles A Staley, Jayme B Stokes, et al. Surgical management of hepatic neuroendocrine tumor metastasis: results from an international multi-institutional analysis. *Annals of surgical oncology*, 17(12):3129–3136, 2010.

[652] Clavien P-A. Dindo D, Demartines N. Classification of surgical complications: a new proposal with evaluation in a cohort of 6336 patients and results of a survey. *Annals of surgery*, 240(2):205–213, 2004.

[653] Lindsay A Bliss, Elan R Witkowski, Catherine J Yang, and Jennifer F Tseng. Outcomes in operative management of pancreatic cancer. *Journal of surgical oncology*, 110(5):592–598, 2014.

[654] Louis Jacob and Karel Kostev. Cancer is associated with intraoperative and postprocedural complications and disorders. *Journal of cancer research and clinical oncology*, 142(4):777–781, 2016.

[655] Dong Wook Shin, Juhee Cho, So Young Kim, Eliseo Guallar, Seung Sik Hwang, Be-Long Cho, Jae Hwan Oh, Ki Wook Jung, Hong Gwan Seo, and Jong Hyock Park. Delay to curative surgery greater than 12 weeks is associated with increased mortality in patients with colorectal and breast cancer but not lung or thyroid cancer. *Annals of surgical oncology*, 20(8):2468–2476, 2013.

[656] George Malietzis, Marco Giacometti, Alan Askari, Subramanian Nachiappan, Robin H Kennedy, Omar D Faiz, Omer Aziz, and John T Jenkins. A preoperative neutrophil to lymphocyte ratio of 3 predicts disease-free survival after curative elective colorectal cancer surgery. *Annals of surgery*, 260(2):287–292, 2014.

[657] Avo Artinyan, Sonia T Orcutt, Daniel A Anaya, Peter Richardson, G John Chen, and David H Berger. Infectious postoperative complications decrease long-term survival in patients undergoing curative surgery for colorectal cancer: a study of 12,075 patients. *Annals of surgery*, 261(3):497–505, 2015.

[658] Matthieu Faron, Jean-Pierre Pignon, David Malka, Abderrahmane Bourredjem, Jean-Yves Douillard, Antoine Adenis, Dominique Elias, Olivier Bouche, and Michel Ducreux. Is primary tumour resection associated with survival improvement in patients with colorectal cancer and unresectable synchronous metastases? a pooled analysis of individual data from four randomised trials. *European Journal of Cancer*, 51(2):166–176, 2015.

[659] Sohei Satoi, Hiroki Yamaue, Kentaro Kato, Shinichiro Takahashi, Seiko Hirono, Shin Takeda, Hidetoshi Eguchi, Masayuki Sho, Keita Wada, Hiroyuki Shinchi, et al. Role of

adjuvant surgery for patients with initially unresectable pancreatic cancer with a long-term favorable response to non-surgical anti-cancer treatments: results of a project study for pancreatic surgery by the japanese society of hepato-biliary-pancreatic surgery. *Journal of hepato-biliary-pancreatic sciences*, 20(6):590–600, 2013.

[660] Vic J Verwaal, Serge van Ruth, Eelco de Bree, Gooike W van Slooten, Harm van Tinteren, Henk Boot, and Frans AN Zoetmulder. Randomized trial of cytoreduction and hyperthermic intraperitoneal chemotherapy versus systemic chemotherapy and palliative surgery in patients with peritoneal carcinomatosis of colorectal cancer. *Journal of clinical oncology*, 21(20):3737–3743, 2003.

[661] S Anwar, MB Peter, J Dent, and NA Scott. Palliative excisional surgery for primary colorectal cancer in patients with incurable metastatic disease. is there a survival benefit? a systematic review. *Colorectal Disease*, 14(8):920–930, 2012.

[662] Yong Sik Yoon, Chan Wook Kim, Seok-Byung Lim, Chang Sik Yu, So Yeon Kim, Tae Won Kim, Min-Ju Kim, and Jin Cheon Kim. Palliative surgery in patients with unresectable colorectal liver metastases: a propensity score matching analysis. *Journal of surgical oncology*, 109(3):239–244, 2014.

[663] Kostan W Reisinger, Jeroen LA van Vugt, Juul JW Tegels, Claire Snijders, Karel WE Hulsewe, Anton GM Hoofwijk, Jan H Stoot, Maarten F Von Meyenfeldt, Geerard L Beets, Joep PM Derikx, et al. Functional compromise reflected by sarcopenia, frailty, and nutritional depletion predicts adverse postoperative outcome after colorectal cancer surgery. *Annals of surgery*, 261(2):345–352, 2015.

[664] Michael Poullis, James McShane, Mathew Shaw, Michael Shackcloth, Richard Page, Neeraj Mediratta, and John Gosney. Smoking status at diagnosis and histology type as determinants of long-term outcomes of lung cancer patients. *European Journal of Cardio-Thoracic Surgery*, 43(5):919–924, 2013.

[665] Harry W Herr, James R Faulkner, H Barton Grossman, Ronald B Natale, Ralph deVere White, Michael F Sarosdy, and E David Crawford. Surgical factors influence bladder cancer outcomes: a cooperative group report. *Journal of clinical oncology*, 22(14):2781–2789, 2004.

[666] F-W Kreth, N Thon, M Simon, M Westphal, G Schackert, G Nikkhah, B Hentschel, G Reifenberger, T Pietsch, M Weller, et al. Gross total but not incomplete resection of glioblastoma prolongs survival in the era of radiochemotherapy. *Annals of oncology*, 24(12):3117–3123, 2013.

[667] Patrice Forget, Jean-Pascal Machiels, Pierre G Coulie, Martine Berliere, Alain J Poncelet, Bertrand Tombal, Annabelle Stainier, Catherine Legrand, Jean-Luc Canon, Yann Kremer, et al. Neutrophil: lymphocyte ratio and intraoperative use of ketorolac or diclofenac are prognostic factors in different cohorts of patients undergoing breast, lung, and kidney cancer surgery. *Annals of surgical oncology*, 20(3):650–660, 2013.

[668] Mark J Truty. The role and techniques of vascular resection. In *Multimodality Management of Borderline Resectable Pancreatic Cancer*, pages 203–222. Springer, 2016.

[669] Oliver Strobel, Viktoria Berens, Ulf Hinz, Werner Hartwig, Thilo Hackert, Frank Bergmann, Jurgen Debus, Dirk Jager, Markus W Buchler, and Jens Werner. Resection after neoadjuvant therapy for locally advanced,unresectable pancreatic cancer. *Surgery*, 152(3):S33–S42, 2012.

[670] Ingo Hartlapp, Justus Muller, Werner Kenn, Ulrich Steger, Christoph Isbert, Michael Scheurlen, C-T Germer, Hermann Einsele, and Volker Kunzmann. Complete pathological remission of locally advanced, unresectable pancreatic cancer (lapc) after intensified neoadjuvant chemotherapy. *Oncology Research and Treatment*, 36(3):123–125, 2013.

[671] Pavlos Papavasiliou, Yun Shin Chun, and John P Hoffman. How to define and manage borderline resectable pancreatic cancer. *Surgical Clinics of North America*, 93(3):663–674, 2013.

[672] Qing-Hua Ke, Shi-Qiong Zhou, Ji-Yuan Yang, Wei Du, Gai Liang, Yong Lei, and Fei Luo. S-1 plus gemcitabine chemotherapy followed by concurrent radiotherapy and maintenance therapy with s-1 for unresectable pancreatic cancer. *World J Gastroenterol*, 20(38):13987–13992, 2014.

[673] Heather L Lewis and Syed A Ahmad. Neoadjuvant therapy for resectable pancreatic adenocarcinoma. In *Difficult Decisions in Hepatobiliary and Pancreatic Surgery*, pages 583–597. Springer, 2016.

[674] Renata D Peixoto, Caroline Speers, Colleen E McGahan, Daniel J Renouf, David F Schaeffer, and Hagen F Kennecke. Prognostic factors and sites of metastasis in unresectable locally advanced pancreatic cancer. *Cancer medicine*, 4(8):1171–1177, 2015.

[675] Caroline S Verbeke. Resection margins in pancreatic cancer. *Surgical Clinics of North America*, 93(3):647–662, 2013.

[676] Chang Moo Kang, Yong Eun Chung, Jeong Youp Park, Jin Sil Sung, Ho Kyoung Hwang, Hye Jin Choi, Hyunki Kim, Si Young Song, and Woo Jung Lee. Potential contribution of preoperative neoadjuvant concurrent chemoradiation therapy on margin-negative resection in borderline resectable pancreatic cancer. *Journal of Gastrointestinal Surgery*, 16(3):509–517, 2012.

[677] James J Lee, Justin Huang, Christopher G England, Lacey R McNally, and Hermann B Frieboes. Predictive modeling of in vivo response to gemcitabine in pancreatic cancer. *PLoS Comput Biol*, 9(9):e1003231, 2013.

[678] James Coates, Luis Souhami, and Issam El Naqa. Big data analytics for prostate radiotherapy. *Frontiers in Oncology*, 6:149, 2016.

[679] Lisette G de Pillis, Weiqing Gu, and Ami E Radunskaya. Mixed immunotherapy and chemotherapy of tumors: modeling, applications and biological interpretations. *Journal of theoretical biology*, 238(4):841–862, 2006.

[680] Yangjin Kim, Hyun Geun Lee, Nina Dmitrieva, Junseok Kim, Balveen Kaur, and Avner Friedman. Choindroitinase abc i-mediated enhancement of oncolytic virus spread and anti tumor efficacy: a mathematical model. *PloS one*, 9(7):e102499, 2014.

[681] Franziska Michor and Kathryn Beal. Improving cancer treatment via mathematical modeling: surmounting the challenges is worth the effort. *Cell*, 163(5):1059–1063, 2015.

[682] Henrica MJ Werner, Gordon B Mills, and Prahlad T Ram. Cancer systems biology: a peek into the future of patient care? *Nature reviews clinical oncology*, 11(3):167–176, 2014.

[683] Arnaud H Chauviere, Haralampos Hatzikirou, John S Lowengrub, Hermann B Frieboes, Alastair M Thompson, and Vittorio Cristini. Mathematical oncology: how are the mathematical and physical sciences contributing to the war on breast cancer? *Current breast cancer reports*, 2(3):121–129, 2010.

[684] Alan T Lefor. Computational oncology. *Japanese journal of clinical oncology*, 41(8):937–947, 2011.

[685] Jean-Charles Nault, Aurelien De Reynies, Augusto Villanueva, Julien Calderaro, Sandra Rebouissou, Gabrielle Couchy, Thomas Decaens, Dominique Franco, Sandrine Imbeaud, Francis Rousseau, et al. A hepatocellular carcinoma 5-gene score associated with survival of patients after liver resection. *Gastroenterology*, 145(1):176–187, 2013.

[686] Eithne Costello, William Greenhalf, and John P Neoptolemos. New biomarkers and targets in pancreatic cancer and their application to treatment. *Nature Reviews Gastroenterology and Hepatology*, 9(8):435–444, 2012.

[687] Eric A Collisson, Anguraj Sadanandam, Peter Olson, William J Gibb, Morgan Truitt, Shenda Gu, Janine Cooc, Jennifer Weinkle, Grace E Kim, Lakshmi Jakkula, et al. Subtypes of pancreatic ductal adenocarcinoma and their differing responses to therapy. *Nature medicine*, 17(4):500–503, 2011.

[688] W Du and O Elemento. Cancer systems biology: embracing complexity to develop better anticancer therapeutic strategies. *Oncogene*, 34(25):3215–3225, 2015.

[689] Fraser M Smith, William M Gallagher, Edward Fox, Richard B Stephens, Elton Rexhepaj, Emanuel F Petricoin, Lance Liotta, M John Kennedy, and John V Reynolds. Combination of seldi-tof-ms and data mining provides early-stage response prediction for rectal tumors undergoing multimodal neoadjuvant therapy. *Annals of surgery*, 245(2):259266, February 2007.

[690] Christopher CL Liao, Anuja Mehta, Nicholas J. Ward, Simon Marsh, Tan Arulampalam, and John D. Norton. Analysis of post-operative changes in serum protein expression profiles from colorectal cancer patients by maldi-tof mass spectrometry: a pilot methodological study. *World Journal of Surgical Oncology*, 8(1):33, 2010.

[691] Aurlien Thomas, Nathan Heath Patterson, Martin M. Marcinkiewicz, Anthoula Lazaris, Peter Metrakos, and Pierre Chaurand.

[692] B. Michael Ghadimi, Marian Grade, Michael J. Difilippantonio, Sudhir Varma, Richard Simon, Cristina Montagna, Laszlo Fzesi, Claus Langer, Heinz Becker, Torsten Liersch, and Thomas Ried. Effectiveness of gene expression profiling for response prediction of rectal adenocarcinomas to preoperative chemoradiotherapy. *Journal of Clinical Oncology*, 23(9):1826–1838, 2005. PMID: 15774776.

[693] Y. Wang, T. Jatkoe, Y. Zhang, M. G. Mutch, D. Talantov, J. Jiang, H. L. McLeod, and D. Atkins. Gene expression profiles and molecular markers to predict recurrence of dukes' b colon cancer. *J Clin Oncol*, 22, 2004.

[694] Y. Jiang, G. Casey, I. C. Lavery, Y. Zhang, D. Talantov, M. Martin-McGreevy, M. Skacel, E. Manilich, A. Mazumder, D. Atkins, C. P. Delaney, and Y. Wang. Development of a clinically feasible molecular assay to predict recurrence of stage ii colon cancer. *J Mol Diagn*, 10, 2008.

[695] E. J. A. Morris, N. J. Maughan, D. Forman, and P. Quirke. Who to treat with adjuvant therapy in dukes b/stage ii colorectal cancer? the need for high quality pathology. *Gut*, 56, 2007.

[696] N Hosseini, M Y Sir, C J Jankowski, and K S Pasupathy. Surgical Duration Estimation via Data Mining and Predictive Modeling: A Case Study. *AMIA ... Annual Symposium proceedings. AMIA Symposium*, 2015:640–8, 2015.

[697] M Jajroudi, T Baniasadi, L Kamkar, F Arbabi, M Sanei, and M Ahmadzade. Prediction of survival in thyroid cancer using data mining technique. *Technology in cancer research & treatment*, 13(4):353–9, aug 2014.

[698] Tomohiro Tanaka, Masayuki Kurosaki, Leslie B. Lilly, Namiki Izumi, and Morris Sherman. Identifying candidates with favorable prognosis following liver transplantation for hepatocellular carcinoma: Data mining analysis. *Journal of Surgical Oncology*, 112(1):72–79, 2015.

[699] Maciej Bobowicz, Marcin Skrzypski, Piotr Czapiewski, Michał Marczyk, Agnieszka Maciejewska, Michał Jankowski, Anna Szulgo-Paczkowska, Wojciech Zegarski, Ryszard Pawłowski, Joanna Polanska, Wojciech Biernat, Janusz Jaskiewicz, and Jacek Jassem. Prognostic value of 5-microrna based signature in t2-t3n0 colon cancer. *Clinical & Experimental Metastasis*, pages 1–9, 2016.

[700] Hari Nathan and Timothy M. Pawlik. Limitations of claims and registry data in surgical oncology research. *Annals of Surgical Oncology*, 15(2):415–423, 2008.

[701] Tina Hernandez-Boussard, Suzanne Tamang, Douglas Blayney, Jim Brooks, and Nigam Shah. New Paradigms for Patient-Centered Outcomes Research in Electronic Medical Records: An example of detecting urinary incontinence following prostatectomy. *eGEMs (Generating Evidence & Methods to improve patient outcomes)*, 4(3), may 2016.

[702] Przemyslaw Waliszewski. Computer-aided image analysis and fractal synthesis in the quantitative evaluation of tumor aggressiveness in prostate carcinomas. *Frontiers in Oncology*, 6:110, 2016.

[703] P Tracqui. Biophysical models of tumour growth. *Reports on Progress in Physics*, 72(5):056701, 2009.

[704] H. M. Byrne and M. A. J. Chaplain. Growth of necrotic tumors in the presence and absence of inhibitors. *Mathematical Biosciences*, 135(2):187–216, 1996.

[705] Mindy D Szeto, Gargi Chakraborty, Jennifer Hadley, Russ Rockne, Mark Muzi, Ellsworth C Alvord, Kenneth A Krohn, Alexander M Spence, and Kristin R Swanson. Quantitative metrics of net proliferation and invasion link biological aggressiveness assessed by mri with hypoxia assessed by fmiso-pet in newly diagnosed glioblastomas. *Cancer research*, 69(10):4502–4509, 2009.

[706] Zuoren Yu, Timothy G Pestell, Michael P Lisanti, and Richard G Pestell. Cancer stem cells. *The international journal of biochemistry & cell biology*, 44(12):2144–2151, 2012.

[707] Piyush B Gupta, Tamer T Onder, Guozhi Jiang, Kai Tao, Charlotte Kuperwasser, Robert A Weinberg, and Eric S Lander. Identification of selective inhibitors of cancer stem cells by high-throughput screening. *Cell*, 138(4):645–659, 2009.

[708] Mihai Tanase and Przemyslaw Waliszewski. On complexity and homogeneity measures in predicting biological aggressiveness of prostate cancer; implication of the cellular automata model of tumor growth. *Journal of surgical oncology*, 112(8):791–801, 2015.

[709] Charles W Kimbrough, Charles R St Hill, Robert CG Martin, Kelly M McMasters, and Charles R Scoggins. Tumor-positive resection margins reflect an aggressive tumor biology in pancreatic cancer. *Journal of surgical oncology*, 107(6):602–607, 2013.

[710] Suzette E Sutherland, Martin I Resnick, Gregory T Maclennan, and Howard B Goldman. Does the size of the surgical margin in partial nephrectomy for renal cell cancer really matter? *The Journal of urology*, 167(1):61–64, 2002.

[711] U.S. National Cancer Institute. Nci dictionary of cancer terms. In *https://www.cancer.gov/publications/dictionaries/cancer-terms*, 2016-09-30.

[712] Jennifer S Fang, Robert D Gillies, and Robert A Gatenby. Adaptation to hypoxia and acidosis in carcinogenesis and tumor progression. In *Seminars in cancer biology*, volume 18, pages 330–337. Elsevier, 2008.

[713] Kai Ruan, Gang Song, and Gaoliang Ouyang. Role of hypoxia in the hallmarks of human cancer. *Journal of cellular biochemistry*, 107(6):1053–1062, 2009.

[714] X. Zheng, S. M. Wise, and V. Cristini. Nonlinear simulation of tumor necrosis, neovascularization and tissue invasion via an adaptive finite-element/level-set method. *Bulletin of Mathematical Biology*, 67(2):211, 2005.

[715] John P. Sinek, Sandeep Sanga, Xiaoming Zheng, Hermann B. Frieboes, Mauro Ferrari, and Vittorio Cristini. Predicting drug pharmacokinetics and effect in vascularized tumors using computer simulation. *Journal of Mathematical Biology*, 58(4):485, 2008.

[716] Douglas Hanahan and Robert A Weinberg. The hallmarks of cancer. *cell*, 100(1):57–70, 2000.

[717] A Giatromanolaki, E Sivridis, MI Koukourakis, A Polychronidis, and C Simopoulos. Prognostic role of angiogenesis in operable carcinoma of the gallbladder. *American journal of clinical oncology*, 25(1):3841, February 2002.

[718] H G Augustin. Translating angiogenesis research into the clinic: the challenges ahead. *The British Journal of Radiology*, 76(suppl'1):S3–S10, 2003. PMID: 15456709.

[719] Michael M. Gottesman. Mechanisms of cancer drug resistance. *Annual Review of Medicine*, 53(1):615–627, 2002. PMID: 11818492.

[720] Nicholas C Turner and Jorge S Reis-Filho. Genetic heterogeneity and cancer drug resistance. *The lancet oncology*, 13(4):e178–e185, 2012.

[721] X Sui, R Chen, Z Wang, Z Huang, N Kong, M Zhang, W Han, F Lou, J Yang, Q Zhang, et al. Autophagy and chemotherapy resistance: a promising therapeutic target for cancer treatment. *Cell death & disease*, 4(10):e838, 2013.

[722] Niko Beerenwinkel, Roland F Schwarz, Moritz Gerstung, and Florian Markowetz. Cancer evolution: mathematical models and computational inference. *Systematic Biology*, 2014.

[723] Kimiyo N Yamamoto, Akira Nakamura, and Hiroshi Haeno. The evolution of tumor metastasis during clonal expansion with alterations in metastasis driver genes. *Scientific reports*, 5, 2015.

[724] Sebastien Benzekry, Amanda Tracz, Michalis Mastri, Ryan Corbelli, Dominique Barbolosi, and John M.L. Ebos. Modeling spontaneous metastasis following surgery: An in vivo-in silico approach. *Cancer Research*, 76(3):535–547, 2016.

[725] Wenzhuo He, Chenxi Yin, Guifang Guo, Chang Jiang, Fang Wang, Huijuan Qiu, Xuxian Chen, Ruming Rong, Bei Zhang, and Liangping Xia. Initial neutrophil lymphocyte ratio is superior to platelet lymphocyte ratio as an adverse prognostic and predictive factor in metastatic colorectal cancer. *Medical Oncology*, 30(1):439, 2013.

[726] Anna Kane Laird. Dynamics of tumour growth. *British journal of cancer*, 18(3):490, 1964.

[727] Robert A Gatenby and Robert J Gillies. A microenvironmental model of carcinogenesis. *Nature Reviews Cancer*, 8(1):56–61, 2008.

[728] Yan Mao, Evan T Keller, David H Garfield, Kunwei Shen, and Jianhua Wang. Stromal cells in tumor microenvironment and breast cancer. *Cancer and Metastasis Reviews*, 32(1-2):303–315, 2013.

[729] Sang Yun Ha, So-Young Yeo, Yan-hiua Xuan, and Seok-Hyung Kim. The prognostic significance of cancer-associated fibroblasts in esophageal squamous cell carcinoma. *PloS one*, 9(6):e99955, 2014.

[730] Etienne Baratchart, Sebastien Benzekry, Andreas Bikfalvi, Thierry Colin, Lindsay S Cooley, Raphael Pineau, Emeline J Ribot, Olivier Saut, and Wilfried Souleyreau. Computational modelling of metastasis development in renal cell carcinoma. *PLoS Comput Biol*, 11(11):e1004626, 2015.

[731] Steffen Erik Eikenberry, T Sankar, MC Preul, EJ Kostelich, CJ Thalhauser, and Yang Kuang. Virtual glioblastoma: growth, migration and treatment in a three-dimensional mathematical model. *Cell proliferation*, 42(4):511–528, 2009.

[732] G Sciume, R Santagiuliana, Mauro Ferrari, P Decuzzi, and BA Schrefler. A tumor growth model with deformable ecm. *Physical biology*, 11(6):065004, 2014.

[733] Natalie Turner, Ben Tran, Phillip V Tran, Mathuranthakan Sinnathamby, Hui-Li Wong, Ian Jones, Matthew Croxford, Jayesh Desai, Jeanne Tie, Kathryn Maree Field, et al. Primary tumor resection in patients with metastatic colorectal cancer is associated with reversal of systemic inflammation and improved survival. *Clinical colorectal cancer*, 14(3):185–191, 2015.

[734] Regina V Tse, Laura A Dawson, Alice Wei, and Malcolm Moore. Neoadjuvant treatment for pancreatic cancer–a review. *Critical reviews in oncology/hematology*, 65(3):263274, March 2008.

[735] Daneng Li and Eileen M OReilly. Adjuvant and neoadjuvant systemic therapy for pancreas adenocarcinoma. In *Seminars in oncology*, volume 42, pages 134–143. Elsevier, 2015.

[736] Alfons Hoekstra, Bastien Chopard, and Peter Coveney. Multiscale modelling and simulation: a position paper. *Philosophical Transactions of the Royal Society of London A: Mathematical, Physical and Engineering Sciences*, 372(2021):20130377, 2014.

[737] Mark Robertson-Tessi, Robert J Gillies, Robert A Gatenby, and Alexander RA Anderson. Non-linear tumor-immune interactions arising from spatial metabolic heterogeneity. *bioRxiv*, page 038273, 2016.

[738] Steffen Eikenberry, Craig Thalhauser, and Yang Kuang. Tumor-immune interaction, surgical treatment, and cancer recurrence in a mathematical model of melanoma. *PLoS Comput Biol*, 5(4):e1000362, 2009.

[739] Philipp M Altrock, Lin L Liu, and Franziska Michor. The mathematics of cancer: integrating quantitative models. *Nature Reviews Cancer*, 15(12):730–745, 2015.

[740] Fabio Raman, Elizabeth Scribner, Olivier Saut, Cornelia Wenger, Thierry Colin, and Hassan M Fathallah-Shaykh. Computational trials: Unraveling motility phenotypes, progression patterns, and treatment options for glioblastoma multiforme. *PloS one*, 11(1):e0146617, 2016.

[741] Andrea Hawkins-Daarud and Kristin R. Swanson. Precision medicine in the clinic: Personalizing a model of glioblastoma through parameterization. *bioRxiv*, 2016.

[742] Anne Baldock, Russell Rockne, Addie Boone, Maxwell Neal, Carly Bridge, Laura Guyman, Maceij Mrugala, Jason Rockhill, Kristin Swanson, Andrew Trister, A Hawkins-Daarud, and David Corwin. From patient-specific mathematical neuro-oncology to precision medicine. *Frontiers in Oncology*, 3:62, 2013.

[743] Andrea Hawkins-Daarud, Russell Rockne, Alexander Anderson, and Kristin Swanson. Modeling tumor-associated edema in gliomas during anti-angiogenic therapy and its impact on imageable tumor. *Frontiers in Oncology*, 3:66, 2013.

[744] Maxwell Lewis Neal, Andrew D Trister, Tyler Cloke, Rita Sodt, Sunyoung Ahn, Anne L Baldock, Carly A Bridge, Albert Lai, Timothy F Cloughesy, Maciej M Mrugala, et al. Discriminating survival outcomes in patients with glioblastoma using a simulation-based, patient-specific response metric. *PLoS One*, 8(1):e51951, 2013.

[745] Karl Y. Bilimoria, Andrew K. Stewart, David P. Winchester, and Clifford Y. Ko. The national cancer data base: A powerful initiative to improve cancer care in the united states. *Annals of Surgical Oncology*, 15(3):683–690, 2008.

[746] SC Ferreira Jr, ML Martins, and MJ Vilela. Reaction-diffusion model for the growth of avascular tumor. *Physical Review E*, 65(2):021907, 2002.

[747] Lisette G de Pillis, Daniel G Mallet, and Ami E Radunskaya. Spatial tumor-immune modeling. *Computational and Mathematical Methods in medicine*, 7(2-3):159–176, 2006.

[748] DG Mallet and LG De Pillis. A cellular automata model of tumor-immune system interactions. *Journal of theoretical biology*, 239(3):334350, April 2006.

[749] Trisilowati Trisilowati, Scott McCue, and Dann Mallet. Numerical solution of an optimal control model of dendritic cell treatment of a growing tumour. *ANZIAM Journal*, 54:664–680, 2013.

[750] Karline Soetaert, Jeff Cash, and Francesca Mazzia. *Solving differential equations in R*. Springer Science & Business Media, 2012.

[751] C Garlanda, C Parravicini, M Sironi, M De Rossi, R Wainstok de Calmanovici, F Carozzi, F Bussolino, F Colotta, A Mantovani, and A Vecchi. Progressive growth in immunodeficient mice and host cell recruitment by mouse endothelial cells transformed by polyoma middle-sized t antigen: implications for the pathogenesis of opportunistic vascular tumors. *Proceedings of the National Academy of Sciences*, 91(15):7291–7295, 1994.

[752] Mary E. Edgerton, Yao-Li Chuang, Paul Macklin, Wei Yang, Elaine L. Bearer, and Vittorio Cristini. A novel, patient-specific mathematical pathology approach for assessment of surgical volume: Application to ductal carcinoma in situ of the breast. *Anal. Cell. Pathol.*, 34(5):247–63, 2011.

[753] Leonid Hanin and Marco Zaider. Effects of surgery and chemotherapy on metastatic progression of prostate cancer: Evidence from the natural history of the disease reconstructed through mathematical modeling. *Cancers*, 3(3):3632, 2011.

[754] Leonid Hanin and Olga Korosteleva. Does extirpation of the primary breast tumor give boost to growth of metastases? evidence revealed by mathematical modeling. *Mathematical biosciences*, 223(2):133–141, 2010.

[755] Feras MO Al-Dweri, Damian Guirado, Antonio M Lallena, and Vicente Pedraza. Effect on tumour control of time interval between surgery and postoperative radiotherapy: an empirical approach using monte carlo simulation. *Physics in medicine and biology*, 49(13):2827, 2004.

[756] J Jesus Naveja, Flavio F Contreras-Torres, Andres Rodrjguez-Galvan, and Erick Martjnez-Loran. Computational simulation of tumor surgical resection coupled with the immune system response to neoplastic cells. *Journal of Computational Medicine*, Article ID 831538, 2014.

[757] Chongming Jiang, Chunyan Cui, Li Li, and Yuanzhi Shao. The anomalous diffusion of a tumor invading with different surrounding tissues. *PloS one*, 9(10):e109784, 2014.

[758] R Santagiuliana, Mauro Ferrari, and BA Schrefler. Simulation of angiogenesis in a multiphase tumor growth model. *Computer Methods in Applied Mechanics and Engineering*, 304:197–216, 2016.

[759] Hiroshi Haeno, Mithat Gonen, Meghan B Davis, Joseph M Herman, Christine A Iacobuzio-Donahue, and Franziska Michor. Computational modeling of pancreatic cancer reveals kinetics of metastasis suggesting optimum treatment strategies. *Cell*, 148(1):362–375, 2012.

[760] Yangjin Kim. Regulation of cell proliferation and migration in glioblastoma: New therapeutic approach. *Frontiers in Oncology*, 3:53, 2013.

[761] Ioannis T Konstantinidis, Andrew L Warshaw, Jill N Allen, Lawrence S Blaszkowsky, Carlos Fernandez-del Castillo, Vikram Deshpande, Theodore S Hong, Eunice L Kwak, Gregory Y Lauwers, David P Ryan, et al. Pancreatic ductal adenocarcinoma: is there a survival difference for r1 resections versus locally advanced unresectable tumors? what is a true r0 resection? *Annals of surgery*, 257(4):731–736, 2013.

[762] Lisette G. de Pillis, Ami E. Radunskaya, and Charles L. Wiseman. A validated mathematical model of cell-mediated immune response to tumor growth. *Cancer Research*, 65(17):7950–7958, 2005.

[763] Bjorn Eiben, Rene Lacher, Vasileios Vavourakis, John H Hipwell, Danail Stoyanov, Norman R Williams, Jorg Sabczynski, Thomas Bulow, Dominik Kutra, Kirsten Meetz, et al. Breast conserving surgery outcome prediction: A patient-specific, integrated multi-modal imaging and mechano-biological modelling framework. In *International Workshop on Digital Mammography*, pages 274–281. Springer, 2016.

[764] Yixun Liu, Samira M Sadowski, Allison B Weisbrod, Electron Kebebew, Ronald M Summers, and Jianhua Yao. Patient specific tumor growth prediction using multi-modal images. *Medical image analysis*, 18(3):555–566, 2014.

[765] Jianjun Paul Tian, Avner Friedman, Jin Wang, and E. Antonio Chiocca. Modeling the effects of resection, radiation and chemotherapy in glioblastoma. *Journal of Neuro-Oncology*, 91(3):287, 2008.

[766] Urszula Ledzewicz, Heinz Schattler, Avner Friedman, and Eugene Kashdan. *Mathematical methods and models in biomedicine*. Springer Science & Business Media, 2012.

[767] F Cornelis, O Saut, P Cumsille, D Lombardi, A Iollo, J Palussiere, and T Colin. In vivo mathematical modeling of tumor growth from imaging data: Soon to come in the future? *Diagnostic and interventional imaging*, 94(6):593–600, 2013.

[768] Katrina K Treloar, Matthew J Simpson, Parvathi Haridas, Kerry J Manton, David I Leavesley, DL Sean McElwain, and Ruth E Baker. Multiple types of data are required to identify the mechanisms influencing the spatial expansion of melanoma cell colonies. *BMC systems biology*, 7(1):1, 2013.

[769] CS Mayo, ML Kessler, A Eisbruch, G Weyburne, M Feng, I El Naqa, JA Hayman, S Jolly, L Holevinski, C Anderson, et al. The big data effort in radiation oncology: data mining or data farming? *Advances in Radiation Oncology*, 2016.

[770] P. W. Lavori and R. Dawson. Dynamic treatment regimes: practical design considerations. *Clin Trials*, 1(1):9–20, 2004.

[771] S. A. Murphy. An experimental design for the development of adaptive treatment strategies. *Stat Med*, 24(10):1455–81, 2005.

[772] K. M. Kidwell. SMART designs in cancer research: Past, present, and future. *Clin Trials*, 11(4):445–456, 2014.

[773] B. Chakraborty and E. E. M. Moodie. *Statistical methods for dynamic treatment regimes : reinforcement learning, causal inference, and personalized medicine*. Statistics for biology and health,. Springer, New York, NY, 2013.

[774] M. Mielcarek, B.E. Storer, M. Boeckh, P.A. Carpenter, G.B. McDonald, H.J. Deeg, and *et al*. Initial therapy of acute graft-versus-host disease with low-dose prednisone does not compromise patient outcomes. *Blood*, 113:2888–94, 2009.

[775] S. Assouline and J. H. Lipton. Monitoring response and resistance to treatment in chronic myeloid leukemia. *Curr Oncol*, 18(2):e71–83, 2011.

[776] J. Cortes, A. Quintas-Cardama, and H. M. Kantarjian. Monitoring molecular response in chronic myeloid leukemia. *Cancer*, 117(6):1113–22, 2011.

[777] M. L. Sorror, M. B. Maris, R. Storb, F. Baron, B. M. Sandmaier, D. G. Maloney, and B. Storer. Hematopoietic cell transplantation (HCT)-specific comorbidity index: a new tool for risk assessment before allogeneic hct. *Blood*, 106(8):2912–9, 2005.

[778] S. A. Murphy. A generalization error for q-learning. *Journal of Machine Learning Research*, 6:1073–1097, 2005.

[779] L. Orellana, A. Rotnitzky, and J. M. Robins. Dynamic regime marginal structural mean models for estimation of optimal dynamic treatment regimes, part i: main content. *Int J Biostat*, 6(2):Article 8, 2010.

[780] J. M. Robins. *Optimal Structural Nested Models for Optimal Sequential Decisions*, pages 189–326. Springer, 2004.

[781] E. Arjas and O. Saarela. Optimal dynamic regimes: presenting a case for predictive inference. *Int J Biostat*, 6(2):Article 10, 2010.

[782] P. J. Martin, J. D. Rizzo, J. R. Wingard, K. Ballen, P. T. Curtin, C. Cutler, M. R. Litzow, Y. Nieto, B. N. Savani, J. R. Schriber, P. J. Shaughnessy, D. A. Wall, and P. A. Carpenter. First- and second-line systemic treatment of acute graft-versus-host disease: recommendations of the american society of blood and marrow transplantation. *Biol Blood Marrow Transplant*, 18(8):1150–63, 2012.

[783] S. Theurich, H. Fischmann, A. Shimabukuro-Vornhagen, J. M. Chemnitz, U. Holtick, C. Scheid, N. Skoetz, and M. von Bergwelt-Baildon. Polyclonal anti-thymocyte globulins for the prophylaxis of graft-versus-host disease after allogeneic stem cell or bone marrow transplantation in adults. *Cochrane Database Syst Rev*, 9:CD009159, 2012.

[784] I. Walker, T. Panzarella, S. Couban, F. Couture, G. Devins, M. Elemary, G. Gallagher, H. Kerr, J. Kuruvilla, S.J. Lee, J. Moore, T. Nevill, G. Popradi, J. Roy, K.R. Schultz, D. Szwajcer, C. Toze, and R. Foley. Pretreatment with anti-thymocyte globulin versus no anti-thymocyte globulin in patients with haematological malignancies undergoing haemopoietic cell transplantation from unrelated donors: a randomised, controlled, open-label, phase 3, multicentre trial. *Lancet Oncology*, 17(2):164–173, 2016.

[785] N. Kroger, C. Solano, C. Wolschke, G. Bandini, F. Patriarca, M. Pini, A. Nagler, C. Selleri, A. Risitano, G. Messina, W. Bethge, J. Perez de Oteiza, R. Duarte, A.M. Carella, M. Cimminiello, S. Guidi, J. Finke, N. Mordini, C. Ferra, J. Sierra, D. Russo, M. Petrini, G. Milone, F. Benedetti, M. Heinzelmann, D. Pastore, M. Jurado, E. Terruzzi, F. Narni, A. Volp, F. Ayuk, T. Ruutu, and F. Bonifazi. Antilymphocyte globulin for prevention of chronic graft-versus-host disease. *New England Journal of Medicine*, 374(1):43–53, 2016.

[786] M. T. Van Lint, C. Uderzo, A. Locasciulli, I. Majolino, R. Scim, F. Locatelli, G. Giorgiani, W. Arcese, A. P. Iori, M. Falda, A. Bosi, R. Miniero, P. Alessandrino, G. Dini, B. Rotoli, and A. Bacigalupo. Early treatment of acute graft-versus-host disease with high- or low-dose 6-methylprednisolone: a multicenter randomized trial from the italian group for bone marrow transplantation. *Blood*, 92(7):2288–93, 1998.

[787] O. Bembom and M. J. van der Laan. Statistical methods for analyzing sequentially randomized trials. *J Natl Cancer Inst*, 99(21):1577–82, 2007.

[788] P. F. Thall, C. Logothetis, L. C. Pagliaro, S. Wen, M. A. Brown, D. Williams, and R. E. Millikan. Adaptive therapy for androgen-independent prostate cancer: a randomized selection trial of four regimens. *J Natl Cancer Inst*, 99(21):1613–22, 2007.

[789] P. F. Thall2005 and J. K. Wathen. Covariate-adjusted adaptive randomization in a sarcoma trial with multi-stage treatments. *Stat Med*, 24(13):1947–64, 2005.

[790] Weiss LR-Mruphy SA. Oetting I, Levy JA. *Statistical methodology for a SMART design in the development of adaptive treatment strategies.* American Psychiatric Publishing, address = Arlington, VA, chapter = 8, pages = 179-205, year = 2011, type = Book Section.

[791] M. Wolbers and J. D. Helterbrand. Two-stage randomization designs in drug development. *Stat Med*, 27(21):4161–74, 2008.

[792] E. E. M Moodie, J. C. Karran, and S. M. Shortreed. A case study of SMART attributes: a qualitative assessment of generalizability, retention rate, and trial quality. *Trials*, 2016.

[793] W. E. Pelham and G. A. Fabiano. Evidence-based psychosocial treatments for attention-deficit/hyperactivity disorder. *J Clin Child Adolesc Psychol*, 37(1):184–214, 2008.

[794] T. S. Stroup, J. A. Lieberman, J. P. McEvoy, M. S. Swartz, S. M. Davis, R. A. Rosenheck, D. O. Perkins, R. S. Keefe, C. E. Davis, J. Severe, J. K. Hsiao, and CATIE Investigators. Effectiveness of olanzapine, quetiapine, risperidone, and ziprasidone in patients with chronic schizophrenia following discontinuation of a previous atypical antipsychotic. *Am J Psychiatry*, 163(4):611–22, 2006.

[795] M. Gunlicks-Stoessel and L. Mufson. Early patterns of symptom change signal remission with interpersonal psychotherapy for depressed adolescents. *Depress Anxiety*, 28(7):525–31, 2011.

[796] A. J. Rush, M. Fava, S. R. Wisniewski, P. W. Lavori, M. H. Trivedi, H. A. Sackeim, M. E. Thase, A. A. Nierenberg, F. M. Quitkin, T. M. Kashner, D. J. Kupfer, J. F. Rosenbaum, J. Alpert, J. W. Stewart, P. J. McGrath, M. M. Biggs, K. Shores-Wilson, B. D. Lebowitz, L. Ritz, G. Niederehe, and STAR*D Investigators Group. Sequenced treatment alternatives to relieve depression (star*d): rationale and design. *Control Clin Trials*, 25(1):119–42, 2004.

[797] L. S. Schneider, P. N. Tariot, C. G. Lyketsos, K. S. Dagerman, K. L. Davis, S. Davis, J. K. Hsiao, D. V. Jeste, I. R. Katz, J. T. Olin, B. G. Pollock, P. V. Rabins, R. A. Rosenheck, G. W. Small, B. Lebowitz, and J. A. Lieberman. National institute of mental health clinical antipsychotic trials of intervention effectiveness (catie): Alzheimer disease trial methodology. *Am J Geriatr Psychiatry*, 9(4):346–60, 2001.

[798] C. P. Belani, J. Barstis, M. C. Perry, R. V. La Rocca, S. R. Nattam, D. Rinaldi, R. Clark, and G. M. Mills. Multicenter, randomized trial for stage IIIb or IV non-small-cell lung cancer using weekly paclitaxel and carboplatin followed by maintenance weekly paclitaxel or observation. *J Clin Oncol*, 21(15):2933–9, 2003.

[799] K. K. Matthay, C. P. Reynolds, R. C. Seeger, H. Shimada, E. S. Adkins, D. Haas-Kogan, R. B. Gerbing, W. B. London, and J. G. Villablanca. Long-term results for

children with high-risk neuroblastoma treated on a randomized trial of myeloablative therapy followed by 13-cis-retinoic acid: a children's oncology group study. *J Clin Oncol*, 27(7):1007–13, 2009.

[800] M. V. Mateos, A. Oriol, J. Martnez-Lpez, N. Gutirrez, A. I. Teruel, R. de Paz, J. Garca-Laraa, E. Bengoechea, A. Martn, J. D. Mediavilla, L. Palomera, F. de Arriba, Y. Gonzlez, J. M. Hernndez, A. Sureda, J. L. Bello, J. Bargay, F. J. Pealver, J. M. Ribera, M. L. Martn-Mateos, R. Garca-Sanz, M. T. Cibeira, M. L. Ramos, M. B. Vidriales, B. Paiva, M. A. Montalbn, J. J. Lahuerta, J. Blad, and J. F. Miguel. Bortezomib, melphalan, and prednisone versus bortezomib, thalidomide, and prednisone as induction therapy followed by maintenance treatment with bortezomib and thalidomide versus bortezomib and prednisone in elderly patients with untreated multiple myeloma: a randomised trial. *Lancet Oncol*, 11(10):934–41, 2010.

[801] John Kuruvilla, C. Tom Kouroukis, Ann Benger, Matthew C. Cheung, Neil L Berinstein, Stephen Couban, Matthew D. Seftel, Kang Howson-Jan, Armand Keating, Massimo Federico, David A Macdonald, Harold J. Olney, Morel Rubinger, Michael Voralia, A. Robert Turner, Tara Baetz, Annette E Hay, Marina Djurfeldt, Ralph M. Meyer, Bingshu Chen, and Lois Shepherd. A randomized trial of rituximab vs observation following autologous stem cell transplantation (asct) for relapsed or refractory cd20-positive b cell lymphoma: Final results of NCIC CTG LY.12. *Blood*, 122:155–155, 2013.

[802] M. Crump, J. Kuruvilla, S. Couban, D. A. MacDonald, V. Kukreti, C. T. Kouroukis, M. Rubinger, R. Buckstein, K. R. Imrie, M. Federico, N. Di Renzo, K. Howson-Jan, T. Baetz, L. Kaizer, M. Voralia, H. J. Olney, A. R. Turner, J. Sussman, A. E. Hay, M. S. Djurfeldt, R. M. Meyer, B. E. Chen, and L. E. Shepherd. Randomized comparison of gemcitabine, dexamethasone, and cisplatin versus dexamethasone, cytarabine, and cisplatin chemotherapy before autologous stem-cell transplantation for relapsed and refractory aggressive lymphomas: NCIC-CTG LY.12. *J Clin Oncol*, 32(31):3490–6, 2014.

[803] T. M. Habermann, E. A. Weller, V. A. Morrison, R. D. Gascoyne, P. A. Cassileth, J. B. Cohn, S. R. Dakhil, B. Woda, R. I. Fisher, B. A. Peterson, and S. J. Horning. Rituximab-chop versus chop alone or with maintenance rituximab in older patients with diffuse large b-cell lymphoma. *J Clin Oncol*, 24(19):3121–7, 2006.

[804] M. H. van Oers, R. Klasa, R. E. Marcus, M. Wolf, E. Kimby, R. D. Gascoyne, A. Jack, M. Van't Veer, A. Vranovsky, H. Holte, M. van Glabbeke, I. Teodorovic, C. Rozewicz, and A. Hagenbeek. Rituximab maintenance improves clinical outcome of relapsed/resistant follicular non-hodgkin lymphoma in patients both with and without rituximab during induction: results of a prospective randomized phase 3 intergroup trial. *Blood*, 108(10):3295–301, 2006.

[805] R. M. Stone, D. T. Berg, S. L. George, R. K. Dodge, P. A. Paciucci, P. P. Schulman, E. J. Lee, J. O. Moore, B. L. Powell, M. R. Baer, C. D. Bloomfield, and C. A. Schiffer. Postremission therapy in older patients with de novo acute myeloid leukemia: a randomized trial comparing mitoxantrone and intermediate-dose cytarabine with standard-dose cytarabine. *Blood*, 98(3):548–53, 2001.

[806] Botton S-Pautas C Arnaud P de Revel T Reman O Terre C Corront B Gardin C Le QH Quesnel B Cordonnier C Bourhis JH Elhamri M Fenaux P Preudhomme

C Michallet M Castaigne S Dombret H. Thomas X, Raffoux E. Effect of priming with granulocyte-macrophage colony-stimulating factor in younger adults with newly diagnosed acute myeloid leukemia: a trial by the acute leukemia french association (alfa) group. *Leukemia*, 21:453–61, 2007.

[807] L. Wang, A. Rotnitzky, X. Lin, R. E. Millikan, and P. F. Thall. Evaluation of viable dynamic treatment regimes in a sequentially randomized trial of advanced prostate cancer. *J Am Stat Assoc*, 107(498):493–508, 2012.

[808] E. E. Moodie, B. Chakraborty, and M. S. Kramer. Q-learning for estimating optimal dynamic treatment rules from observational data. *Can J Stat*, 40(4):629–645, 2012.

[809] E. F. Krakow, M. Hemmer, T. Wang, B. Logan, M. Aurora, S. Spellman, D. Couriel, A. Alousi, J. Pidala, M. Last, S. Lachance, and E. E. M. Moodie. Tools for the precision medicine era: How to develop highly personalized treatment recommendations from cohort and registry data using Q-learning. *American Journal of Epidemiology*, (Accepted), 2017.

Index

Milton Keynes UK
Ingram Content Group UK Ltd.
UKHW051945071024
449327UK00026B/2168